AF348293

HISTORY AND DESCRIPTION

OF THE

ROYAL MUSEUM

OF

NATURAL HISTORY.

HISTORY AND DESCRIPTION

OF THE

ROYAL MUSEUM

OF

NATURAL HISTORY,

PUBLISHED BY ORDER OF THE ADMINISTRATION OF THAT
ESTABLISHMENT.

TRANSLATED FROM THE FRENCH OF M. DELEUZE.

With three Plans and fourteen Views of the Galleries, Gardens,
and Menagerie.

PARIS:

PRINTED FOR A. ROYER, AT THE JARDIN DU ROI,

BY L. T. CELLOT, RUE DU COLOMBIER, N° 30.

1823.

HISTORY AND DESCRIPTION

OF THE

ROYAL MUSEUM

OF

NATURAL HISTORY.

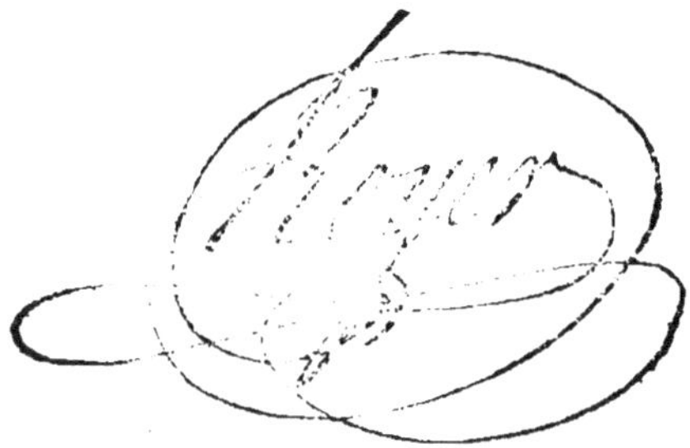

HISTORY AND DESCRIPTION

OF THE

ROYAL MUSEUM

OF

NATURAL HISTORY,

PUBLISHED BY ORDER OF THE ADMINISTRATION OF THAT
ESTABLISHMENT.

TRANSLATED FROM THE FRENCH OF M. DELEUZE.

With three Plans and fourteen Views of the Galleries, Gardens.
and Menagerie.

PARIS:

PRINTED FOR A. ROYER, AT THE JARDIN DU ROI,

BY L. T. CELLOT, RUE DU COLOMBIER, N° 30.

1823.

THE Editor begs leave, in this place, to render his most
grateful acknowledgments to those Friends who have so
kindly aided him in the prosecution of this present work;
had they not expressed their wishes to the contrary, he
would have made a point of mentioning their names.
As it is, the kindness and goodwill with which they
undertook to assist him, in translating his native into a
foreign tongue, will ever remain deeply impressed upon
his mind.

JARDIN DU ROI, 1st January 1823.

PREFACE.

Having been intrusted by the Administrators of the Museum, with the publication of this work, it was my duty to comply with their request; but I should have found myself unequal to the task, had not the Professors themselves furnished me with the means. From them I received much assistance, both by verbal and written instructions; more especially in those parts of the science with which I was less acquainted, in consequence of their having less reference to that branch which is the particular object of my attention.

I have been also considerably assisted in the descriptive part of this work, by several of the gentlemen employed as assistant naturalists in the Museum, and whose care it is to prepare and arrange the collections for the galleries. M. Valenciennes furnished me with a list and description of all the vertebrated animals, as well as of the mollusca and zoophytes; to which M. Dufresne has added an account of the shells. M. Latreille took upon himself to trace out the article for the

crustacea, the arachnides, and the insects. To
M. Delafosse, who lectured for the eminent professor whose recent loss is so much felt, is due
the article on mineralogy; and M. Laurillard has
supplied me with the details I wanted for the
description of the cabinet of comparative anatomy, and for the collection of fossil bones.
I beg leave here to present these Gentlemen
with my acknowledgments and gratitude for
their kind assistance. Lastly, Mr. Royer, one of
the clerks in the office of the administration,
who has been at the expence of this edition, has
had drawn and engraved for it the accompanying Plans and Views.

Although this work can be considered but as
a mere sketch, it will nevertheless give a tole-
rable idea of the riches the Museum contains;
and it will impress the reader with the utility
and importance of that establishment. May it
remind every Frenchman of the obligations the
natural sciences owe to our Kings, who uninterruptedly, from Louis XIII down to our actual
Sovereign, have favoured their study and facilitated their progress.

HISTORY AND DESCRIPTION

OF THE MUSEUM

OF NATURAL HISTORY.

INTRODUCTION.

Of all our institutions for public instruction the Museum of natural history is, without doubt, one of the most useful and the most worthy of admiration. All the natural sciences are there taught, and illustrated by extensive collections, so disposed as to facilitate study and afford the means of immediate comparison. Professors chosen from among the most learned men of France, assist the student

to observe, to distinguish and to class the objects
of nature; not simply by exposing their theories,
or by descriptions, which a well written book
might supply, but by placing before him the objects themselves, and pointing out the particulars to
which his attention should principally be directed.
On quitting the lectures he may proceed to the
botanical garden, where an immense number of
plants are cultivated; to the menagerie, where
many rare animals are kept; or to the zoological
and botanical galleries, where are assembled preserved specimens, the accumulated treasure of
ages, and where he may study families, genera
and species, examine the minutest details, and
familiarise himself with every scientific character.
For the courses of mineralogy and geology, rich
collections methodically arranged afford him the
same resources. In fine, in the course of comparative anatomy he is taught the organisation of
animals; in those of chemistry, the composition
of bodies and their use in the arts; in that of agriculture, the best methods of naturalizing and multiplying useful vegetables; and in that of *iconography*, the art of delineating objects of natural
history, so as to express their distinctive charac-

ters. As those who cultivate a science have frequent occasion to seek in books the developement of the principles exposed by the professor, there is a library exclusively devoted to natural history, to which they have daily access, and in which are found descriptions and drawings of objects, with the history of science, and the actual state of human knowledge.

From the connexion that exists between the several parts of natural history, the professors are enabled, by mutual communications, to extend their views beyond the branch which immediately occupies them, and by the union of their observations and discoveries, to present what may properly be denominated the philosophy of nature.

The man of science visits the Museum to augment his knowledge and fix his ideas; the meditative man contemplates an assemblage of wonders, which exalts his admiration of the riches of nature and the power of that being who has assigned its invariable laws; the studious youth is inspired by the recollection of the illustrious

men who have succeeded each other in the establishment, and whose labours and fame are every
where recalled to his remembrance; the curious
are delighted with the splendour of the collections, and with the view of prodigious numbers
of foreign plants and animals; and amid this variety of objects, the man of letters and the artist
find an inexhaustible fund of ideas, and models of
every description.

In the midst of this imposing spectacle, the activity of peaceful industry, the variety of the plantations, the balmy air, the over-arching walks,
the view which, from a rising ground covered
with ever-green trees, extends over the city and
the surrounding country; the number of persons
who forget in these retreats the cares and agita-
tions of the world; all conspire to render the Museum an abode of tranquillity and intellectual
delight.

Those who devote themselves to the study of a
single branch of natural history, soon learn to
appreciate that part of the establishment which
particularly interests them, but even after a lapse

of years are hardly able to understand the value
of the rest. Foreigners, and those amongst the
curious who have other duties to fulfil, other occupations to pursue and little time at their disposal, require to be told what is most worthy of
their attention, and directed where to find it.
With this view, we shall offer a brief description of every part of the establishment : but
before we conduct our readers to what is most
interesting or valuable, it may be proper to
make them acquainted with the establishment
itself ; its foundation; its successive additions;
its ancient administration ; the names of the learned men who have adorned it ; its reorganisation, and its present state. This notice will be
succinct, and those who are desirous of further
details are referred to the Memoirs of M. de Jussieu, inserted in the annals of the Museum, volumes 1, 2, 3, 4, 6 and 11; and to the History of
the Museum, written in german by Fischer.

The name of *Museum of Natural History*, is of
recent date; it was given at the period when the
garden assumed its present form, and was employed to designate the union of three former

establishments, the King's Garden, Cabinet and Menagerie. How these three establishments were successively founded and united, we shall summarily explain.

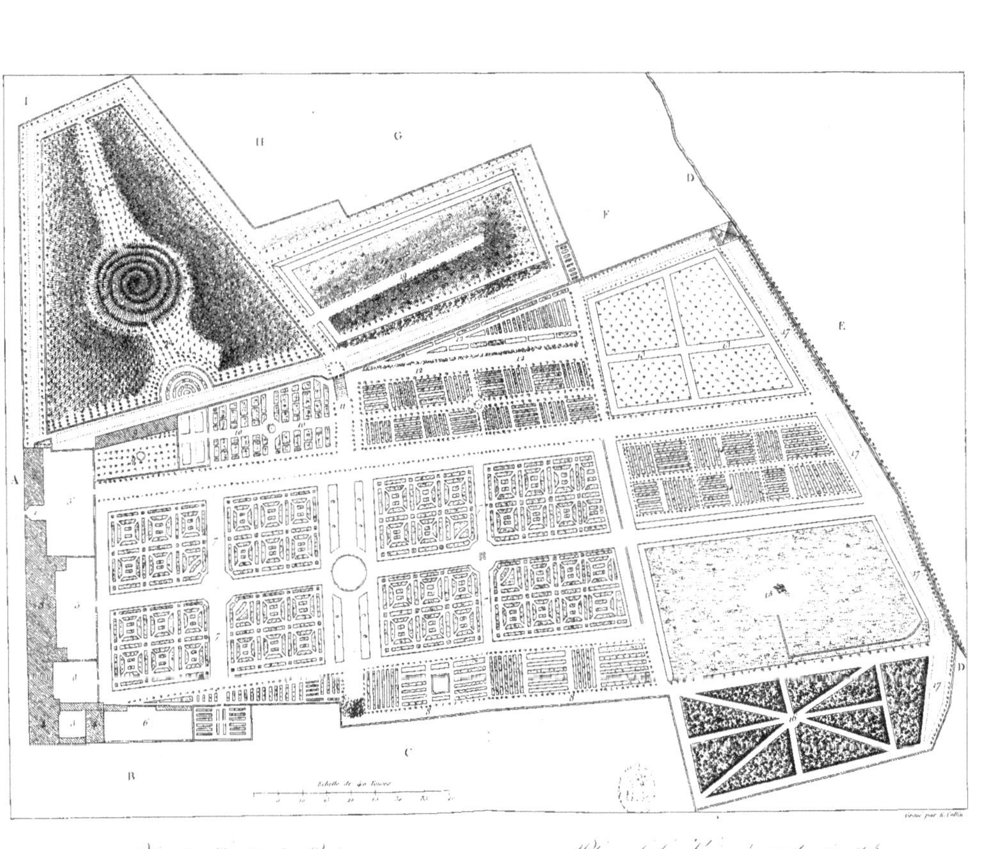

Plan du Jardin du Roi en 1640.

Plan of the King's garden in 1640.

HISTORY

OF THE MUSEUM.

FIRST PERIOD.

FROM THE FOUNDATION OF THE GARDEN TILL 1759.

The King's Garden was founded by Louis XIII, by an edict given, and registered by the parliament, in the month of May 1635.

At the entreaty of Herouard his first physician, and Guy de la Brosse, physician in ordinary, he had nine years before, by letters patent, authorised the foundation, and the acquisition in the *faubourg Saint Victor* of a house and twenty four acres of land, which now form a part of the garden. He gave the superintendance of this establishment to the first physician and his successors, with the power of choosing an intendant , who should reside in the garden and exercise the direction. Herouard chose Guy de la Brosse , and this choice was confirmed by the king. But the death of Herouard and other circumstances having retarded the execution of these letters patent, the foundation cannot in strictness be dated before 1635.

The edict given at this period confirmed the gift of the house and land, named Bouvard , then first physician to the king , superintendant, and Guy de la Brosse intendant of the garden , and contained in short all necessary provisions relative to the objects of the foundation , to its expenditure , and to the nomination and duties of those to be employed in it. The following were its most essential regulations : « Considering that the ope-
» rations of pharmacy are not taught in the
» schools of medicine , we have been humbly
» entreated by the *sieur* Bouvard , that three
» doctors, chosen by him from the faculty of
» Paris , may be provided to *demonstrate the in-*
» *terior of plants* , and all medicines, and to employ themselves in the composition of drugs
» by simple and chemical methods ; for these
» reasons we have appointed three of our counsellors, physicians of the faculty of Paris , to
» demonstrate the interior of plants , and to per-

» form all necessary pharmaceutic operations ;
» for the instruction of students ; and have named
» to this office MM. Jacques Cousinot , Urbain
» Boudineau and Cureau de la Chambre.

» It is also our pleasure that in a cabinet, set
» a part for that purpose, specimens should be
» kept of all drugs either simple or compound ,
» together with every thing rare in nature that
» may be met with ; of which cabinet the said la
» Brosse shall have the key, and regulate its
» opening on the days of demonstration.

» And as the said la Brosse , on whom will
» devolve the care of directing and cultivating
» the garden , may not always be able to attend
» to the *external demonstration of plants ,* we
» have also created a subdemonstrator to assist
» him in the external demonstration in the garden,
» which office will be filled by Vespasian Robin
» our arborist; each of which officers shall attend,
» in the exercise of his employment , at the days
» and hours appointed by our superintendant......
» To all of whom we have appointed the follow-
» ing salaries, viz : to our first physician super-
» intendant of the whole establishment 3,000 li-
» vres ; to each of the three demonstrators, 1,500
» livres; to la Brosse and those who may succeed
» him in the office of intendant 6,000 livres; to the
» subdemonstrator , 1,200 livres. It is further our

» pleasure that the said la Brosse should dispose
» of the buildings, with the exception of the labo-
» ratory and cabinet for the preservation of the
» specimens and rarities, and of those which shall
» be hereafter erected for the facility of instruction,
» and should select the gardeners, porters, etc. For
» the expences of which garden, we have granted
» to our intendant the sum of 4,000 l., besides his
» salary. We farther grant, 400 livres, to the sub-
» demonstrators for the purchase of drugs, and
» 400 l., for the salaries of assistants in the labora-
» tory. For the payment of which sums a fund
» of 21,000 livres, will Es. provided by us, etc.
» Given at Saint-Quentin, in the month of May
» 1635. Registered 15 of May. »

It is important to remark that the *livre tournois*
of that period was worth two livres and a half
of the present money ; thus the sum of 21,000 l.,
would now represent 52,500 francs. The increase
of the expenditure has by no means been proportioned to that of the establishment.

The medical faculty opposed the registering of
this edict, and demanded that the professors should
be chosen on their presentation, and not on that
of the superintendant; they also complained of
the appointment of la Brosse, and especially
desired that chemistry might not be taught. Their
opposition however was unavailing.

Guy de la Brosse, established in his office, repaired and disposed the buildings, and in the
first year formed a parterre 292 feet long and
227 broad, composed of such plants as he could
procure, the greater number of which were given him by John Robin, the father of Vespasian
arborist or botanist to the king. The number
of these plants, comprehending the varieties,
amounted to 1800. He then prepared the ground,
procured new plants by correspondance, traced
the plan of the garden to the extent of ten acres,
and opened it in 1640 (1). The ensuing year he
published a catalogue of the plants cultivated,
the number of which, comprising varieties,
was 2,360. He caused those to be drawn which
he was apprehensive of losing, and even had
some of them engraved.

Unhappily de la Brosse terminated his career
at the moment when, having surmounted the
obstacles that opposed his progress, he began to
witness the success of his labours. He was, properly speaking, the founder of the King's Garden,
and died there in 1643. His remains were deposited in the chapel which formed a part of the
building; and when it became necessary to de-

(1) Over the principal gate was this inscription : *Royal Garden Of
Medicinal Plants ;* which remained till the gate itself was demolished
by Buffon to enlarge the galleries.

molish this chapel to build the stair case to the galleries, they were placed in a private vault.

Such is the origin of an establishment which has since attained so high a degree of prosperity. The limits of this notice oblige us to trace with rapidity the history of its progress and vicissitudes to the end of the last century; we will enter further into particulars in describing the changes effected by the reorganisation, in order to give an exact idea of its present state.

The death of de la Brosse was a great misfortune : those who succeeded him, had neither the same zeal nor the same activity. The professors, depending on the will of the superintendants, were repetedly changed; nor were the lectures given with punctuality. Fourqueux, counsellor of the parliament, was named to replace de la Brosse, and it was impossible for him to bestow the requisite attention on the business of cultivation and instruction. Vespasian Robin alone gave lessons in botany. The superintendant however obtained funds for the construction of a green house and a large basin.

Bouvard, on quitting the place of first physician, wished to retain that of superintendant of the king's garden. This arrangement was defeated by Vautier his successor, on the ground that by the terms of the edict the superintendance of the gar-

den was annexed to the appointment of first physician. Vautier, who was the personal enemy of
Fourqueux, wished to appoint an intendant of
his own choice, and experiencing some opposition, he took no further interest in the garden.
From that time the establishment declined, the
plants perished for want of cultivation, and the
lectures were neglected. Vautier however performed an essential service, by substituting a
course of anatomy for that bearing the name of
the *interior of plants*, which was intended to give
a general knowledge of their properties and uses.

The place of superintendant, again vacant by
the death of Vautier in 1652, was given to Vallot.
At this period *Gaston d'Orléans*, brother of Louis
the thirteenth, had established a botanical garden
at his palace of Blois, which had acquired much
celebrity by the works of Morison, and by drawings of the most remarkable plants. This circumstance awakened the attention of Vallot, and he
appointed Denys Jonquet, a physician who cultivated plants at Saint-Germain-des-Prés, successor
to Robin. Jonquet was seconded by Fagon great
nephew of Guy de la Brosse, who, having been
brought up in the garden, had there acquired a
taste for botany, and who was much attached to
the scene of his infancy. This young man, since
become celebrated, travelled at his own expence

through several provinces of France , and among the Alps and Pyrenees, and sent the fruit of his researches to the garden. At the same time he procured plants from foreign countries ; so that in 1665, the number of species and varieties amounted to 4,000. To reward the zeal of Fagon , Vallot first named him professor of chemistry, and then professor of botany, after the death of Jonquet, in 1671 , thus uniting the two chairs of botany and chemistry.

Gaston of Orleans, not satisfied with assembling plants of every country in his garden at Blois, had them described by learned botanists, and the most remarkable species drawn on vellum by the painter Robert eminent for his skill in that branch of the art. After the death of Gaston in 1660. Colbert persuaded the king to purchase these drawings, and to attach to the Museum a painter who should be obliged to add a certain number every year. Robert was appointed to this place ; and his works, continued till his death in 1684 , possess a degree of truth and finish which none of his successors have been able to surpass. He was followed by J. Joubert , a landscape painter destitute of the talent suited to subjects of natural history, who called in the assistance of Aubriet; and this very able artist , thus attached to the garden in a subordinate capacity, afterwards succeeded

to the title. It is thus that the magnificent collection of drawings of plants and animals has
been formed, which was at first deposited in the
king's library, and which is now the most valuable part of that of the Museum.

Vallot dying in 1671 , Colbert united the superintendance of the garden to that of the king's
buildings, already held by himself, leaving to the
first physician the title of intendant only, with the
direction of the cultivation. In the month of December he procured a declaration from the king,
regulating the administration of the garden ,
and gave commissions to the professors defining
their duties. From this moment the establishment
assumed increasing importance, and it would have
advanced still more rapidly had the principal administration not been united with other offices.

Dacquin first physician and intendant of the
garden in 1672, favoured exclusively the study of
anatomy, which was therefore taught with signal
success , especially by the celebrated Guichard
Joseph Duverney, named professor in 1679. The
lectures of this learned man attracted a great concourse of pupils.

Fagon who had for several years filled the botanical and chemical chairs with applause , being
encumbered with other duties, meditated the re-
signation of his place , and wishing to appoint a

successor worthy of himself, called from a remote part of France, Joseph Pitton de Tournefort, who was then only twenty six years of age, but who already announced what he was one day to become, to whom he yielded the chair of botany in 1683. Ten years after, Fagon became first physician. This place gave him the intendance of the garden, and from the singular respect in which he was held, the title of superintendant was reestablished in his favour.

From the same cause he was enabled to obtain the patronage of government for the establishment to which he was so much attached, and during the fifteen years it was under his direction, every thing resumed new life. This increasing prosperity might have gone on in uninterrupted progression if Fagon, reflecting that it was due to his own zeal and merit, and that his successors might not be animated by the same views, had taken advantage of his influence to procure a settled administration, independant of the caprices of a superior.

We have already mentioned that Duverney filled the chair of anatomy, and Tournefort that of botany : the names of these two professors shed a lustre over the establishment. Tournefort previous to his appointment enjoyed a brilliant reputation, which was heightened in 1693, by the

publication of his Elements of Botany. This work,
by presenting a new method for the classification of plants, gave a solid basis to the science,
and rendered the study of it equally easy and
agreeable.

Tournefort made several journeys to procure
plants, and in 1700, went to the Levant, accompanied by the painter Aubriet. During his absence
his chair was filled by his colleague Morin, and
at his return in 1702, he introduced into the garden the plants of which he had collected the seeds
in his travels : several of them, before unknown,
have since been propagated. He died in 1708,
and left his collection of natural history and
herbarium, to the garden. This herbarium is not
extensive, but it is rendered valuable by the
plants gathered in the Levant, and indicated in
the *corollarium* of the *Institutiones Rei Herbariæ* (1). Tournefort's chair was given to Danty
d'Isnard, who had been named to the botanical
section of the academy of sciences, in 1716.

Fagon, who like his predecessors might have
vested his authority in an intendant, judged it
better simply to confide the oversight of the cul-

(1) The *Institutiones Rei Herbariæ*, which is a translation with additions of the elements of Botany, was published by Tournefort in 1700.
The name of each plant is accompanied by a descriptive phrase. The
corollarium was added after his return from the Levant.

ture to a man of information, exclusively occupied with this charge. He made choice of
Sebastian Vaillant, who, after listening to Tournefort in 1691, had renounced his place of surgeon at the *Hôtel-Dieu*, to devote himself to
the study of plants. This young man, who was
afterwards elected to the academy of sciences,
zealously seconded the views of Fagon; and in
1708 was appointed by him subdemonstrator,
to supply the place of the professor, in case of
absence, and to conduct the herborisations of the
pupils in the country, for the purpose of making
them acquainted with indigenous plants. He also
undertook to increase the collection of drugs,
which was confided to his care and arranged for
study, and obtained funds for the construction
of two hot-houses.

Vaillant formed a very considerable herbarium,
the genera of which were methodically arranged,
and the species, accompanied by tickets indicating all the synonyms then known. This herbarium, which at his death in 1722 was purchased
by order of the king, forms the basis of that of
the Museum.

What above all should save the name of
Vaillant from oblivion, is his first public dis-
course on assuming the functions of assistant professor, in the absence of the principal; in which

he demonstrates the existence of two sexes, and
of the phenomena of fecundation, in vegetables:
thus it was in the King's Garden that this great
discovery, which had been only hinted at before
and was not generally admitted, was first positively announced, and supported by irrefragable
proofs.

Danty d'Isnard retired after delivering a single
course of lectures. Fagon then occupied himself
in seeking among the pupils of Tournefort, a person capable of disseminating his doctrine, and of
filling a chair which his fame had rendered so
conspicuous. A young man of Lyons who, after
studying medecine at Montpellier, had come to
Paris to perfect himself in botany under that celebrated master, appeared in every respect suited to his purpose. Antony de Jussieu, whose
name has become celebrated by the impulse
which himself, his two brothers, and his nephew have given to botanical science, was
named professor in 1709, though only twenty
three years of age : he fully justified this choice
by the ability displayed in his lectures, and by several memoirs, which soon after procured him
admittance to the academy of sciences. In 1716
he visited Spain and Portugal, and the next year
brought back a great accession of plants to the
garden. He was accompanied by his brother

Bernard, then seventeen years old, who formed the resolution of renouncing the study of medecine for that of botany, to which he afterwards gave a new direction.

It was Antony de Jussieu who, in 1720, entrusted Declieux, lieutenant of the royal navy, with a young coffee tree, which, transported to Martinique, became the parent of the immense culture of the West-Indies.

Fagon, who had resigned his botanical chair in 1683 to Tournefort, had also been replaced in the duties of that of chemistry, by a succession of learned professors; Saint Yon, Louis Lemery, Berger and Geoffroy : the reputation gained by the latter pointing him out to Fagon as a successor worthy of himself, he gave him up the title in 1712.

This professor taught chemistry and the *materia medica* with brillant success. In the chemical and pharmaceutic experiments, he was assisted by demonstrators who were generally men of distinguished merit : it was not however until 1695, that this place of demonstrator, which had till then been held by commission, was given to a regular professor.

After Duverney had been appointed professor of anatomy in 1679, that science continued to be much better taught in the garden than in any

other school of France ; and the progress it has
since made , is due to the method he introduced ,
and to the ardour his lectures had excited. Having
reached an advanced age, and feeling his strength
unequal to the continuance of his duties , he was
replaced by his ablest pupils, among whom was
his nephew ; who , as well as his successors , seconded the impulse given to the science.

The indefatigable Fagon thus enjoyed the fruit
of his labours. The three chairs were filled by
celebrated professors, aided by able demonstrators : a large amphitheatre had been constructed
capable of containing six hundred auditors, and
it was usually filled : the cultivation was carefully directed by Vaillant , and additional plants
were procured from his correspondants in America : interesting objects were added to the collection of drugs , and new illustrations , to that of osteology; and the painter Aubriet continued to swell
the number of drawings of plants and animals.

Such was the establishment at the death of
Louis XIV, in 1715. Fagon , aged and infirm ,
resigned the place of first physician , which
was given to Poirier, and retired into the garden , where he was born , and where he died
in 1718.

« His memory, says M. de Jussieu , will
» always be venerated in the Museum , for

» which he procured numerous collections, pro-
» per places to receive and preserve them, and,
» above all, professors who reflected honour on
» the establishment. »

Poirier, superintendant at the death of Fagon, survived him but a few days, and was succeeded by Dodart: but the administration of the
garden, which had been detached from the place
of first physician by a declaration of the king in
1718, was given to Chirac, physician to the duke
of Orleans. Chirac, solely occupied with the practice of medicine, took little interest in the natural sciences: he would allow nothing however
to be done without his orders; and wished to
subject the professors to a dependance on himself, and a conformity to impracticable regulations. At last, seeing the impossibility of directing
every thing in person, he placed in the garden,
with the title of inspector, a man destitute of
knowledge, whom he was soon obliged by superior authority to dismiss. The correspondance
was shackled, the cultivation neglected, and the
funds destined for the establishment diverted to
other uses. The power of Chirac being increased
by his succeeding to the place of first physician
on the death of Dodart, the professors could no
longer make known their grievances; and the
establishment, notwithstanding the exertions of

Anthony and Bernard de Jussieu, and the sacrifices they unceasingly made to procure manure,
ustensils, and other necessary objects, was gradually falling to decay. Chirac was displeased at
their zeal, and Bernard de Jussieu who, besides
succeeding Vaillant in 1722 as subdemonstrator,
was also charged with the care of the collection of drugs, to which the name of *Cabinet
of Natural History* now began to be given, was
deprived of that place. Aubriet, guided by his
own choice and the orders of the intendant, delineated chiefly such medicinal plants as were
already well known.

The lectures in anatomy and chemistry met
with fewer obstacles, either because they demanded no expenditure, or because Chirac took a
greater interest in those sciences than in the other
branches of natural history. The celebrated Duverney, finding himself too weak to continue
his lectures, devolved his duties upon Winslow.
Among his able assistants must be reckoned his
brother, Peter Duverney, and his nephew, James-
Francis-Maria Duverney, for whom the place of
demonstrator was afterwards created, and who
was the master of Daubenton. Duverney died in
1730, in the eighty second year of his age, and
the fifty first of his professorship. Hunaud, a pupil
of Winslow who united to much knowledge

an uncommon facility of elocution, supported the reputation of the anatomical chair, and was constantly attended by a great number of pupils.

Geoffroy had been professor of chemistry since 1710; on his death in 1731, his chair was given to Louis Lemery, who greatly improved the science, by a degree of clearness which it had not before possessed, and extended it beyond the *materia medica*, to which it had till then been confined.

It is to be remarked that the demonstrators and assistants of the professors of chemistry and anatomy, were generally members of the academy of sciences. Simon Boulduc, a very distinguished man, was the first who bore the title of demonstrator, and at his death in 1729, the place was given to his son Giles-Francis Boulduc.

Chirac having ended his unprofitable days in 1732, his son in law Chicoisneau succeeded him as first physician; but the administration of the garden was finally separated from that place. The necessity of confiding it to a person exclusively devoted to its duties was now felt. The king appointed with the title of intendant Charles-Francis de Cysternay du Fay, of an ancient family which, since the fifteenth century, had followed the profession of arms. Young du Fay had himself served with distinction in the war with Spain, but his leisure hours were consecrated to

science, for which he had imbibed a taste from
his father; and with such success, that he was admitted into the academy of sciences in 1723.
After the war, he abandoned the service, and on
being appointed to the intendance of the King's
Garden, devoted his whole time and attention
to the restoration of the establishment. He began
by repairing the disorders occasioned by the negligence of the preceeding administration. His
attention was principally directed to botany, and
he restored the place of keeper of the cabinet
to Bernard de Jussieu. He obtained from the government funds for the necessary reparations;
travelled into England and Holland, to establish
correspondances and collect specimens; rendered permanent the office of demonstrator of
anatomy, and procured the appointment of J.-
F.-M. Duverney, nephew of the professor, about
the year 1736 : he also made considerable additions of rare and useful objects to the cabinet,
and augmented it by the gift of his own collection of precious stones.

In 1739, he was attacked by the small-pox:
feeling his death approach, and wishing to complete the services he had rendered to the establishment, by ensuring the continuance of its
prosperity, he penned a request to the ministry
that Buffon might be appointed his successor.

SECOND PERIOD.

BUFFON was already distinguished by several memoirs on mathematics, natural philosophy and rural economy, which had gained him admittance to the academy of sciences; but he was not yet known as a naturalist. Endowed with that power of attention which discovers the most distant relations of thought, and that brillancy of imagination which commands the attention of others to the result of laborious investigations, he was equally fitted to succeed in different walks of genius. He had not yet decided to what object he should devote his talents and acquirements, when his nomination to the place of intendant of the King's Garden determined him to attach himself to natural history; and as the progress of the science depended very much upon the prosperity of the institution, he used his utmost endeavours to render it worthy of its destination. As his reputation increased, he employed the advantages afforded by his credit and celebrity, to enrich the establishment to which

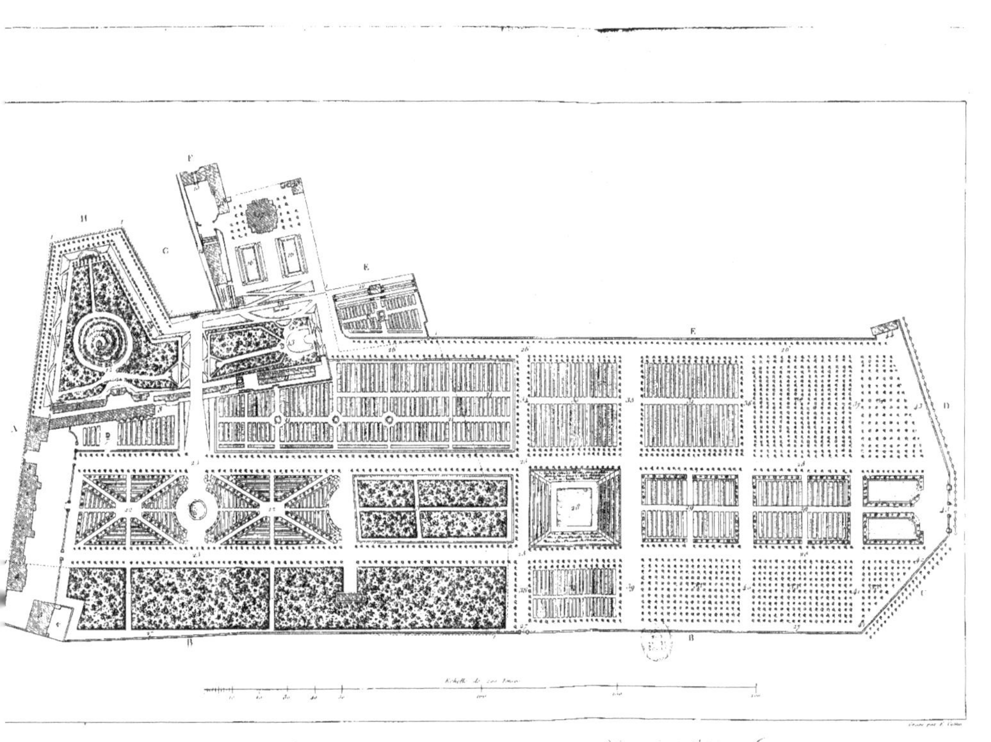

Plan du Jardin du Roi en 1788.
Plan of the King's Garden in 1788.

he had allied himself; and to him are owing its growth and improvement, till the period of its reorganisation, and that extension and variety which rendered a new organisation necessary.

In the construction of a vast edifice, a single architect projects the plan, orders the distribution, and directs the first operations; but different artists are employed to perfect the details, and to form a harmonious whole by labouring after a common conception: when the attention of an individual is distracted by a variety of objects, essential things are likely to be neglected for others of less importance. If the Museum owes its splendour to Buffon, to this magnificent establishment Buffon owes his fame. If he had not been placed in the midst of collections, furnished by government with the means of augmenting them, and thus enabled by extensive correspondance to elicit information from all the naturalists of his day, he would never have conceived the plan of his Natural History, or would never have been able to execute it: that genius which embraces a great variety of facts, in order to deduce from them general conclusions, is continually exposed to err, if it has not at hand all the elements of its speculations.

When Buffon entered upon his office, the cabinet consisted of two small rooms, and a third, containing the preparations of anatomy which were not exposed to public view : the herbarium was in the apartment of the demonstrator of botany : the garden, which was limited to the present nursery on the eastern side, to the green-house on the north, and the galleries of natural history on the west, still presented empty spaces, and contained neither avenues nor regular plantations.

Buffon first directed his attention to the increasing of the collections, and to the providing of more commodious places for their reception. They were arranged in two large rooms of the building which contains the present galleries, which was formerly the dwelling house of the intendant; and soon after, were opened to the public on appointed days.

He next occupied himsef in the embellishment of the garden. Having cut down an old avenue which did not correspond with the principal gate, he replaced it in 1740 by one of lime trees in the proper direction; and planted another parallel on the other side of the parterre. These avenues, which have existed for eighty years, terminate towards the extremity of the nursery, and mark the limits of the garden at that period : they were afterwards continued when new ground was taken in.

From the moment when the charge of the cabinet was given to Bernard de Jussieu, he had bestowed unceasing care upon its arrangement and preservation. The extent of his knowledge, and the facility with which he seized the affinities of bodies, and classed them in their natural order, qualified him particularly for this task, rendered more difficult by the increase of the collections; but being diverted by other occupations, and residing at some distance from the garden, he expressed a desire to be replaced in an office which required unwearied activity and ceaseless assiduity. Buffon also felt that his researches in natural history, needed the assistance of a man who had still all the ardour of youth, and who possessed in a high degree, both the spirit of method and a talent for observation. Gifted with that genius which seizes the principal characters of objects, and unites them in splendid combinations, he had neither time nor patience for the examination of details; to which the weakness of his sight was also an obstacle. He made choice of his countryman Daubenton, who was then twenty nine years of age, and who, after studying botany under de Jussieu, and anatomy under Winslow and Duverney, had retired to Montbard, the place of his birth, to practice medicine. Buffon invited him to Paris, and

in 1745 procured him the place of keeper of the cabinet, with a lodging in the garden, and appointments which soon rose from 5oo to 4ooo fr. He
charged him with the arrangement of the cabinet,
and associated him to his own studies, in the descriptive part of his Natural History, especially in
the anatomy.

In 1749, on the publication of the first volumes
of the Natural History, which attracted the attention of all Europe, Buffon appealed to the liberality of naturalists, whom he invited to send
him whatever remarkable objects they might
meet with in their researches. These he placed in the Museum; Daubenton arranged them,
and the building soon became too confined for
its contents. Buffon, who had already resigned a
part of his habitation, now thought proper to
abandon it entirely, and in 1766, removed to
n° 13 *rue des fossés Saint-Victor*. The collection was then disposed in four large saloons,
which alone formed the cabinet till the reorganisation. The two first contained the animals;
the third, the minerals; the fourth, the herbarium,
the ancient drugs, and different productions of
the vegetable kingdom (1).

(1) The skeletons were still crowded in a separate apartment; for
as comparative anatomy then occupied but a small number of the
learned, those objects were not exhibited, which were not likely to
interest public curiosity.

The saloons were open to the public two days in each week, and the pupils had hours set a part for study. Daubenton was always present to give the necessary explanation, and foreign naturalists often resorted to him for instruction.

He thus divided his attention between the minute care of naming and arranging the objects, and that of directing the students. His patience was inexhaustible ; but the duties of his place became too laborious for the exertions of an individual, and Buffon, perceiving the necessity of naming an assistant, obtained the creation of a place of subdemonstrator, to which he appointed the younger Daubenton, cousin of the former, with a salary of 2,400 francs.

Whilst the collections yearly increased, either by the contributions of naturalists and travellers, or by acquisitions made by the king at the solicitation of Buffon , the lectures were continued with activity and zeal : not that Buffon himself attached the same importance to all the sciences, but because they were professed by able men, who made personal sacrifices for the prosperity of the establishment.

Antony de Jussieu subserved its interests not merely by his lectures , but by sending young men at his own expence to travel through the provinces, to collect seeds and plants. He formed a library of natural history and a considerable herbarium, which were of eminent service to his illustrious brother and nephew, and which have been always as much at the disposal of those who cultivate the science, as if they belonged to the establishment; with this advantage, that desired explanations are never withheld by the courtesy of the possessors.

The chair of botany, vacant by his death in 1758, was filled by Lemonnier, a member of the academy of sciences, and physician in chief to the army in Germany.

Lemonnier, like his master Bernard de Jussieu, was attached by inclination to the science. In 1770, being appointed first physician to the king, and consequently obliged to reside at Versailles, he was replaced by the present professor, Antony-Laurence de Jussieu, the nephew of Bernard; but he still continued to employ his leisure in rearing and naturalising plants, and, when stript of place and fortune in his old age, by the revolution, found a source of consolation and enjoyment in this favourite pursuit.

Bernard de Jussieu had, at first, alone directed the cultivation; but the number of plants which required peculiar care augmenting every day, he instructed for that purpose a gardener named Bertamboise: this individual dying in 1745, Buffon

named to succeed him, J.-A. Thouin, gardener
at Stord near l'Ile-Adam, who distinguished
himself alike by his knowledge and his zeal,
and who was the head of a family, since become so respectable, by its services to the King's
Garden and other establishments.

At his death in 1768, Bernard de Jussieu made
choice of his son, Andrew Thouin, then twenty
years of age, as the person best fitted to repair
the loss. This young man, brought up in the garden, formed by the lessons of the greatest masters, and passionately fond of study, soon acquired that extensive knowledge which opened
his way to the academy of sciences, and procured him the title of professor of agriculture,
which he still adorns. By the united care of the
professor, the demonstrator, and the young gardener, the collection of living plants in a short
time became more numerous and interesting
than could have been hoped; as the intendant
solely occupied with the cabinet, which he had
in a manner created, concerned himself less
about the other parts of the establishment.

We shall presently see that the solicitations
of M. A.-L. de Jussieu, determined Buffon to direct his attention to botany, for which the establishment was originally formed; and when,
in 1772, he resolved on giving it all possible

extent and regularity , he commenced the execution of his plan with the botanical garden :
but before we speak of what was done after
1772 , we must trace the progress of chemistry
and anatomy from 1739 until that period.

Boulduc, demonstrator of chemistry, who had
succeeded his father in 1729 , died in 1742 ;
his chair was given to Rouelle, whose reputation was already established by private lectures,
and by memoirs, which two years after procured his election to the academy of sciences.

Rouelle was devotedly attached to chemistry ;
being bent upon discovery , he did not consider it as the art of operating singular trans-
formations , or of making pharmaceutic preparations , but as a means of penetrating the
secrets of nature , and the laws of the composition of bodies. Equally an enemy to the occult
qualities , vague explanations and obscure language of the ancient chemists, he attached himself to the doctrine of Stahl , which, though
resting on incomplete observations, had the advantage of combining facts , and referring secondary phenomena to general principles. This
magnificent theory had for some years given a
new direction to chemistry; but it was understood in France by a few men of science only, and
had still many adversaries. From the manner in

which it was professed by Rouelle, it soon became generally known : his method was first
to excite the attention by surprising experiments,
and then direct it to general views. His elocution was often incorrect and coarse , but
always animated and picturesque, and his enthusiasm communicated itself to his auditors. His
lectures , by exciting a taste for chemistry , diffused the ideas of Stahl , as those of Fourcroy
have since disseminated the principles of Lavoisier, and the new nomenclature.

When forced by ill health to retire in 1768,
he was replaced by his brother, who became
titular professor in 1770. His successor had not
the same talent for striking the imagination ,
but he was thoroughly acquainted with the practice , and rendered his lectures interesting by the
choice of experiments.

While Rouelle zealously inculcated the new
theory , Bourdelin continued to repeat the ideas
of Lemery; a most shocking discrepancy hence
resulted, between the doctrines of the professor
and those of the demonstrator; which was the
more apparent, as the latter declaimed with
vehemence against the old opinions. But Bourdelin, a man of correct mind and inoffensive character , and moreover wholly devoted to the
practice of medicine , was neither incensed at

the attacks of his colleague, nor jealous of his reputation; he acknowledged the superiority of the new theory, though he had not time to study it sufficiently for the purpose of instruction, and in 1770, when his advanced age obliged him to discontinue his lectures, he desired to be replaced by Macquer, a pupil of Rouelle.

During the first years of his residence in the garden, Buffon was unfortunate enough to lose several of the professors whom he had found there on his arrival. Sensible of their services, which ought never to be forgotten, he endeavoured to replace them by men capable of imparting new lustre to the establishment. Science having been deprived of Hunaud in the same year with Boulduc, his place was bestowed upon Winslow, the most celebrated anatomist in Europe.

Winslow had long lectured for Duverney, of whom he ought to have been the immediate successor, and though seventy four years of age, he resumed his functions with the zeal of youth. He was happy amid a circle of pupils, who heard him with the respect due to his age, and the admiration excited by his genius; but after eight years of laborious exertion, feeling the want of repose, he demanded an assistant, and the reversion of his place was given to Ferrein, already a member of the academy of sciences, and professor at the college of France.

Ferrein, become titular professor at the death of Winslow in 1760, filled the chair of anatomy with distinction : sometime before his death, he was assisted by M. Portal, then very young, but already favourably known by his private lectures. He died in 1769, and his chair was given to Antony Petit, who enjoyed a splendid reputation as a physician and anatomist, and especially as a teacher. This reputation was still heightened by the brilliancy of his lectures at the King's Garden, to which not only his medical pupils, but persons indifferent to the science, flocked in such crowds that the amphitheatre was unable to contain them. He possessed the art of giving interest to the driest subject, and those who heard him fifty years since, still retain the impression of his eloquence.

In 1743, the establishment lost the painter Aubriet, who had accompanied Tournefort in his travels, and who was the oldest inhabitant at the garden. His drawings on vellum are numerous, and though not equally finished with those of Robert, are very exact : during the last years of his life, he was aided by mademoiselle de Basseporte, who succeeded to his title and employment.

Buffon had now attained the meridian of his glory : his works, which assigned him the first rank amongst the authors of his time, had diffused an universal taste for natural history, while the collections he had formed, facilitated the study of this science. In foreign countries also, he enjoyed the highest reputation; and the authors of new observations or discoveries, eagerly communicated them to a man of genius, by whom to be mentioned was a sort of passport to immortality.

The moment was arrived for realising his long meditated projects, when, in 1771, he was attacked by an illness which caused the most serious apprehensions for his life. During his convalescence, he learned that the count d'Angiviller had obtained the reversion of his place of intendant of the King's Garden. Buffon was deeply wounded by this proceeding : M. d'Angiviller however succeeded in regaining his friendship : as his place of director of the king's buildings, and chief of the academies of painting and sculpture, required him to point out the great men whose statues were to be executed in marble at the public expence, he asked permission of the king to erect one to Buffon. This was the most flattering distinction that could be conferred on a living man, as it had

till then been reserved for the memory of those,
who had rendered the most eminent services
to their country. But the king , reading the
judgment of posterity on the merits of Buffon,
in that of his contemporaries, assented to the
proposal; and the celebrated Pajou was charged
with the execution. The statue was finished in
1776, and placed on the staircase leading to the
galleries : it is now in the library of the Museum.

The health of Buffon being perfectly reestablished in the beginning of 1772, he resolved to
fix his residence once more in the garden , and
to employ his influence for the benefit of the
establishment. With the aid of government, he
purchased two houses adjoining the Museum,
one of which he destined for the dwelling of
the intendant, and removed into it accordingly :
the first floor was appropriated to his household,
and the others, to such objects as had not yet
found their place in the Museum.

The return of Buffon forms an epoch in the
history of the garden : from that moment , every
branch of the establishment rapidly increased ,
and the way was prepared for the improvements which have taken place since the new
organisation. To give an idea of his services during the sixteen years of his administration , we

must consider them under three points of view; the buildings, the collections, and the instruction.

The botanical garden was in the same state in which it had been left by Tournefort. The ground appropriated to the scientific arrangement of plants was so inadequate to the object, that it became necessary to cultivate them wherever a vacant spot could be found; and as they were thus distributed without regard to classification or natural affinities, the professor was frequently obliged to go from one extremity of the garden to the other, to finish his lecture : the soil also was exhausted, and the delicate plants could be preserved only by the utmost care.

Buffon, yielding to the reiterated instances of M. de Jussieu, exposed the necessities of the institution to the minister, the duke de la Vrilliere; and obtained, in 1773, the sum of 36,000 francs, for the formation of a new scientific garden. The ground was prepared, and the plants, taken up in the autumn with suitable precautions, were transplanted at the end of winter. M. de Jussieu took this occasion to dispose them according to the new method, of which his uncle Bernard had conceived the idea, fifteen years before, in arranging the garden at Trianon. The nomenclature of Linnœus was substituted for that of Tournefort, as being more convenient, and already generally received throughout Europe; and the botanical lectures were from that moment regularly given. The botanic garden and orangery were enclosed with an iron railing. A building, begun at the corner of the orangery for a green-house, of which the construction did not answer the desired purpose, was demolished, and only the cellars and such portions of the walls preserved as were necessary to support the terrace of the upper green-house; under which convenient shelters were formed for delicate plants. With the ruins of this building and the bad soil from the botanic garden, a gentle slope was formed from the lower alleys to the rising ground, ornamented on each side by rows of dwarf elms and an iron balustrade.

Some years after, Buffon undertook to prolong the garden, and to double its extent, by the addition of the land which separated it from the *rue de Seine*, and by demolishing the houses near the residence of the intendant and the great southern alley.

This space, occupied by plants of domestic use, belonged almost exclusively to the monks of Saint-Victor's Abbey : some woodyards near the quay, were municipal property, let by the city to timber merchants. The municipal administration were ready to give up such parts of the ground as pertained to the city; but the monks could not sell the property of the abbey : Buffon laboured to surmount this obstacle, and succeeded in the following manner.

Between the garden, the boulevard, and the *rue Poliveau*, was a vast enclosure, traversed by the rivulet *Bièvre*, and belonging, with the buildings and improvements, to a single individual : Buffon bought it in 1779, for the sum of 142,000 francs. He then proposed to the monks of Saint-Victor, to exchange their lands for others of the same value, forming a part of his recent purchase : being assured of their consent, he obtained for them the necessary authorisation from the king.

When in possession of the tract to be annexed to the garden, he easily procured the purchase of it by the government; and after indemnifying the former residents, he began his operations in 1782. In the first place he demolished the buildings which interfered with the regularity of his plan , and laid out a street, parallel to the great alley which bounded the garden on the south. The inhabitants of that quarter gave it the name of *rue de Buffon*, which was confirmed by the municipal body. A wall was raised round that part of the garden which was not on a level with the street, and the remainder was enclosed by an iron railing.

The limits of the garden being thus fixed, he continued the principal alley to the quay, and formed two others parallel, along the walls which formed its two sides. That on the south, which borders the *rue de Buffon*, was left without trees, from the fear of throwing too much shade on the cultivated ground; the corresponding one, leading from the foot of the small hill to the quay, was planted with horse-chesnuts. The terrace wall, which bordered the establishment on the east, was pulled down, and the botanic garden and nursery were both enlarged. Between the two principal alleys a large square basin was sunk to the level of the river, and garnished on the sides with various shrubs. A parterre, appropriated to the most interesting plants that flourish in our climate, occupied the remainder of the ground towards the river.

On each side of the garden, four large squares were enclosed by a trellis; of which, that next the *rue de Buffon* was planted in quincunx, with trees of the four seasons; those adjoining the botanic garden were destined, the first, for an assemblage of fruit trees, and the second for the sowing of economical plants; and the others, which are now a *school of agriculture*. for a short

time supplied the place of a nursery. The transverse alleys which separate them, were planted with trees of different species.

All these operations were confided to the superintendance of M. Thouin, and finished in 1784.

Still the enclosure extended, on the north,
only to the botanic garden and the adjacent hill
planted with ever-greens. The ground opposite the scientific garden, at the entrance of the
alley of horse-chesnuts, belonged to a private
company, who destined it for building; but Buffon determined the government to purchase it.
As it was below the adjacent level, bordered
by terraces, and thus sheltered towards the north
and west, the seed-beds were removed thither,
as well as those plants which required peculiar
care ; and a subterranean passage was formed,
under the avenue, to the botanic garden. These
changes being effected in 1786, the terrace on
which the seed-beds had been placed in 1774,
was destined for a large hot-house; which was
built two years after, and which bears the name
of Buffon.

The garden had been doubled in extent by the
late acquisitions, the distribution of it was regular and beautiful, and every possible advantage was offered for the culture and study of
vegetables; but the perfection of one part of

the establishment only rendered more sensible the deficiencies of the rest. The cabinet was not spacious enough to contain the vast accession of objects, and the amphitheatre was both too small , and in other respects inconvenient. As every part of the garden had already an appropriate destination, farther additions became necessary; and these were made in 1787 , when the king purchased, and annexed to the establishment, the hotel de Magny, with its courts and gardens, situated between the hill of ever greens and the *rue de Seine.*

On this ground Buffon constructed the amphitheatre, which now serves for the lectures of botany and chemistry, and removed the lodging of MM. Daubenton and Lacépède to the hotel de Magny. The second floor of the cabinet, which was thus left vacant, was fitted up for the reception of the collections , and permission obtained from the government to erect an addition to the former galleries : the work was immediately begun, and continued without intermission; but it was not completed till after the death of Buffon.

As the buildings became more extensive and the objects were disposed in a more striking manner , more value was attached to the collections, and the celebrity of the establishment increased. Individuals offered specimens to the cabinet, where they were seen inscribed with the name of the donor, in preference to retaining them at home; learned societies eagerly contributed to the progress of knowledge, by enriching a public deposit; and sovereigns, as an agreeable present to the king, sent to his Museum duplicates of the curiosities in their own. The academy of sciences, for instance, having acquired Hunaud's anatomical collection, added it to that of Duverney in the garden; the count d'Angiviller gave Buffon his private cabinet; the missionaries in China, sent him whatever interesting objects they could procure in a country where they alone could penetrate; the king of Poland presented a very considerable collection of minerals; and the empress of Russia, not being able to induce Buffon to visit Saint-Petersburgh, invited his son, and on his return presented him with several animals from the north, which were wanting to the cabinet, and with various objects of natural history collected in her dominions.

The government neglected nothing for the perfecting of an establishment which did honour to the nation, as a repository of light and a centre of communication. More considerable funds than had before been granted, were placed

at the disposal of M. Daubenton, for the purchase of objects, interesting from their rarity, or their utility to science; foreign trees
were transplanted ; the cabinet of zoology
was enriched by the collection of Sonnerat in
India, by that of Commerson, made in Bougainville's voyage round the world, and by a part of
that brought by Dombey from Peru and Chili,
of which half the objects were detained by the
Spanish government, who even prevented the
publication of his narrative : commissions of correspondence, accompanied by a salary, were also
given to learned travellers, who engaged to collect objects for the botanical garden and the
cabinet.

Nevertheless, it must be owned that all these
collections were not at that moment of much utility, and that it is only at a later period, and
since the new organisation of the establishment, that their importance has been felt, and
their end attained. Buffon was not a friend to
method : he described the exterior form, the
habits and economy and animals, and soared
to the most elevated general views; but he dis-
liked the labour of distinguishing characters,
and settling principles of classification. In the
arrangement of the cabinet, he wished to excite
curiosity by striking contrasts, so that, like his

own writings, it should present a picture of the most remarkable things in nature, independant of system, which he regarded as the artifice of man. This manner of considering natural history was particularly pleasing, to a mind that delighted in contemplating the universe of things as a whole : and indeed, in nature where all is harmony, the most different beings are placed side by side, and the imagination seizes at once the links which unite, and the characters which separate them.

According to Buffon, the end of a general collection was attained, when it captivated the attention, and led the beholder to seek in living nature what was thus imperfectly represented; it was even deemed a useful exercise to separate what related to a peculiar study, from the crowd of objects that surrounded it.

We do not pretend to defend this view, which would not be at all admissible, now that we have objects enough to form a series and room for their exhibition, and that each branch of natural history is studied as a separate science. One of its necessary consequences was the neglect of whatever was not calculated to interest the public; when a collection arrived the most remarkable objects were selected to fill the empty spaces, and the rest were preserved in boxes.

Daubenton had prepared the skeletons of such quadrupeds as he could procure, which were crowded into a narrow space, and concealed from view. To remedy this inconvenience, more extensive galleries were necessary ; and as the Museum united the productions of the three kingdoms, it was indispensable that each collection should be arranged by a professor of the science which it was intended to illustrate, so as to accord with his own lectures. As there was at that period no professor of zoology nor of mineralogy, the botanical garden was the only part of the establishment methodically distributed throughout.

Far from reproaching Buffon with not having effected what it was perhaps impossible, at that time, to perform, we should acknowledge our obligations to him, for having assembled not only the numerous collection of birds contained in his work, and that of fishes described by M. de Lacépède, but also a multitude of objects of all kinds, which have since been properly arranged, and have eminently contributed to the progress of natural history.

In 1784, Daubenton the younger being obliged by bad health to resign his place of keeper and demonstrator of the cabinet, Buffon appointed as his successor M. de Lacépède : who was thus fixed in the pursuit of natural history, in which he has since made so eminent a figure both as a professor and an author.

The connections of the King's Garden with learned societies, with travellers, and with the naturalists of foreign countries, requiring a continual correspondence, for which the intendant and keeper of the cabinet had not time, and which they could not confide to ordinary secretaries, Buffon obtained the creation of a new place of assistant keeper of the cabinet, to which the correspondence was attached. To this place he appointed his friend Faujas de Saint-Fond, who, by his extraordinary activity and varied knowledge, was of all men the best calculated to fill it. M. Faujas did not limit himself to epistolary intercourse, but made several journeys, and came back laden with treasures for the Museum, collected by himself, or obtained by solicitations in the name of Buffon.

Mademoiselle Basseporte, during thirty years, had added to the collection of drawings on vellum such objects as were pointed out to her by Bernard de Jussieu: her zeal was unabated, but her talent, which had never equalled that of Aubriet, was enfeebled by age. Buffon, desirous of attaching to the garden the ablest artist in this line, obtained the reversion of her place for

M. Van Spaendonck, who immediately assumed
its active duties, and succeeded Mademoiselle
Basseporte in 1780. This choice had the unfore-
seen and happy consequence of founding a chair
of *iconography* in the Museum.

The garden and cabinet being open to the public, it became necessary to maintain an exact
police, and the requisite authority for this purpose
was vested in an inspector, with a body of guards
at his disposal, and a salary of 4000 fr; but the
place has since been suppressed, and the service
of the Museum confided to a company of invalids.

Thus, in the space of sixteen years, Buffon accomplished the extensive plans of improvement
which he had conceived on fixing his residence
in the garden : it now remains for us to trace the
progress of instruction, which depended not di-
rectly upon him, but which was principally
owing to his judicious choice of professors.

There were chairs at that time for botany,
chemistry, and anatomy only ; but as Daubenton
and his assistant repaired daily to the cabinet,
naturalists were enabled to obtain explanations
of the objects before them; and these private
lessons were the more useful, as they were
adapted to the capacity and knowledge of the
hearer.

Lemonnier had been professor of botany since

1758, and Bernard de Jussieu demonstrator, since 1722; but the former being obliged to reside at Versailles, and the latter finding himself weakened through age, M. de Jussieu, his nephew, was chosen to supply the places of both, and was thus charged with the lectures in the garden, and the botanical excursions in the country. During the last years of his life, Bernard de Jussieu intrusted the details of cultivation wholly to M. Andrew Thouin: it was a signal satisfaction to him to witness the replanting of the botanic garden. When he walked in the establishment, his former pupils crowded around him, listening to him with eagerness, and treasuring up with veneration his slightest words. Among his services to the garden must be reckoned the education of his nephew, who has made of botany a regular science, by developing and perfecting the natural method.

Bernard de Jussieu terminated his honourable career in 1777, and was succeeded by his nephew. He has written only a few memoirs, inserted in the transactions of the academy of sciences, of which he was made a member at the age of twenty six years. More occupied with the progress of science than his own reputation, he imparted his discoveries without reserve to his pupils, and by his liberality in com-

municating knowledge, as well as by the introduction of his method, he has acquired a lasting
title to the admiration and gratitude of naturalists.

Though Lemonnier was fully impressed with
the ability of M. de Jussieu, his interest in the
establishment, led him to wish that the two
places of professor and demonstrator should be
kept distinct; he therefore resolved on retiring
with the title of honorary professor, as soon as he
could find a second person adapted to his purpose.
With this view he had cast his eyes on M. Desfontaines, whose merit he appreciated, and whose
future reputation he foresaw; but as he was not at
that time sufficiently known, Lemonnier thought
better to defer his project, and it was not until
1786 that he took steps for its accomplishment.
M. Desfontaines had then just returned from his
travels in Barbary, with the plants of which he
has since published the history; he had also attracted the attention of botanists by several excellent memoirs, and had been admitted into the
academy of sciences. Lemonnier proposed him to
Buffon, who approved the choice. M. Desfontaines
wished to take the place of demonstrator, and
leave that of professor to M. de Jussieu, but the
latter preferred retaining the functions exercised
by his uncle during fifty five years.

At the appointment of M. Desfontaines, the bo-

lanic garden was already very rich : the instruction was no longer limited to the demonstration
of medicinal plants, and the progress of the
science since Tournefort, by the labours of
Linnœus, Adanson, and de Jussieu, authorised
and required a more philosophic plan. M. Desfontaines was the first to perceive the importance of a general knowledge of the nature of
vegetables, the functions peculiar to each organ, and the phenomena of the different periods
of their developement, in order duly to understand their generic and specific characters : he
therefore divided his course into two parts; the
first devoted to the anatomy and physiology of
vegetables ; the second, to the classification and
description of the genera and species. From that
period botanical instruction was no longer confined to the exterior forms of plants, but comprised their affinities, uses, and modifications. To
the method of teaching adopted in the King's
Garden since 1788, are to be ascribed those
works which have made of vegetable physiology
the basis of botany, and led to the applications
of this science in agriculture and the arts.

M. Desfontaines delivered his lectures for some
years in the botanic garden, but as the number
of pupils increased, he assembled them in the amphitheatre, where the necessary specimens for

illustration were exhibited, the remainder of the
family under consideration being exposed in the
garden.

Whilst M. Desfontaines thus taught the whole
of the science, by describing successively all the families , M. de Jussieu made a botanical excursion
every week into the country. In these walks, besides naming the species that offered themselves
as he passed, he took every opportunity of remarking the characters of the different groups
of vegetables, and thus exercising his pupils in the
application of the principles which he has since
made public, in a work that became classical immediately on its appearence.

The method adopted in the King's Garden.
and since followed in the different schools of
France and in foreign countries, having shewn
the order and dependancy of the vegetable kingdom, the prodigious number of plants daily discovered, no longer present themselves as insulated beings, but as parts of a general series.

Chemistry was still confided to two professors, the first of whom exposed the theory and
the second performed the experiments. This separation ought never to have taken place ; no
farther at least than it coincided with that since
adopted, by which the task of exposing the principles of the science, and the experiments neces-

sary to illustrate them, is assigned to one professor, and its applications in the arts to another.

Macquer, who had filled the place of Bourdelin since 1770, and enjoyed the title since
1777, differed entirely from his predecessor:
he resumed the theory of Becker and Stahl,
which Rouelle had first rendered popular, and
it is evident from his Chemical Dictionary, that a
man of so just and clear an understanding must
have treated his subject with superior method
and exactness. He did not aim at striking the imagination of his hearers, but endeavoured to fix
their attention on important phenomena, and to
lead them, by their own inferences, to the explanation. Though a partisan of the doctrine of Stahl, as
well as the demonstrators, he was struck with
the discoveries of Lavoisier; he even anticipated
the changes they must produce, and, far from rejecting their consequences, prepared his pupils
to receive them. At his death, in 1784, Buffon
named to succeed him M. Fourcroy, who had not
attained his thirtieth year, and was not even a
member of the academy, but who had acquired
a name by his lectures, as the substitute of professor Bucquet.

From Fourcroy's entrance into the garden must
be dated the propagation of the new theory,
which has changed the face of the science.

The discoveries of Cavendish on the decomposition of water, had rendered necessary a new nomenclature, distinguishing simple from compound bodies, and indicating the elements of the latter. This project, conceived by Guiton-Morveaux, was adopted by Lavoisier and other chemists, among whom was Fourcroy, and executed with an exactness of method, of which the sciences offered no example. But though this nomenclature, in which every word was the expression of a fact, greatly facilitated the labour of becoming acquainted with the existing and possible combinations of matter, it was still a new language; which of all innovations is the most difficult to introduce in practice. The advantages of the new nomenclature were too obvious to be denied; but if it had been learned only in books, it would never perhaps have become universal. Fourcroy adopted it in his lectures, and as his eloquence, in which the clearest ideas were conveyed in the most precise and harmonious diction, captivated the attention of his hearers, his expressions remained imprinted on the memory, and chemistry was no longer spoken of but in the terms he employed. The same language being used by the demonstrator, soon became familiar; and to the lectures in the King's Garden is truly owing the rapid progress of chemistry.

Rouelle inflamed the imagination of his pupils by surprising phenomena; but he had not the skill of classing facts, and establishing principles: Fourcroy excited the same enthusiasm, and directed the attention of a multitude of able men to chemical studies, by exposing the discoveries of his contemporaries, by proclaiming the names of their authors, and announcing their consequent results. He acquired a reputation, which soon extended beyond the land which gave him birth: foreign princes maintained young men at Paris, to attend his lectures, and to enrich their native country with his doctrine. Men of taste listened to his eloquence, and became devoted to a science presented under such fascinating forms. The amphitheatre could not contain the crowd who resorted to it, and it twice became necessary to enlarge it. The language he taught has been enriched and perfected, as the discovery of a greater number of simple substances has modified the names of their different compounds: but its principles are the same. The taste which he diffused for the science has increased; several distinguished chemists have been formed in his school; and the professors who now occupy the two chairs, are his pupils.

Rouelle the younger, who had been titular démonstrator since 1770, dying in 1779, his place was given to M. Brongniart, professor at the school of pharmacy, author of an Essay on Chemistry, and one of the writers in the Journal of Sciences, Arts and Trades. He at first repeated the language of Macquer, but soon adopted the new nomenclature.

Antony Petit maintained the reputation which Duverney and Winslow had given to the chair of anatomy, and saw the number of his pupils every day increase ; but he at length became less exact in delivering his lectures, and this negligence excited the louder murmurs from the universal eagerness to hear him. He was therefore assisted by Vic-d'Azir, a man in every respect worthy to fill his place, and to whom he wished it should be given at his death ; but Buffon thought it due to M. Portal, who, ten years before, had lectured for Ferrein. In 1778, M. Portal was named to the chair, which he still occupies ; uniting this place to that of first physician to the king.

Though Vic-d'Air assisted Petit only two years, he conferred a lasting benefit upon science by diffusing just notions on comparative anatomy, for which a separate chair has since been erected in the Museum, and which has become the basis of zoology.

This sketch of what was done for the establishment from 1772 to 1778, is only a recital of its obligations to Buffon. That great man conceived at the same time the plan of his Natural History, and the enlargement, or rather the creation, of the Museum. His project, which he had meditated for several years, was to form a complete whole susceptible of improvement in all its parts. It was not until he had ensured the means of realizing it, that he commenced its execution: from that time, no obstacle could arrest his progress, and no labour exhaust his patience. His enterprise was crowned with the most brilliant success, and, at the close of his life, he had every reason to congratulate himself on the sacrifices it had cost. The King's Garden and Museum were justly cited as his work, and at the same time as the most splendid institution ever formed for the progress of natural history, and the point of union for all the students of nature. If the French gloried in so beautiful a monument, foreigners admired it without jealousy, because it was equally useful to the scientific men of every nation.

The establishment however was far from being what it became a few years after, and the collections of that period would appear insignificant, if compared with their present splendor; but the plan was formed, the impulse given, and every thing prepared for further progress. If this progress has taken place only since the death of Buffon, and since several men of learning, animated with the same zeal, have been charged with separate branches of the administration, it must be avowed to his glory, that without him the new organisation would not have become necessary, nor probably the Museum of natural history have been at this time in existence.

THIRD AND LAST PERIOD.

FROM THE DEATH OF BUFFON TO THE PRESENT TIME,
1788 TO 1821.

NEW ORGANISATION.

At the death of Buffon, which happened on the 16th. of April 1788, the place of intendant of the garden, was given, not to the count d'Angiviller who had obtained the reversion, but to his brother, the marquis de la Billarderie. That nobleman carried on the works begun by Buffon, and constructed the small hot-house of the *ficoideæ*. It was he who attached to the institution the chevalier de la Marck, already celebrated for his french Flora, by an appointment of botanist of the cabinet, with the charge of the herbariums. His conciliating character rendered him dear to the professors, and he neglected nothing to promote the interest, and maintain the order of the institution; but his credit was small, compared with that of his predecessor. As the systematic reduction of the public expenses was already begun at court, he could not with the least prospect of success demand extraordinary funds; and the establishment had become too extensive for the form of administration in use in the time of Buffon.

On the 20th of August 1790, M. Lebrun made a report in the name of the committee of finances of the constitutant assembly, on the state of the King's Garden, in which its expenses were estimated at 92,222 francs; 12,777 francs being for necessary repairs. This report, which was the signal for a new organisation, was followed by the draught of a decree proposing the reduction of the intendant's salary from 12,000 to 8,000 francs; the suppression of several places, particularly that of commandant of the police of the garden; an increased stipend to some of the professors; the creation of a chair of natural history, etc., etc.

During the discussion, the president of the assembly received an address from the officers of the King's Garden — for thus the keepers and demonstrators of the cabinet, the professors, and the painter were denominated; the reasonable views of which being approved by the assembly, it was submitted to the comittee of finances, and the definitive report adjourned, in order to receive a plan of organisation for the institution. In consequence of this decision, M. de la Billarderie convoked the officers, who, naming M. Daubenton president and M. de Lacépède secretary, drew up a report, which was signed by the persons then in office, and by MM. Petit and Lemonnier, as honorary professors.

The regulations proposed were very similar to those adopted three years after. During that interval, the officers ceased to assemble, or to occupy themselves about the administration; but continued to discharge their professional duties with so much zeal, that several of them delivered twice the number of lectures prescribed, and, after the public lessons, gave private instructions to such pupils as thought proper to consult them. This was all that in existing circumstances could be done, to keep alive the ardour for science, and to aid the progress of the student.

The disorders of the revolution beginning at this period, M. de la Billarderie withdrew from France, and his place of intendant was filled by the appointment of M. de Saint-Pierre in 1792.

M. de Saint-Pierre undertook the direction of the King's Garden at a difficult conjuncture. That distinguished writer was gifted with eminent talents as a painter of nature, and a master of the milder affections ; he knew how at once to awaken both the heart and the imagination ;

but he wanted exact notions in science, and his timid and melancholy character deprived him of that knowledge of the world and that energy of purpose, which are alike requisite for the exertion of authority. Nevertheless, he was precisely the man for the crisis : his quiet and retired life shielded him from persecution, and his prudence was a safeguard to the establishment. He busied himself with the details of the garden, and was careful not to take any step nor to make any proposition, without consulting its principal inhabitants. He presented several memoirs to the ministry containing some very sound regulations, concieved in a spirit of economy which circumstances rendered necessary. In each of these memoirs, of which we have a copy, we notice the following words : After consulting the elders ; by which term he designated the persons who had been long attached to the establishment, though without an official share in its administration.

By retrenching every needless expenditure, he provided funds for objects of avowed utility; and built a hot-house, which now bears his name, in continuation of that of the Peruvian cactus, behind the labyrinth, and adjoining the terrace of the *rue du Jardin-du-Roi.*

The menagerie at Versailles being abandoned and the animals likely to perish of hunger, M. Couturier, intendant of the king's domains in that city, offered them by order of the minister to M. de Saint-Pierre : but as he had neither convenient places for their reception, nor means of providing for their subsistence, he prevailed on M. Couturier to keep them, and immediately addressed a memoir to the government, on the importance of establishing a menagerie in the garden. This address had the desired effect, and proper measures were ordered to be taken for the preservation of the animals, and their removal to the Museum ; which however was deferred till eighteen months after.

A decree of the legislative assembly, of the 18th of August 1792, having suppressed the universities, the faculties of medecine, etc., there was reason to fear that the King's Garden would be involved in the proscription ; but as it was considered as national property, and as visitors of all classes were equally well received ; as the people believed it to be destined for the culture of medicinal plants, and the laboratory of chemistry to be a manufactory of saltpetre, it was respected.

Still a disorderly faction, rendered formidable by its triumph of the 31st of May, threatened every vestige of the monarchy. An institution of which the officers had been appointed by the king, was naturally the object of its jealousy. The peril was imminent: and it would have been impossible to escape it, if there had not been found in the convention some men of courage who saw the tendency of these measures, and sought to arrest their progress. Among them must be particularly distinguished M. Lakanal, president of the committee of public instruction, who, when informed of the danger, repaired secretly to the garden to confer what MM. Daubenton, Thouin and Desfontaines, on the means of averting it. He demanded of them a copy of the regulations that had been submitted to the constituant assembly; and the next day, the 10th of June 1793, obtained a decree for the organisation of the establishment, of which we shall cite the most essential articles.

« The establishment shall henceforth be called » the *Museum of Natural History.*

» Its object shall be the teaching of natural his-» tory in all its branches.

» All the officers of the Museum shall have the » title of professor, and enjoy the same rights.

» The place of intendant shall be suppressed, » and the salary equally divided amongst the pro-» fessors.

» The professors shall choose a director and a » treasurer every year, from among themselves.

» The director shall preside in the assemblies of
» the officers, and be charged with the execution
» of their deliberations: the same person shall not
» be continued in office more than two years in
» succession.

» The vacancies in their own body shall be fill-
» ed by the professors (1).

» Twelve courses of lectures shall be given in
» the Museum.

» 1. A course of mineralogy;
» 2.   of general chemistry;
» 3.   of chemistry applied to the arts;
» 4.   of botany;
» 5.   of rural botany;
» 6.   of agriculture;
» 7 and 8. two courses of zoology;
» 9. a course of human anatomy;
» 10.   of comparative anatomy;
» 11.   of geology;
» 12.   of *iconography*.

» The subjects to be treated of in the courses,
» and the details relative to the organisation of
» the Museum, shall be specified in a regulation to
» be drawn up by the professors, and communi-
» cated to the committee of public instruction. »

(1) This article was abrogated by a law of May 1802: at present, the
body of professors and the academy of sciences each name a candi-
date, for the acceptance of the king; but the voice of the professors is
usually seconded by the academy.

The third section provides for the formation
of a library, where all the books on natural history in the public repositories, and the duplicates
of those in the national library, shall be assembled; and also the drawings of plants and animals, taken from nature in the Museum. The
fourth clause insists on a correspondance with all
similar institutions in France.

By this decree, twelve chairs were established, without naming the professors; the distribution of their functions being left to the officers themselves. These officers were :

MM. Daubenton, keeper of the cabinet, and
professor of mineralogy at the college of France;

Fourcroy, professor of chemistry ;

Brongniart, demonstrator ;

Desfontaines, professor of botany;

De Jussieu, demonstrator ;

Portal, professor of anatomy;

Mertrud, demonstrator;

Lamarck, botanist of the cabinet, with the
care of the herbarium;

Faujas Saint-Fond, assistant keeper of the cabinet, and corresponding secretary;

Geoffroy, sub - demonstrator of the cabinet ;

Vanspaendonck, painter ;

Thouin, first gardener.

No difficulty occurred respecting those officers,

who were already professors or demonstrators : but MM. Faujas and Lamarck were otherwise situated : the correspondance thenceforth pertaining to the assembly, and the herbarium being placed under the direction of the professor of botany, they were left without employment. In this embarrassment, M. Faujas, well known by his travels and his beautiful work on the volcanoes of the Vivarais, was made professor of geology ; and M. de Lamarck, equally versed in zoology and botany, and reputed the best conchyologist in France, was appointed to teach the history of invertebrated animals. The determinate direction thus given to his labours, has procured the scientific world some distinguished works.

The administration was aware of the importance of dividing the zoological instruction into three parts ; but as M. de Lacépède had a few months before resigned the place of sub-demonstrator and keeper of the cabinet, the third chair, to which he had unquestionable claims, was left vacant, in the hope that, at a more favorable moment, he would be called to fill it: which accordingly took place. In the mean time, M. Geoffroy, who had succeeded him in the cabinet, undertook alone to teach the history of quadrupeds, birds, fishes, and reptiles.

On the 9th of July 1793, the professors having

received notice of this decree, met, and appointed M. Daubenton president, M. Desfontaines secretary, and M. Thouin treasurer. From that
time, they assembled on stated days, and
planned the supplementary regulations enjoined
by the legislative body.

Their first care was to obtain the creation of
certain places, which the recent changes rendered necessary.

The general administration of the cabinet belonged to the assembly, and the care of the collections, to the several professors; the places cf
keeper and assistant keepers of the cabinet, were
therefore suppressed. But as it was necessary to
have some person charged with the key of the
galleries, the preservation of the objects, and the
reception of visitors, these duties were devolved
on M. Lucas, who had passed his life in the establishment, and enjoyed the confidence of Buffon.

M. Andrew Thouin, being made professor of
agriculture, M. John Thouin was appointed first
gardener. Four places of assistant naturalist (1),
were created, for the arrangement and preparation of objects under the direction of the professors, and three painters were attached to the Museum, viz : M. Marechal and the brothers Henry

---

(1) Filled by MM. Desmoulins, Dufresne, Valenciennes and Deleuze ; the two first for zoology, the others for mineralogy and botany.

and Joseph Redouté. These regulations and appointments were approved by the government.

At the same time the library was disposed for the reception of the books and drawings; which last were contained in sixty-four portfolios.

In 1794, M. Toscan was appointed librarian, and M. Mordant Delaunay adjunct; and the library was opened to the public on the 7th of September 1794.

The animals from the menagerie at Versailles, those from Rincy, and others belonging to individuals who made a trade of exhibiting them for gain, having been removed to the Museum in 1794, dens were formed under the galleries of the cabinet for those which it was necessary to confine, and the others were placed in stables, or among the forest-trees along the *rue de Buffon*. In the mean time, a small building at the extremity of the avenue of horse chesnuts was arranged as a temporary menagerie for ferocious beasts.

The house of the intendant was disposed for the lodging of two professors; the saloons of the cabinet were more perfectly arranged ; and it was decided that new galleries should be constructed on the second floor : in fine a decision of the committee of public instruction of September 1794, ordered the acquisition of the house and lands adjoining the Museum on the north-west;

which had already been deemed necessary by the constitutant assembly.

The report of the committee of public instruction, approved the regulations of the professors, and fixed the organisation of the Museum in its present form, with the exception of slight modifications exacted by the change of circumstances. A law in conformity, of the 11th, of December 1794, created a third chair of zoology, to which M. de Lacépède was appointed ; gave the whole administration of the establishment to the professors; increased their salary from 2,800 to 5,000 francs; fixed the expences of the following year at 194,000 fr. ; and ordained that the land between the *rue Poliveau*, the *rue de Seine*, the river, the *boulevard*, and the *rue Saint-Victor*, should be united to the Museum. A still more vast but impracticable plan had been presented, which was withdrawn at the solicitation of the professors.

The aspect of the ground, and the object of the establishment both indicate the fitness of these limits; but as the opportunity had been neglected of procuring the lands by exchange, it was only by degrees that the portion was obtained which forms the new menagerie : several spots are still in the hands of individuals, and are an obstacle to the completion and regularity of the plan.

The wretchedness of the times was now sen-
sibly felt ; the reduced state of the finances, the
depreciation of the funds, the cessation of foreign
commerce, and the employment of every species
of revenue and industry for the prosecution of
the war, were serious hindrances to the project-
ed improvements.

And indeed, not only during the first years of
terror and destruction, but from 1795 to the end
of the century, the establishment presented astonishing contrasts. Houses and lands of great value
were annexed to the garden, magnificent collections were acquired, and the most useful buildings were commenced; yet every thing languish-
ed within : much was undertaken, and nothing
completed. Funds were wanting to pay the
workmen, to provide nourishment for the animals, and to defray the expence of the collections. Potatoes were cultivated in the beds destined for the rarest plants, and the establishment
was threatened with a decay the more irre-
parable as it affected all its parts. One obstacle
being surmounted others started up : the funds
received were bestowed on the object of most
immediate necessity, and others scarcely less important were neglected. However, when the
public distress had attained its utmost height, not
a moment of discouragement was felt by the ad-

ministrators; they deliberated on the best means
of meeting the exigency, and made themselves
respected by an exemple of zeal, moderation and
disinterestedness. Some of them being called to
employments connected with the government,
used their influence in favour of the establishment to which they were more particularly attached.

At the end of the year 1794, the amphitheatre
was finished in its present state, and in it was
opened, on the 25th of January 1795, the *Normal*
school; an extraordinary institution, but founded
on an unfeasable and visionary plan.

It was fancied that men already ripe in years,
by a few lectures from eminent masters, might be
rendered capable of extending instruction, and
diffusing through the provinces the elements of
sciences, which very few of them had been prepared by previous education to understand. Every
reasonable man, felt the impossibility of realizing
such a scheme, and the institution fell of itself
soon after. It had the effect of exciting the public
attention, and fixing it upon an establishment,
become as it were the paramount of all institutions that might be formed for the study of
nature.

From its re-organisation to the close of the
century, the Museum received a considerable in-

crease of land, and large additions to the collections; of which all possible advantage was subsequently taken. We have remarked that in 1788,
it was extended to the *rue de Seine*, by the acquisition of the hotel de Magny ; the houses and
gardens towards the *labyrinth*, on the west, the
property of an individual named Leger, were
added in virtue of a resolution of the committee of
finance of the 10th of June 1795. The principal
building was reserved for the administration and a
part of the collections ; the others, which before
the revolution had been occupied by a religious
community called the New-Converts, were assigned for the lodgings of the professors. Towards
the east, a vast tenement belonging to the nation,
which had served for the administration of hackney coaches, and had since been converted into
a warehouse for flour, was annexed by a law of
August in the same year. The buildings and courts
were destined for a green-house and a cabinet
of anatomy, which were begun after a plan of
M. Molinos, architect of the Museum: but the
work proceeded slowly from the causes already
mentioned : the green-house was not finished
till five years after, and the gallery of anatomy
not before 1817.

It had long since been decreed that the land
situated to the east and north of the green-house

should be added to the menagerie; but it was necessary to purchase it an expence which appeared too considerable. However some wood yards were acquired, and served for the first parks of the ruminating animals. The two elephants from the menagerie of the statdholder, which arrived in April 1798, were lodged in a stable and yard of the newly acquired buildings.

The fitting up of the galleries on the second floor of the cabinet was continued, but delays in the procuring of glass plates and in regulating the admission of light, retarded the display of the zoological collections.

All the funds that could be disposed of were meanwhile employed in the construction of an additional hot-house, for the plants which captain Baudin was bringing from America. As his voyage was undertaken solely on account of the Museum, it deserves in this place a more particular mention.

In 1796, captain Baudin informed the officers of the Museum that, during a long residence in Trinidad, he had formed a rich collection of natural history, which he was unable to bring away, but which he would return in quest of, if they would procure him a vessel. The proposition was acceded to by the government, with the injunction that captain Baudin should take with him four naturalists : the persons appointed to accompany him, were, Maugé and Levillain for zoology, Ledru for botany, and Riedley, gardener of the Museum, a man of active and indefatigable zeal.

Captain Baudin weighed anchor from Havre on the 30th of september 1796. He was wrecked off the Canary isles, but was furnished with another vessel by the Spanish government, and shaped his course towards Trinidad. That Island having fallen into the hands of the English, so that it was impossible to land, he repaired to Saint-Thomas; and thence, in a larger vessel, to Porto-Rico. Having remained about a year in those two islands, he returned to Europe, and entered the port of Fecamp the 12th of June 1798. His collections, forwarded by the Seine, arrived at the Museum on the 12th of July following.

Never had so great a number of living plants, and especially of trees, from the West Indies, been received at once : there were one hundred large tubs, several of which contained stocks from six to ten feet high. They had been so skilfully taken care of during the passage that they arrived in full vegetation, and succeeded perfectly in the hot-houses.

The result of this voyage was not confined to augmenting the store of living plants, but added greatly to the riches of the cabinets. The herbarium was increased by a vast number of specimens carefully gathered and dried by MM. Ledou and Riedley. Riedley had besides made a collection of all the different kinds of wood of Saint-Thomas and Porto-Rico, with numbers affixed referring to the flower in the herbarium; which enabled the professor of botany to determine the species of the tree. The two zoologists brought back a numerous collection of quadrupeds, birds and insects. That of birds made by Maugé was particularly interesting, from their perfect preservation, and from the fact that the greater part of them were new to the Museum. The success of this voyage occasionned the planning of another, of which we shall soon have occasion to speak, and which was attended with results still more beneficial.

In 1798 the professors presented a memoir to the government exposing the wants of the Museum. The magnificent collections which had been received were still in their cases, liable to be destroyed by insects and comparatively useless, for want of room to display them. There were no means of nourishing the animals, because the contractors who were not paid refused to make further advances. The same distress existed in 1799, and it was the more to be regretted from the value of the recent collections. We will not stop to enumerate them here, but barely name the most important. In June 1795 arrived the cabinet of the statholder, rich in every branch of natural history and especially in zoology. In February 1796, M. Desfontaines gave the Museum his collection of insects from the coast of Barbary. In November of the same year a collection was received from the low countries, and that of precious stones was removed from the mint to the Museum. In February 1797, the minister procured the African birds, which had served for the drawings of Levaillant's celebrated work. In 1798 the collection formed by Brocheton in Guyana, and the numerous objects of animated and vegetable nature, collected under the tropics by captain Baudin and his indefatigable associates, filled the hot-houses and the galleries of the Museum.

The government manifested the most unceasing and lively concern for the establishment, and did every thing in its power to promote its interests; but the penury of the finances rendered it impossible to furnish the necessary funds for the arrangement of the collections, the repairs of the buildings, the payment of the salaries, and the nourishment of the animals. Petitions were useless: the funds were absorbed by the armies whose courage remained unabated amidst the disasters that overwhelmed them. The state of exhaustion was equally evident at home and abroad; when the events of November 1799, by displacing and concentrating power, established a new order of things, whose chief by degrees rendered himself absolute, and by his astonishing achievements cast a dazzling lustre on the nation, and suddenly created great resources.

Embarassment was still felt during the first months of 1800, and so small were the pecuniary supplies of the establishment, that it was necessary to authorise M. Delaunay, superintendant of the menagerie, to kill the least valuable animals, to provide food for the remainder. The face of things however speedily changed.

The extraordinary man who was placed at the head of affairs, felt that his power could not be secured by victory alone ; and that having made himself formidable abroad, it was necessary to gain admiration at home, by favouring the progress of knowledge, by encouraging the arts and sciences, and by erecting monuments which should contribute to the glory and prosperity of the nation.

Among other objects he turned his attention to the Museum, to which he furnished funds for continuing the works that were begun, acquiring land for its enlargement, and still further augmenting the collections.

No foreign animals had for some years been added to the menagerie, and if we except the lions, which had produced young, and the elephants from Holland, it contained few that were remarkable. Several were said to exist in London which the owner, M. Penbrock, wished to dispose of, and in July 1800 M. Chaptal, then minister of the interior, sent M. Delaunay to England on this errand. For the sum of 17,500 francs he bought eight quadrupeds, viz. two tigers, male and female; a male and female lynx, a mandrill, a leopard, a panther, and a hyena, with a number of birds. These animals arrived in safety, and were placed in lodges constructed at the bottom of the horse-chesnut avenue. Sir Joseph Banks took the opportunity of presenting to the Museum several curious plants.

In the mean time the cameleopard and other valuable preparations were placed in the galleries; the skeletons, which had long been inaccessible to the public, were put in order; from the numerous specimens of minerals which had been stored away or had remained in cases, such as were judged proper for exhibition in the galleries were selected; the collection of insects was classed by M. Latreille, appointed assistant naturalist in 1794 ; and series of specimens in the several branches were composed from among the duplicates in the cabinet, for the schools of the departments. All the parts of the establishment were conducted with equal judgement and zeal, because each was confided to a separate chief; and its progressive movement was no longer retarded.

Nevertheless in October 1800, the professors had reason to apprehend its ruin, from a measure which the minister of the interior, brother of the first consul, wished to extend to this in common with other public institutions, viz. of appointing, under the title of accountable administrator, a director general, or intendant, charged with the general administration , and the correspondence with the government; thus reducing the officers of the Museum to the simple function of delivering lectures and preserving the collections.

The professors made the strongest representations to the minister on this subject: they proved that each part of the establishment required a separate director; that the administration was essentially linked with the instruction; that intendants were always inclined to favour particular branches, and that they could not be acquainted with all the parts of so vast a whole :

that all those intrusted with the direction of the
garden, excepting Guy de la Brosse, Dufay and
Fagon, who were in fact its founders, had neglected it, and that several had checked its progress; that Buffon, the only person who had
since taken pride in the institution and employed
his credit for its advancement, had felt the necessity of a different system; that Daubenton had
refused the title of perpetual director, offered
him by his colleagues through respect for his age
and gratitude for his services; that since the new
organisation the general order had not been an
instant troubled, notwithstanding the vicissitudes of politics, and the public misfortunes; that
the Museum being immediately dependant on the
minister it was sufficient that an account should
be rendered by the annual director, and that no
extraordinary expenditure should be made without permission; that the place of intendant, given
at first to some person distinguished in the natural
sciences, might at length be bestowed on a man
destitute of any just idea of their utility; that the
funds destined for the Museum might be converted to other uses; that the professors would
be placed in a state of subordination which
would damp their zeal and paralyse their efforts;
and that some amongst them, who held eminent
posts under government, could no longer pre-

serve their chairs when subjected to the controul
of a perpetual chief.

The minister turned a deaf ear to these remon-
strances : he wished to appoint to the place of
director M. de Jussieu, who used his credit only
to enforce the reasons of the professors, and to
prevent the execution of a plan fraught with irre-
parable mischief. Happily nothing was decided
when in the month of November M. Chaptal,
minister of the interior *ad interim,* determined
the first consul to yield to their representations.

The steady progress and harmonious concur-
rence of all the parts of the Museum demonstrate the utility of the present form of administration, and it is to be hoped that the project
of concentring an authority which has no connexion with politics, will not again be brought
forward under a more enlightened and paternal
government. At its foundation the garden was
of so small an extent, that a single man sufficed
for its administration and improvement ; and
at that time, though botany, anatomy and chemistry only were taught, with a view to medecine, it was often necessary to solicit the favour
of the court. Its funds are now fixed by the
budget, and it is for the administrators to consider
how they may be the most usefully employed.
Each proposes improvements in his own de-

partment, and all unite to justify the confidence of the government, and to ensure the prosperity of an establishment, the glory of which is their common property. A succeeding professor may present a science under a different form, but the administrative assembly is constantly animated by the same spirit : its progress is more or less rapid according to circumstances, but its motion is never retrograde, being always directed towards the same end.

These reflections are not unconnected with the history of the Museum, and their propriety will be fully evinced, by the continuation of the picture which we have undertaken to sketch.

In 1801, during the ministry of M. Chaptal, to whom the Museum is under great obligations, the botanical garden, which had been filling up since 1773, was increased in extent one third, and the two parterres opposite the cabinet were planted as we now see them ; the upper gallery of the cabinet was finished and glazed, and the principal objects were methodically arranged; the green-house was terminated and filled with magnificent shrubs; the plan of the menagerie was finally settled and adopted; several wood-yards were purchased and transformed into parks; the first floor of Leger's house, mentioned in a preceding page, received the herbarium, and

the second was arranged for a zoological laboratory; the skeletons and anatomical preparations
were temporarily placed in rooms of the building called *la Régie,* or the office of hackney
coaches; and in 1802, the Museum had become
so organised as to admit of teaching, with the
requisite illustrations, all the natural sciences.
In fine the institution was in a very flourishing
state; the works were carried on with surprising activity, the lectures attracted a concourse
of foreigners as well as natives, and the amphitheatre was often filled with pupils : the greatest
order reigned throughout, and the checks received since the new organisation were no
longer felt. Thus when we contemplate nature
herself, she is always seen greater and more fertile after the rudest storms.

We must here speak of an enterprise which
more than any other contributed to spread the
fame of the Museum, and to diffuse the knowledge of which it is the source, viz. the publication of the *Annals;* for the conception and execution of which a tribute should be paid to the
memory of Fourcroy. When this learned man
saw the Museum fixed upon a stable basis, he
persuaded his colleagues to unite in publishing
their observations, with a design principally to
make known the riches of the collections. The

proposal being adopted by the professors, they determined on publishing ten sheets every month, with five or six engravings executed by the ablest artists under the inspection of M. Vanspaen-donck. As a mutual intelligence was necessary in order to vary the subjects, the professors assembled weekly for that purpose, and charged M. Deleuze, one of the assistant naturalists, with the collection of the memoirs and the oversight of the press. The first volume, consisting of six numbers, was published in 1802, and the work immediately acquired a reputation which it has constantly sustained. During the cessation of foreign intercourse its publication was retarded, but the composition was not the less attended to. To the 20th volume it bore the title of *Annals of the Museum*, and was continued under that of *Memoirs :* it now forms twenty-six quarto volumes. Communications of other naturalists are sometimes admitted.

Though M. Daubenton had been forty years assembling minerals, and since the new organisation had received many specimens from foreign countries, the collection was still incomplete, and even inferior to that of several individuals. It might long have remained so, if an opportunity had not presented itself of at once rendering it worthy of the establishment.

A German named Weiss had formed a superb cabinet at Paris, which he wished to sell entire. The professors represented to the government the prejudice to science of allowing an assemblage of objects to leave the country, which it would be difficult to procure in detail, and begged authority to treat with M. Weiss, offering to give in exchange for his collection, that of precious stones belonging to the Museum. The minister listened to the proposal, and demanded a report from the council of mines on the merits of the collection. It consisted of 1676 choice specimens, and was valued at 150,000 francs. After much discussion, M. Weiss accepted the offer of this sum: the precious stones were estimated by jewellers, and as they were not equal to the price agreed on, the remainder was paid by the government. From that moment (1802) the museum has possessed a regular series of mineralogical specimens, with very few intervals.

The same year M. Geoffroy presented to the cabinet a collection of objects of natural history, formed during a residence of four years in Egypt: in which were found several of the sacred animals preserved for thousands of years in the tombs of Thebes and Memphis (1).

(1) See the Annals of the Museum, vol. 1. p. 254.

Soon after the definitive organisation of the Museum, the plan of a vast menagerie had been formed, in which animals of all descriptions might be placed in a manner suited to their habits. This project, conceived at a time when the means of execution were not calculated, was soon abandoned. In 1802, several pieces of land were obtained for the herbivorous animals, and the architect M. Molinos presented the plan of the great rotundo in the centre of the menagerie, as a lodging for ferocious beasts: the first stone was laid in 1804, but when raised a few feet it was found to be ill adapted to its object, and the work was suspended: we shall presently see to what use it was afterwards appropriated.

In 1804 a case of drawers was placed in the gallery of the second floor of the cabinet, occupying its whole extent, above which were displayed in glazed cases the insects, shells, crustaceous animals, madrepores, etc.: in the drawers beneath were deposited the duplicates of the collection of entomology, and some rare insects, whose colours it was feared might be injured by the light.

About the same time, the Museum was enriched with very precious geological collections. The emperor Napoleon presented that of fossil fishes obtained from the count Gazola, that offered him by the city of Verona, and that of Corsican rocks received from M. Barral, an officer of the island; which fill one of the largest room of the cabinet.

The anatomical preparations were continued with such activity, that in 1805 one hundred and one quadrupeds, five hundred birds, and as many reptiles and fishes were placed in the cabinet. The male elephant from Holland having died the preceding year, M. Cuvier undertook its dissection, assisted by his pupils in zoology and anatomy, and by the painter Marechal. His researches however were still imperfect, and many parts required further examination; for which an opportunity was afforded eighteen months after by the death of another elephant, bought to replace the first. On this occasion he examined and delineated all the organs not perfectly ascertained : and some years after, the female dying also, he confirmed his former observations and compared the sexes. The anatomy of the elephant, of which we had only the skeleton before, is now as well known as that of the horse; a fresh proof of the utility of a menagerie for the progress of natural history.

The inconveniences experienced in the dissection of the elephant at the hottest season of the year, rendered obvious the necessity of an anatomical laboratory, and that now existing was constructed.

In the same year 1804, the Museum was suddenly enriched by the most considerable accessions in zoology and botany that it had ever received. In the beginning of 1800, the Institute proposed to the first consul to send two vessels to Australasia, for the purpose of discovery in geography and the natural sciences. The project was embraced, and twenty three persons were named by the Institute and the Museum to accompany the expedition. The two ships the Geographer and the Naturalist, the first commanded by captain Baudin, and the second by captain Hamelin, sailed from Havre on the 19th of October 1800. They touched at the Isle of France, where the greater part of the persons embarked with scientific views remained; reconnoitred the western shore of New-Holland, and repaired to Timor, where they lay six weeks. They then revisited the same coast, made the circuit of Vandiemen's land, and steering northwards to Port Jackson, lay by in that harbour for five months: thence they resumed their course to Timor, by Bass's straits, and returning to France entered the port of Lorient the 25th of March 1804.

Of the five zoologists who went out in this

expedition, two remained in the Isle of France, and two, Maugé and Levillain, died on the passage; Peron, the only survivor, attached himself intimately to Lesueur the painter of natural history, an excellent observer, and these two indefatigable men amassed an infinite variety of objects. They were occupied a fortnight in disembarking the collection at Lorient, of which we cannot better give an idea than in the words of M. Cuvier's report to the Institute.

« Every day, he says, affords new proofs of the value of this collection, consisting of more than one hundred thousand specimens of animals of all classes. It has already furnished several important genera, and the number of new species, according to the report of the professors of the Museum, exceeds two thousand five hundred. Every thing that it was possible to preserve has been brought home, either dried, carefully stuffed, or in spirits; nor has the preparation of skeletons been neglected whenever it was practicable; of which that of the crocodile of the Moluccas is sufficient proof. »

The same voyage procured us several living animals, among which were the zebra and the gnu, presented by M. Janson, governor of the Cape, to the empress Josephine, and by her, to the Museum.

The botanical collection was not less important. The vegetation of New-Holland differs remarkably from that of other parts of the globe :
some plants from that region were already
known through the English, and by the voyage
of M. de la Billardiere, but the number was comparatively small. In 1804 were received several
living shrubs which have been easily multiplied;
and a great number of seeds and preserved species, of which three fourths were new, and of
which several have escaped the learned researches
of Mr. Robert Brown : some of them have been
published in our Annals. It is worthy of remark
that the plants of New-Holland, from Port-
Jackson to the Straits of Entrecasteaux, do not require to be placed in hot-houses, like those of the
tropics, but pass the winter in the open air in
the southern parts of France, and many of them
even in Paris. Thus the metrosideros, the melaleuca and the leptospermum, which at first excited so much admiration by the beauty of their
flowers, have been introduced into our gardens.
The magnificent eucalyptus, which is one hundred and fifty feet in height and seven or
eight in diameter, is also beginning to be propagated in the southern departments. The season
at which they bloom requires that they should be
preserved in the orangery : but their habits, in

this respect, may be changed, by raising them from the seed. From the Museum, the beautiful species of myrtle have spread in our nurseries, and thence throughout France.

Whilst the green-house was finishing and additions were making to the cabinet, the menagerie was not neglected; every year a few acres were added along the *rue de Seine*. Enclosures and stalls were formed for the deer, the Ganges-stags, the gnu, the kanguroos, the zebra, etc.: but two very necessary accommodations were still wanting; an aviary, and a separate place for the monkies, of which rare species were often received, and which it was desirable to unite in order to compare them. The aquatic birds, such as the stork, the pelican etc, were fitly placed in the basins; the peacocks paraded about the menagerie, or in an enclosure in the centre of the garden; the ostriches and casuaries had a residence to themselves; but the birds of prey, and such as it was necessary to confine in cages, were dispersed.

As the extraordinary funds had been absorbed for the cabinet and green-house, it was impossible to construct a suitable edifice; a building near the *rue de Seine* was therefore fitted up, as well as circumstances would permit. A number of cells with a western exposure, were prepared for the monkies, and large cages, for the birds of prey.

These cells are but temporary, and do not accord with the rest of the menagerie: an adjoining wood yard might be properly occupied by an aviary, and a lodge for the smaller quadrupeds. Contiguous to the birds of prey is a court for domestic fowls.

But if the design of the menagerie had been completed, it would not have had the desired influence on zoology, without a peculiar organisation connecting it with the different branches of that science. The zoological professor was very properly at its head; but it was impossible for him, occupied with his lectures, correspondence, and arrangements in the cabinet, to superintend its details. It was necessary therefore to appoint a keeper devoted to this duty, and qualified to make the requisite observations. On the 21st of December 1805, the administration made choice of M. Frederic Cuvier, brother to the professor, favourably known by his memoirs in the Annals of the Museum; and at the same time framed a set regulations, in consequence of which the animals are not only properly placed and taken care of, but are observed in every circumstance of their habits, gestation etc.; those that arrive are immediately compared with the analogous species, and such as are wanting for the purposes of science or of rural economy are demanded in foreign countries. If an animal dies, which is not in the galleries of zoology and anatomy, its skin is stuffed, the skeleton is prepared, and the soft parts are preserved in spirits: nor is the precaution neglected of searching the body for intestinal worms: thus, besides the advantages for studying living nature, from the menagerie the cabinet and the collection of drawings are daily enriched.

In 1806, the cabinet of comparative anatomy was temporarily disposed for the admission of the public; who saw methodically arranged, not only the skeletons of numerous animals, but a series of all their organs, prepared by M. Cuvier, or under his direction.

While occupied in forming the cabinet, M. Cuvier discovered that the greater part of fossil bones have no specific identity with existing animals; and wishing to pursue his researches he neglected no means to assemble a collection of remains. Some very remarkable ones were found in the quarries of Montmartre; others were sent him from Germany and other countries. In a series of memoirs in the Annals of the Museum, he made known several species of quadrupeds, that existed before the last revolution that changed the surface of the globe, far more ancient than those found amongst the mummies of Egypt, and differing from those that now inhabit the earth in proportion to the remoteness of the periods at which they lived.

After this publication, M. Cuvier gave his collection, the more valuable because singular in its kind, to the Museum, accepting in exchange only the duplicates of books on natural history in the library. This collection, with that of fishes from mount Bolca, fills one of the saloons of the cabinet.

In 1802, the professors had determined on fitting up the first floor of Leger's house for the botanical collection, but the arrangements were not completed till 1807. In the interval the professors of botany assembled and examined all that the museum possessed in this kind, and in 1808 it was disposed in its present order.

The herbarium was formerly contained in boxes or portfolios, and the collection of fruits grains, gums etc., was bottled up along with Geoffroy's ancient drugs: the whole occupied the room now appropriated to the fishes. M. de Lamarck, who before the new organisation was charged with the herbarium, had several times proposed reducing it to order, but he could not obtain a suitable place for its reception. The different herbariums were not classed, nor even kept together; that of Commerson had been deposited at M. de Jussieu's; that of Dombey was lent, by order of the minister, to l'Heritier, who was writing a description of the plants; and the whole was thus rendered useless to study. In the new botanical galleries a large room was destined for the general herbarium, composed by the union of the several collections, with the exception of such as served, like that of Tournefort, as the type of a classical work; a second room was appropriated to fruits, and other productions of the vegetable kingdom; a third, to specimens of wood; so that all the parts of a vegetable might be compared, and every new acquisition find its place.

The opening of the botanical galleries was an important event in the history of the science. The public are not yet indiscriminately admitted; but persons desirous of instruction receive every necessary communication from the professor. This collection, the most complete in existence, is resorted to for the purpose of determining species and of settling the nomenclature. The authors of monographies, as MM. de Laroche and Duval; of the history of the plants of a given country, as MM. de Humboldt, Bonpland and Kunth; or of general histories, like M. de Candolle; could never have given the same precision to their labours, without the resources afforded by the Museum.

Indeed the importance of such a repository is so generally felt, that many botanists have enriched it with the plants which they had themselves discovered or described : to the generosity of M. de Humboldt we owe the most considerable present of this kind, viz. the herbarium of his travels in the equinoctial regions of America , consisting of 5,600 species, 3,000 of which were new. Among them, are the specimens which served for the engravings of his history of tropical plants.

The formation of separate cabinets of anatomy and botany, left the whole of the former galleries for zoology and mineralogy ; but the space was still too confined, and the building no longer corresponded with the more modern edifices, when the government resolved to perfect the design, and give it at once the requisite extension.

To evince the importance of this service, and complete the history of a monument, which is now the most beautiful of its kind in Europe, it may be proper to enter into more particular details, and to resume the narrative a little further back.

We have seen that in 1766, the cabinet consisted of only two saloons, and that Buffon doubled its extent, by giving up his own dwelling : after the new organisation a gallery lighted from above, was ordered to be built on the second floor. On its completion in 1801, the second floor was found to contain twice as much room as the first, as it had no windows on the sides, and comprised the extent of the library. This cabinet, magnificently fitted up, at first appeared sufficient; but since 1801, so great a number of objects had been received, and so many quadrupeds furnished by the menagerie, that further additions soon became necessary. At the intimation of government, a plan was submitted by the professors to the minister of the interior, in December 1807, which being adopted, and the necessary funds assigned, three new galleries were added, by prolonging those of the first and second floors, as far as the terrace behind the labyrinth. The principal door and the stair-case were placed at the extremity of the edifice, and the entrance to the garden from the street, between the library and the building of the *intendance.*

These important works being terminated in 1810, the interior arrangements were made with such celerity, that the new saloons were occupied in March 1811: one of them was appropriated to rocks, and two others to volcanic productions and fossils. The addition on the second floor was devoted to the quadrupeds and monkies, several of which had remained in the laboratory, or been preserved in cases, for want of room in the cabinet.

The building of the rotundo, which had been suspended for four years, was next resumed. The exterior plan was unchanged, but the interior was altered for the accommodation of herbivorous animals, which, like the elephant, the camel, etc., require warmth and care in the winter, and while suckling their young. It was finished 1812, and forms a picturesque decoration to the managerie, but it is ill adapted to its present use, being detached, and open on five sides, so that it is impossible to maintain an equal temperature and to prevent currents of air. It would be desirable to construct a building for the herbivorous animals on the same plan with that just finished for the beasts of prey, consisting of a suite of stables with a southern exposure. Should the government think proper to be at this expence, which would doubtless contribute to the preservation of costly animals and to the prosperity of the establishment, the rotundo might be converted into a library, for which it is peculiarly fitted, by its form and situation: the library cannot long remain where it is, as the room is too confined, and is besides wanted as an addition to the cabinet of reptiles and fishes.

The necessity of these additions to the buildings must be obvious from the enumeration of those made to the contents of the cabinet. Besides the collections already mentioned, the Corsican rocks of M. Rampasse, were purchased by the Emperor, to complete the series of M. de Barral. In 1808, M. Geoffroy brought from Lisbon a very beautiful collection in every branch of natural history. In 1809, the minister procured the samples of North American wood, collected by M. Michaux, author of a valuable history of the forest-trees of that country; and also a herbarium, containing the original specimens for the Flora of his father, who died in Madagascar. In 1810, twenty-four animals arrived from the menagerie of the King of Holland; minerals were sent from Italy and Germany, by M. Marcel de Serres; and presents of several animals, and a beautiful herbarium from Cayenne, by M. Martin, superintendant of the nurseries in that colony (1).

In the disastrous year of 1813 the budget of the Museum was reduced, and important enterprises were deferred till better times. Even the

(1) M. Martin has introduced the culture of the bread-fruit, by slips of a stock brought from the Friendly Islands by MM. la Billardière and de la Haye, and sent him after being kept a year in the hot-house of the Museum: he had several years before carried from the Isle of France to Cayenne the clove, nutmeg, and pepper trees, which at present yield abundantly.

expences of the menagerie were curtailed; all correspondence with foreign countries was interrupted, and the number of students was diminished by the calls of the army. Nevertheless the most essential operations were regularly continued, and if no new acquisitions were made, means were found to preserve what we already possessed.

In 1814, when the allied troops entered Paris, a body of Prussians were about taking up their quarters in the garden: the moment was critical, and the professors had no means of approaching the competent authorities: the commander consented to wait two hours, and this interval was so employed as to relieve them from all further apprehension. An illustrious son of science, whose name does honour to the country which gave him birth, and to that which he has chosen for the publication of his works, obtained from the Prussian general a safeguard for the Museum, and an exemption from all military requisitions; and though no person was refused admittance it sustained not the slightest injury. The Emperors of Austria and Russia, and the King of Prussia, visited it to admire its riches, and to request duplicates of objects in exchange, and information for the founding of similar institutions in their own dominions

In 1815, when we were condemned a second time to receive the visit of those strangers, returning with more hostile intentions, there was reason to fear that the cabinet would be emptied of a great part of its contents ; and that the Museum of natural history, like that of the fine arts, would be obliged to restore most of the objects obtained by contributions from conquered countries. In fact the magnificent cabinet of the Statholder was reclaimed ; and M. Brugmann was sent to Paris, to receive and transport it. This mission caused the liveliest solicitude to the administrators of the Museum : by the restoration of those objects the series would have been interrupted, and the collection left incomplete. M. Brugmann was too enlightened a man not to perceive, that they would no longer possess the same value when detached; and that in the galleries of Paris they would be more useful even to foreign naturalists. But he was obliged to execute the orders of his sovereign, and could only observe the utmost delicacy in his proceedings, listen to every plan of conciliation, and plead the cause of science, in defending that of the Museum. In this dilemma the professors addressed themselves to M. De Gagern, minister plenipotentiary of Holland, who alone could suspend M. Brugmann's operations, and obtain a revocation of his orders. The application succeeded to their wish: it was agreed that an equivalent should be furnished from the duplicates of the Museum ; and this new collection, consisting of a series of 18,000 specimens, was in the opinion of M. Brugmann himself more precious than the cabinet of the Statholder.

We cannot omit expressing our gratitude to the Emperor of Austria, who caused M. Boose, his gardener at Schœnbrun, to transport to Paris such plants as were wanting in the King's garden ; presented to the Museum two beautiful collections, one of Fungi, modelled in wax, with the greatest accuracy of form and colour, and the other of intestinal worms, formed by M. Bremser ; and directed M. Schribers to send the professors a catalogue of the duplicates of his cabinet for selection, in consequence of which exchanges mutually advantageous took place.

Several wrought stones of price were returned to the Pope ; and objects of natural history and books belonging to individuals, which had been sent to the Museum in the time of the emigration, and which were considered as a deposit, were restored with the permission of the government.

After the peace, the King continued to promote the interests of the Museum ; but the finances were exhausted by the public misfortunes, and it was at first impossible to afford the requisite supplies. As it had suffered less than other establishments, there was less to repair, and during the two first years, only 275,000 francs, instead of 300,000, were granted for its expenditure: but every thing had been subsequently replaced on the former footing, and since the administration of M. Lainé, extraordinary funds have been granted for essential purposes.

The cabinet of anatomy, which was too confined for the collection, has been trebled in extent by the addition of the neighbouring buildings: a hall on the ground floor now contains the larger skeletons, and different classes of objects occupy the divisions of the gallery above.

These arrangements being finished in 1817, the lodge for the beasts of prey, which had been long since planned, was begun by order of M. Lainé in March 1818, and the animals were removed into it in the spring of 1821.

This edifice, of regular but simple architecture, forms a beautiful decoration at one extremity of the menagerie, corresponding with the greenhouse at the other. More ground has since been acquired for the herbivorous animals, and we have hopes that the remainder on the *rue de Seine* will soon be added, agreeably to the long meditated plan. New hot-houses are to be commenced, as those in existence, besides being too small, are imperfectly constructed, exact continual repairs, and are unworthy of the general magnificence of the establishment (1).

Buffon had obtained permission from the King to send naturalists into foreign countries; and the travels of Commerson, Somerat, Dombey, and Michaux, had procured considerable accessions to the garden and the cabinet. Since the new organisation, the two expeditions commanded by capt. Baudin, had doubled the collections. At the restoration the government continued the same advantages, and ordered travellers to be sent into regions little known, to examine their natural productions. Considerable remittances have already been made from Calcutta and Sumatra, by MM. Diart and Duvaucel; from Pondicherry and Chandernagor, by M. Leschenault; from Brazil, by M. St. Hilaire; and from North America, by M. Milbert. M. Lalande, who visited the Cape, and penetrated to a considerable distance into the country, has lately brought back the most numerous zoological collection since that of Peron.

Other travellers without a special mission have

(1) A fine collection of plants from India and Cayenne, received in August 1821, necessitated an addition to that in the Botanic Garden; but this hasty structure can be considered only as a temporary resource.

emulously proved their zeal for science: M. Dussumier Fonbrune has sent home a variety of objects from the Philippine Isles; M. Steven, a learned naturalist in the service of Russia, who had passed twelve years in the Crimea and the government of Caucasus, has enriched the botanical cabinet with a great number of plants from those regions; and M. Dumont Durville, lieutenant of the royal navy, with a herbarium of the shores of the Euxine and the islands of the Archipelago. M. Freycinet has returned from a voyage to the southern ocean, with a general collection, made by the naturalists of the expedition (1); and captain Philibert, recently commanded by the government to traverse the Asiatic seas and visit Guyana, afforded such facilities to M. Perrottet, gardener of the Museum, who accompanied him, that he brought back 158 species of shrubs and trees, from six inches to five feet high, the greater part of which are not found in any garden of Europe (2). To this invaluable collection were added several rare birds, and the celebrated gymnotus or electric eel. A number of living animals, and other objects, have been

(1) M. Guadichaud for botany, and MM. Quay and Gaimard for zoology and mineralogy.

(2) The vegetables of Cayenne were furnished by M. Poiteau, director of the establishment for naturalising foreign plants in that island.

presented by M. Milius, late governor of the Isle of Bourbon.

Hitherto these instances of good fortune have happened at indeterminate periods, and when favourable circumstances induced us to solicit them; but a measure lately adopted by the government assures us in future of their regular annual recurrence.

According to a plan submitted to the King by M. de Cazes, a yearly sum of 20,000 francs has been appropriated to the support of travelling pupils of the Museum, to be appointed by the professors. During the first year they are to prepare themselves under the direction of the professors, and are then to be sent into countries that promise the most abundant harvest of discoveries in natural history. They are required to keep up a constant correspondence with the Museum, and to transport the natural productions of Europe to other quarters of the globe.

Unfortunately the first use of this munificence has been productive only of regret. Of the four travellers commissioned in 1820, two fell victims to their zeal on arriving at the place of destination. M. Godefroy, from whose extensive knowledge important services were expected, perished in a fray with the natives, on landing at Manilla; and M. Havet, a young man distinguished by sound erudition and nobleness of character, died of fatigue at Madagascar. He had studied the language of that island, and was recommended to one of the Kings, whose two sons were residing in Paris for their education. It was expected that he would make known the productions of a country, the interior parts of which have never been explored by any naturalist (1).

These fatal accidents, however, are far from damping the ardour of those engaged in similar pursuits; and, young men are every day soliciting the favour of being sent into distant and barbarous climes. The success of more fortunate adventurers, and the hope of attaching their names to some important discovery, render them insensible to the most painful sacrifices. To representations of the dangers and privations to which they will be exposed, they answer with Euryalus:

> Mene igitur socium summis adjungere rebus,
> Nise, fugis?
> Est hic, est animus lucis contemptor, et istum
> Qui vita bene credat emi, quo tendis honorem.

Having thus traced the successive improvements of the Museum, it only remains to notice the progress of instruction and the professors

(5) The third, M. Plee, is now in the West Indies.

to whom it was confided subsequently to the new organisation.

The mineralogical chair was at first filled by M. Daubenton, who had professed that science during twenty years at the college of France, and who, notwithstanding his great age, delivered his annual course with regularity, without discontinuing his private instructions.

From all that we have said, it must be evident how much the Museum, and science in general, are indebted to his co-operation with Buffon. He assembled and disposed almost all the contents of the former cabinet; and when specially intrusted with the mineralogical collection, he bestowed the utmost pains upon its arrangement, passing his mornings in the gallery in examining specimens, answering questions, and attending to the observations of his pupils. Every person listened with respect to this patriarch of natural history, who at the age of 84 years, his hands and feet deformed by the gout, retained all the force and clearness of his intellect, and that freedom from prejudice which rendered him always accessible to truth.

The professors at their first meeting appointed him director of the establishment, and wished him, at the expiration of two years, to retain the title and functions; which he declined as an infraction of rule, and a dangerous precedent: every thing however continued to be done by his advice. A stranger to political dissensions, he had never been a moment diverted from his scientific labours, and the moderation of his character ensured the tranquillity of his mind. The government had just given him a flattering proof of esteem, by appointing him a member of the first political body of the state, and the growing prosperity of the Museum seemed likely to realise those projects which were the height of his ambition, when he was seized with an apoplectic fit, and expired on the 31st of December 1799. He was buried in the scene where he had spent his life, and where every object recalls the memory of his services.

On the 6th of the following month, the professors, availing themselves for the first time of the right of naming their colleagues, chose for his successor M. Dolomieu, who had been long celebrated as a mineralogist, and as the founder of geology in France. This learned man, whom the love of science had determined to join the expedition to Egypt, had been thrown into prison at Messina, on his return, on a groundless suspicion of having been accessary to the invasion of Malta. The powers that interfered in his behalf had been unable to loose his chains, or to soften his captivity, and the professors were ignorant of the probable period of his deliverance ; but they preferred leaving the chair vacant for a time, to foregoing an opportunity of rendering justice to a man, whose elevated character and devotion to science had not shielded him from the most absurd calumnies, and the most odious persecution. M. Dolomieu was not informed of this step till his liberation, on the 15th of March 1801, by an article in the treaty between France and Naples. He hastened to Paris, and on his first appearance in the amphitheatre was received by the audience with an enthusiasm which manifested their opinion of his merit, and their interest in his sufferings. After finishing his course, he wished to take advantage of the remainder of the summer to visit the Alps, Switzerland, and Dauphiny, to collect minerals for the cabinet; but his health, impaired by the hardships he had undergone, yielded to the fatigues of the journey. On his return he stopped at Neuchatel in the Charolois, at the house of his brother in law, and was there seized with an illness, of which he died on the 26th of November 1801.

The ingenious observations of Bergman and Romé de Lille had for several years fixed the attention of mineralogists on the regular and con-

stant forms of crystals; but they had presented
only detached facts, of which M. Haüy divined
the cause, and by the aid of geometry attained
the general results, which have changed the basis
of the science. Having demonstrated the principles of the discovery, he applied it in his treatise
of mineralogy, published in 1800. This work,
by fixing the classification of minerals in characters depending on the nature of their primitive
molecules, gave an impulse to the science, similar
to that which the discoveries of Lavoisier had
given to chemistry, and in the same manner subjected it to an uniform course and a regular nomenclature; but with this difference, that the
theory and nomenclature of Lavoisier are modified by successive experiments, while the laws of
crystalization are invariable, rigorously determined, and certain in their application from the
measurement of the angles. M. Haüy was called,
on the 18th of December 1801, to fill a chair for
which there could be no competition; and from
that time the instruction has been conformed to
the crystallographic method.

It was at first feared, that this method would
embarrass students not prepared to understand
it; but M. Haüy found means to smooth its aspe-
rities, and to render sensible the laws of decre-
ment and transformation, by models; while, by

presenting the minerals in their pure state, he taught the pupil to distinguish the variations produced by a mixture of different substances. He repeated the most curious experiments on the action of electricity, magnetism, and light; phenomena then recently observed, which offer precise characters depending on the chemical composition of the body. The collection of the cabinet affords the means of tracing the transitions and differences of form, and above and below the specimens for study, are exhibited larger masses of the same species, by which the general aspect may be better ascertained. M. Haüy himself possesses a suite of crystals almost complete, which serve as examples in the lectures, and save the trouble of displacing those in the gallery. He does not confine his instructions to the lecture-room, but willingly affords his pupils more particular explanations at his own house.

The influence of this method has been felt in foreign countries: the Germans associate the new characters with their own classification, and several works have been published uniting the principles of Werner and Haüy, or those of the German and French schools.

Since the new organisation M. Desfontaines has had no occasion to change the method introduced by him in 1786. His lectures are given three times a week during the months of May, June, July, and August, and are generally attended by five or six hundred pupils.

Of all the branches of natural history, botany is the best suited to the female sex ; it presents nothing to offend their delicacy ; it furnishes them amusement in retirement, and lends interest to their walks ; attaches them to the cultivation of their gardens ; assists them to develope a habit of observation in their children ; and affords an opportunity of gratifying their benevolence, by making the poorer inhabitants of the country acquainted with useful plants. The letters of Rousseau first excited a taste for this science in the ladies of France, which has increased with the facility of obtaining instruction. A considerable number repair to the garden at an early hour to attend the lectures, and a separate space has been reserved for them in the amphitheatre.

Since 1770 M. de Jussieu has continued his herborisations during the summer.

The course of agriculture is delivered by M. Thouin, with such illustrations as are possible from the practice in the garden and the collection of models. M. Thouin is charged with the correspondence with all the public gardens of France and other countries, and with the yearly distribution of more than 80,000 parcels of seeds.

the produce of the garden, or collected by travellers.

After the suppression of the universities, the Museum being the only remaining institution of science, M. de Fourcroy redoubled his efforts to confirm the favourable impression made, at the opening of his career, and his activity seemed to augment with the sphere of his exertions. Though called by his celebrity to different political posts, he continued his lectures with undeviating regularity; but when appointed counsellor of state, and charged with the ministry of public instruction, he found it necessary to call in the aid of an assistant. For this purpose he selected his pupil and relative M. Laugier, who performed the duty for several years, and succeeded him as titular professor at his death, which took place in 1809 at the age of fifty-five years. M. Laugier recalls the method of his master, by expounding with clearness the whole science, as augmented by the discoveries of the last twenty years (1).

When a chair of chemical arts was substituted for the office of demonstrator, it was given of right to M. Brongniart, who had succeeded Rouelle the younger in 1779. He was the better qualified

(1) M. Laugier's place of assistant naturalist was bestowed upon M. Chevreul, author of several memoirs in the Annals of the Museum, and of the chemical part of the Dictionary of Natural History.

to fill it, as in his lectures at the King's Garden,
at the school of pharmacy, and the lyceum of
arts, he had always preferred the exhibition of
useful processes to surprising and brilliant experiments.

At his death, in February 1804, he was succeeded by M. Vauquelin, who, having made
practical chemistry his peculiar study, was enabled to give greater scope to this important
part of the science : by the improvement of analytic chemistry and the art of essaying, by the
discovery of chrome and other substances, and
by the introduction of more scientific methods
into common practice, he is allowed to have
exerted a great and beneficial influence on our
manufactures.

As early as the beginning of last century botany
was cultivated with success. A great number of
plants were assembled in the King's Garden, rich
herbariums had been formed, and Tournefort,
from the examination of all the plants then
known, had deduced a method, which in general
preserved the natural relations. The progress of
zoology was less rapid, not from a neglect of
that science, but from the want of resources.
Separate descriptions of animals were published,
curious observations were made upon insects,
and Linnæus had presented in systematic order

and described in precise and picturesque language, the varieties of animated nature. Nevertheless the greater part of the animals of the old and new world were imperfectly known, for want of opportunities of comparing them, and of observing the differences produced by age and other circumstances in the same species.

To the collections of the King's Garden, and to the works of which they facilitated the execution, are owing the wider range and greater exactness of zoology at the present day. The history of quadrupeds by Buffon and Daubenton, that of birds by Buffon and Montbeliard, and that of the cetaceous animals and fishes by M. de Lacépède, made known with accuracy the species which Linnæus had only indicated, and many others whose existence he had not suspected. The galleries of the Museum furnished M. de Lamarck with materials for his history of invertebrated animals, and enabled M. Latreille to perfect his great work on insects. M. Cuvier soon after accomplished in favour of zoology what M. de Jussieu had done for botany, by founding, upon natural relations and invariable characters, a classification now generally adopted.

The three chairs of zoology are still occupied by the professors first appointed to fill them, and the number of their pupils is yearly increasing,

as a taste for the science becomes more generally
diffused, and the collections afford means of more
positive and varied instruction.

M. Geoffroy de St. Hilaire resumed his lectures at his return from Egypt, where he was
employed during four years. In his annual
course, after describing the animals by their apparent characters, he presents zoology under a
general view, embracing and connecting all its
parts. This method reposes on four considerations, which may be termed the four primor-
dial views of anatomical philosophy : viz. the
theory of analogies; the principle of connexions;
the balance of dimensions; and the elective affinities of the organic elements. According to
this plan he no longer confines himself to the
description of external forms, but shews the
cause of these forms in the modifications of the
interior organisation ; thus seeking to link the
parts to the whole, and to present the science
under a larger aspect.

M. Geoffroy had taught the history of all the
vertebrated animals for eighteen months, when
the law of the 7th of December 1794, at the
request of the professors, erected a separate chair
for oviparous quadrupeds, reptiles and fishes :
to which M. de Lacépède, who had left the garden two years before, was called in January

1795. Not contented with completing his course of lectures, M. de Lacépède resumed his former labours in the cabinet, and soon after, on M. Geoffroy's departure for Egypt, took charge of the birds and quadrupeds, in addition to the objects especially committed to his care. By him the collection of birds, the most magnificent that had ever been assembled, was arranged in beautiful order for exhibition, and rendered classical for the study of ornithology. The celebrity which he had acquired by his works, and by his connexion with Buffon, attracted crowds of young men to his lectures, whom he induced to attach themselves to a branch of natural history which had been little cultivated in France. During ten years his whole time was employed in facilitating the study of a science which owed much of its progress to himself; and when called to a post under government, which left him no leisure for these pursuits, he ensured the solid instruction of his pupils by choosing for his assistant M. Duméril, author of the Analytic Zoology, and the co-operator of M. Cuvier in the first volumes of his Comparative Anatomy.

The chevalier de Lamarck, so highly distinguished by his works on invertebrated animals, has for twenty-five years taught the history of mollusca, crustacea, insects, worms and zoophites.

He has also classed the shells and polypuses of the cabinet after a more scientific and exact method, and has characterised all the genera, and determined a great number of living and fossil species. His impaired sight not permitting him to continue his demonstrations, he is replaced by M. Latreille, whose numerous writings, and especially his great work on the classification and generic characters of crustaceous animals and insects, rank him among the first entomologists of Europe.

The three courses just mentioned are delivered in the summer, and continue three or four months.

The chair for human anatomy has always been filled by professors of distinguished merit, and for many years it afforded a more complete body of instruction than any other in the kingdom. In later times, as anatomical courses have been multiplied, though it no longer boasts the same superiority, it has not lost its ancient reputation: since 1778 it has been occupied by M. Portal, first physician to the king and president of the academy of medicine.

M. Mertrud had for several years studied comparative anatomy under Daubenton; yet he did not consider that science in its most elevated point of view. M. Cuvier, appointed to assist

him on the 15th of November 1795, and named
professor after his death on the 1st of November
1802, has taught it in its generality and in its
details, embracing the analogies of all classes of
animals, from the polypus to the elephant, by
the comparison of their essential organs. He
has also formed the cabinet of comparative anatomy, from materials furnished by the menagerie, or contributed by travellers and foreign
naturalists.

The establishment of a course of geology, distinct from that of mineralogy, was a most judicious innovation (1). Without the precise characters afforded by mineralogy, the geologist cannot ascertain the genera and species in their pure
state, nor discern the elements of an aggregate
body, and the alteration of the primitive forms
by the mixture of different substances ; but the
history of the great masses which cover the globe,
of the relative situation and different formation
of rocks, of subterranean fires and volcanic productions, of thermal waters, of fossil bones and
shells found at different depths, forms a peculiar
science, founded on innumerable observations,

(1) Geology was formerly so little attended to that even the name
was known only to men of learning. The word *geology* was not found
in the dictionary of the academy, although the analogous terms *zoology*
and *zoography* were inserted.

and exempt from the systematic absurdities that have disgraced the theory of the earth.

M. Faujas de St. Fond first occupied the chair of geology in the Museum. If the science, notwithstanding the facts with which he had enriched it, was not sufficiently advanced for the establishment of positive laws, he at least had the merit of rendering it popular, and of contributing to its progress since the beginning of the century. The impaired state of his health during the last years of his life, obliged him to reside chiefly in the country, though attached to Paris by the duties of his office and the friendship of his colleagues: he terminated his career at his estate of St. Fond, near Montelimar, the 18th July 1819, at the age of seventy-eight.

M. Cordier, an inspector of the mines, and the pupil and travelling companion of Dolomieu, was named by the professors of the Museum and by the academy of sciences to succeed M. Faujas, and appointed by an ordinance of the 13th of September 1819. At his entrance into the garden he lost no time in reorganising the cabinet of geology, by distributing the rocks into three series, according to their nature, their position, and their locality. In his lectures he contents himself with exposing the actual state of the globe, by a connected view of facts ascertained by observation; and insists particularly on the riches of our own mineral kingdom, and the means of rendering them subservient to the progress of the arts and to the wants of society.

Natural history cannot dispense with the aid of drawings, and the most exact descriptions leave but a vague impression on the mind if unaccompanied by figures: language suffices to express essential characters, but cannot give an idea of the physiognomy and general appearance of objects; it was a fortunate conception therefore to attach a professor of the art to the Museum. This institution has both diffused a taste for drawing, and given it a more useful direction. It is easy to see by comparison, how much the figures in works of natural history are superior at the present day to those of the last century. M. Vanspaendonck, since his appointment in 1774, has formed numerous artists. Though the primary object of his lectures is the imitation of scientific characters, beauty and effect are not neglected; and to this source may perhaps be traced the perfection to which the art of painting flowers is carried in France, and its influence on several of our manufactures. His lessons of *iconography*, which are attended by a great number of young ladies, are given in the library three times a week during four months. The library on these oc-

casions is open only to the pupils, who are at
liberty to continue their work on the interven-
ing days, and are often assisted with the advice
of the professor.

As it is necessary to adapt the instructions to
the greater number of pupils, the professors
cannot in their courses enter into minute details,
nor expose discoveries and principles which
would be understood only by men versed in
science; for these objects the Annals of the Museum offer an appropriate medium of communication. In this work M. Haüy has fixed the characters of different minerals recently added to
his cabinet, and shewn the simplicity of the laws
of crystallography, and the advantage of analytic
formulas; MM. Fourcroy, Vauquelin and Laugier
have communicated the most important results
of their experiments in the chemical laboratory;
M. Desfontaines has described new genera of
plants, that have bloomed in the garden or been
found in the herbarium; M. de Jussieu has defined
the characters of the principal natural families,
with such additions and corrections as the progress
of the science has necessitated in his work;
M. Thouin has explained in detail the management
of the seed-beds and plantations, and the processes
of grafting; MM. Geoffroy and Lacépède have
published new genera of quadrupeds, bats, rep-

tiles and fishes; M. de Lamarck has described the fossils of the environs of Paris; M. Cuvier has made known the anatomy of mollusca, and the skeletons of extinct animals, whose bones he had collected; and the professors in general have contributed extracts from their correspondence with other establishments, or with travellers and foreign naturalists.

As in each branch of instruction the understanding is led from the first rudiments of knowledge to the profoundest principles; so all the parts are aptly connected by a mutual dependance, and unite in the same result of promoting the progress of natural science, and rendering it conducive to the well-being of society.

Two thousand pupils yearly attend the lectures of the Museum, of whom a few only become distinguished naturalists; but all acquire a share of useful knowledge, and a talent for observation. It has been said by Bacon, that ignorance in philosophy is preferable to superficial knowledge; and it cannot be denied that shallow notions of history and philosophy are often employed to sap the foundations of morality and politics. But it is otherwise with the knowledge of nature; in this unbounded science every acquisition is useful, from the simplest perception the deepest researches, and from the minutest details to the most general views: the study of it accords with every age, with every disposition of mind, and every profession in life: it yields assistance to agriculture, medicine and the arts, and powerfully contributes to the wealth of nations. As its object is to ascertain and connect facts, and not to investigate causes, it is free from the uncertainty of hypothesis; and if observation is sometimes incomplete, nature is always at hand to dissipate doubts and to rectify errors.

But to obtain the results that may be hoped from it, and spare the student the laborious researches of his predecessors, there must exist a repository of knowledge, from which he may borrow to enrich it in his turn. This repository is the Museum: founded by our monarchs, adorned by men of genius, and governed by enlightened administrators, it has hitherto resisted every shock, escaped amid every scene of devastation, and excited the admiration of rival nations. The warrant of its duration is its utility, and the protection of a Sovereign, whose glory can only increase as the progress of knowledge shall render more evident the wisdom of his institutions.

COMPARATIVE VIEW

OF

THE ESTABLISHMENT IN 1789 AND 1821.

Before entering on the description of the Museum, we shall sum up its history, in a comparison of its state at the present time and at the death of Buffon; from which it will appear, that the expence has by no means increased in proportion to the increase of the establishment, and to the advantages of the new organisation; and that under any other form of administration the instruction would be more costly, and less complete.

The expences of the King's Garden in 1789 were 104,269 francs, and those of the menagerie of Versailles, since transported to the Museum, 100,000 frs.; making a sum of 204,269 frs.: at present the current expences of the Museum are 300,000 francs.

In 1789 the Garden contained 43 acres; the

Museum contains 79. The galleries of natural history have been raised one story, and nearly doubled in length; and a library of more than 12,000 volumes has been added to the establishment. Two hot-houses, a large green-house, the rotundo in the centre of the menagerie, and the lodge of the carnivorous beasts, have been constructed: two buildings connected with the establishment have been converted, the one into botanical galleries, a zoological laboratory, and a hall of administration, the other, into a cabinet of comparative anatomy and an anatomical laboratory; and additional dwellings have been acquired for the professors, assistant naturalists, etc.: the buildings at present are to those of the former period in the proportion of 7 to 1.

The botanical garden has been augmented one-third; those of agriculture, fruit-trees, and economical plants, and several parterres, have been formed; the grounds have been more richly adorned with shrubs and flowers, and the menagerie has been planted with trees yielding seed for the propagation of useful species: the extent of the culture is as 9 to 1.

Of the collections, that of living plants has been doubled; that in the herbarium is six times as great; and that of fruits and other productions of the vegetable kingdom has increased in the same proportion. The collection of birds and quadrupeds is 20 times as numerous; that of fishes, now the most extensive with which we are acquainted, was formerly insignificant; that of insects, which consists of 40,000 individuals of 22,000 different species, contained only 1,500 specimens; that of shells also has been greatly increased. The rich cabinets of comparative anatomy, geology, and fossil bones were not then in existence. The menagerie of Versailles contained a small number of animals, and was of little use to zoology; that of the Museum has presented successively more than 500 species, and has given rise to many important observations.

The garden, the buildings and the collections, form a magnificent monument, but it is the extent given to the instruction which has infused new life into the institution, and rendered it of general utility. In the King's Garden there were three professors and three demonstrators: in the Museum there are thirteen professors, with aid-naturalists attached to such as need assistance; and twice as many lectures are delivered from each chair.

The Museum employs one hundred and sixty-one persons, of whom ninety-nine are paid by the month, and sixty-two by the year. A correspondence is kept up with all similar establishments, and a prodigious quantity of seeds, slips, etc. are annually distributed.

From their comparative extent the expences of the Museum should be four times as great as those of the King's Garden and menagerie, instead of exceeding them by one third. This surprising economy is due to its organisation, and to a careful, provident, and accountable administration, attentive to every detail, and immediately inspecting the execution of every undertaking. It must be granted, however, that as the Museum is every day receiving living animals, foreign plants and new collections, the preparation and preservation of which necessitate expence, it sometimes requires extraordinary supplies; and the government is too enlightened not to proportion its encouragement to the utility of the institution.

We shall terminate this summary by a reflexion which will not fail to affect those to whom the spectacle of social harmony and domestic felicity is not less interesting than that of nature: how pleasing amid the agitation of a great city to behold an establishment, in which are united fifty families, living in peace, usefully occupied, contented with their lot, attached to the place of their abode and priding themselves in its prosperity, strangers to professional rivalry and political dissensions, and grateful at once to the government which supports and to the administration which directs them. The philosophers employed in investigating nature, communicate with the workmen, who catch the reflexion of knowledge, and learn to delight in the results of their labour.—Here are united the sources of happiness spoken of by Virgil:

Felix qui potuit rerum cognoscere causas . . . .
Fortunatus et ille Deos qui novit agrestes.

DESCRIPTION

OF THE MUSEUM.

CHAPTER FIRST.

GARDENS, GREEN-HOUSE AND HOT-HOUSES.

§ I. TOUR OF THE GARDEN.

PERSONS desirous of surveying all the parts of the garden, without retracing their steps, should proceed in the order we are going to point out.

The visitor is supposed to arrive by the gate on the quay, from which at one glance he embraces the whole establishment. At the opposite extremity of the garden is seen the cabinet of natural history, occupying its whole breadth, and rising above the growth of two enclosures, one of which is the nursery, and the other a square basin, hollowed to the level of the river, and adorned with shrubs. On the right and left are two large avenues of linden-trees; and beyond these, on the right, several cultivated squares, and the menagerie extending to the *rue de Seine;* and on the left, groves of forest-trees bordering the *rue de Buffon.* By the great avenue on the right we arrive at the court of the cabinet, and following the iron railing which separates it from the garden, find ourselves at the entrance of the parallel avenue, with the cabinet in the rear, and a little to the right the house called the *intendance,* which was the residence of Buffon from 1773 till his death (1). From this point we shall begin the circuit of the garden, noticing every thing in our way; first in the lower part, which extends from the cabinet to the river; and next on the high grounds, called the labyrinth and the little hill. We shall then glance over the menagerie, and return by the avenue of horse-chesnuts, which separates it from the botanic garden, etc. This tour alone would occupy nearly an hour; but it requires a much longer space, and must be several times repeated, if we would examine all the curious and instructive objects which invite the attention. We shall not speak particularly at this moment of such parts of the garden as have a special destination; as the squares appropriated to scientific botany, agriculture, and fruit-trees, the green-houses, the mena-

(1) It is at present occupied by MM. de Lamarck and Vanspaendonck, professors; and MM. Lucas, keepers of the galleries.

geric considered as the abode of animals, etc.; each of which must be the object of a separate visit.

Proceeding then from the head of the great avenue of lime-trees planted by Buffon in 1740, on the southern side of the garden, we see on the right, plantations of forest-trees and a cultivated square; and on the left, two enclosures separated by a circular basin, the nursery, the square basin already mentioned, and several parterres: reserving the last for our return, we shall at present notice only what lies between the avenue and the *rue de Buffon*.

The four first squares are composed of trees of every species and every country, which pass the winter in our climate: among them are a gleditschia without thorns, sent from Canada by M. de la Galissonnière in 1748, which is one of the largest trees in the garden; a sophora of Japan, *sophora Japonica*, the first received in Europe; and the first acacia obtained from North America, which its possessor Vespasian Robin planted in the King's Garden in 1635. A few years ago this tree was more than 60 feet in height, but the summit beginning to decay it was lopped, that it might sprout anew from the trunk. From this stock has been propagated one of the most useful and agreeable among trees, to which the name of Robinia was given by

Linnæus, in memory of the person by whom it was introduced. Near the same spot there was formerly a superb horse-chesnut, planted in 1656, and at that time the only individual, except one, in France; it perished in the winter 1766-67, after furnishing the seeds of almost all that now adorn our parks.

The two squares opposite the basin and a little above it, were in Tournefort's time a systematic plantation of trees and shrubs: when the botanic garden was renewed by M. de Jussieu, such trees as could not easily be procured were transplanted, and others were substituted in their room; but those which remain still recall the method of Tournefort. We may particularly notice a juniper (*juniperus excelsa,* Marsch.) 40 feet in height, and 15 from the ground to the first ramification, which was brought from the Levant and planted by Tournefort. This male stock is probably the only example of the species in France. Its leaves crushed between the fingers emit a pungent odour.

At the extremity of this square is a coffee-house, where refreshments are tasted beneath the shade. Beyond are three squares enclosed by a trellis, which during the life of Buffon were a neglected copse, and which have been cultivated within these few years by M. Thouin.

The first is appropriated to annual plants in request for the beauty of their flowers, and is divided into four sections, in which the various species are grouped according to the season at which they bloom: the seeds of these flowers are carefully gathered for distribution. The second is destined for ornamental, vivacious plants, and towards the end of autumn roots and slips of such as can be multiplied in this way are given to amateurs. From these two parterres are frequently selected the models for the lessons of M. Vanspaendonck. The third square is occupied by the seed-beds of trees and shrubs that support our winter : the infant stocks are transplanted to the nursery, and afterwards employed in embellishing the garden. Here is seen a pretty clump of Ispahan peach-trees, the seeds of which were brought from Persia by M. Olivier in 1780 (1).

At the extremity of this square a transverse alley of Virginian tulip-trees (2), *liriodendrum*

---

(1) A history of these trees by M. Thouin is found in the 8th vol. of the Annals of the Museum.

(2) The tulip-tree was introduced into France about 1748, by the marquis de la Galissonnière, governor of Canada. It is one of the most beautiful trees of North America. In its native soil it reaches 150 feet in height, and its trunk is sometimes more than 20 feet in circumference. The stocks in the King's Garden were planted 25 years ago by M. Thouin : they are covered with flowers in the month of June, and for several years they have yielded productive seeds.

tulipifera, marks the original limits of the garden; which are also indicated by the inferior growth of the main avenue, prolonged, as we have already seen, in 1783. Next follow four other squares, of which the first is enclosed by a trellis, and planted with ever-green trees, in gradation for the benefit of the sun, the largest towards the north upon the avenue, and the more diminutive towards the south upon the *rue de Buffon*: among them are some very large silver firs, *pinus picea;* Jerusalem pines; beautiful red cedars, *juniperus Virginiana;* an oak with sweet acorns, *quercus ballota* (1); the variegated and the Mahon hollies, *ilex variegata* and *I. balearica,* etc.

This square is separated by an alley of larches (2) from the following, which is also enclosed by a trellis, and contains a variety of trees whose fruit or foliage arrives at perfection in the autumn. Among them are a pacane-nut hickory, *juglans olivæformis* (3), the largest in France; two stocks

(1) In Spain the glands are eaten like chesnuts.

(2) These trees grow to the height of 120 feet on mountains and in cool and shady situations; but they do not succeed in the King's Garden. The larches and the deciduous cypress are the only trees in the family of the *coniferæ* which lose their leaves in the winter.

(3) This tree, originally from Upper Louisiana, grows abundantly on the borders of rivers and marshes. Its nuts are an object of commerce in the United States. It has not yet bloomed in France.

of the *gingko biloba* (1), the first seen in this climate; a beautiful persimon, or Virginian plumtree, *diospyros virginiana;* a *mespylus linearis,* a species of medlar, with a summit in the form of an umbrella (2); a superb mulberry, *morus rubra,* from North America, whose fruit equals that of the black mulberry, and whose foliage, more compact than that of any other tree, persists till the end of autumn, and is secure from the attack of insects. The square is terminated by an avenue of white maples, *acer eriocarpum,* a species first accurately distinguished by M. Desfontaines, in the 7th volume of the Annals of the Museum.

The fourth square is a thicket of ornamental summer-trees, so distributed as to present agreeable contrasts in their foliage, form and flowers. The Carolinian and the flowering ash, the American black walnut, *juglans nigra,* and the coffee-

(1) A large tree from Japan, called also the Japan walnut, of very singular foliage. It was introduced into England in 1754, and thence brought by M. Petigny to France a few years after. Its fruit is as large as an apple, and contains an excellent kernel. One of the two stocks has borne male flowers.

(2) This tree, originally from North America, was unknown to botanists until the seeds were brought to the Museum about twenty-five years ago. It has multiplied so rapidly, that a small avenue has been formed of it, and it will soon be generally cultivated in pleasure grounds. The form is totally different from that of the other species of the same genus.

tree or Canadian bonduc, *gymnocladus canadensis* (1) are more particularly worthy of notice. This plantation is terminated by an avenue of the *ailanthus* or Japan varnish (2), a superb tree first described by M. Desfontaines in the Memoirs of the Academy of Sciences for 1786. All the trunks bear on one side the impression of the frost by which they were injured in the spring of 1787.

The last square is planted with trees which bloom in the spring. It was of twice its present extent before the building of the bridge, when a part of it was sacrificed to the area on the quay, and the trees were transplanted to a square at the bottom of the corresponding avenue. There still remain the yellow pavia, *pavia lutea;* the Ohio buck eye, *P. Ohioensis;* and the red flowering horse chesnut, *æsculus flore rubro,* a very

(1) This is a diœcious tree, remarkable for its double winged leaves, three feet long, and twenty inches broad: its branches being few, it has in the winter the appearance of a dead stock, and hence it has been named by the French Canadians *chicot,* or stump-tree. We have only the male stock.

(2) M. Desfontaines, who observed its fructification for the first time at M. Lemonnier's at Versailles, determined it to be a new genus of the family of the terebinthaceæ, and published a description of it in the Memoirs of the Academy of Sciences in 1786. He gave it the name of *aylanthus,* which it bears at Amboyna, and which signifies *tree of heaven:* it was first designated under the name of Japan varnish, because it was believed to furnish the fine Japanese varnish. This denomination has been preserved, though improper; it agrees better with the *rhus vernix* of Linnæus.

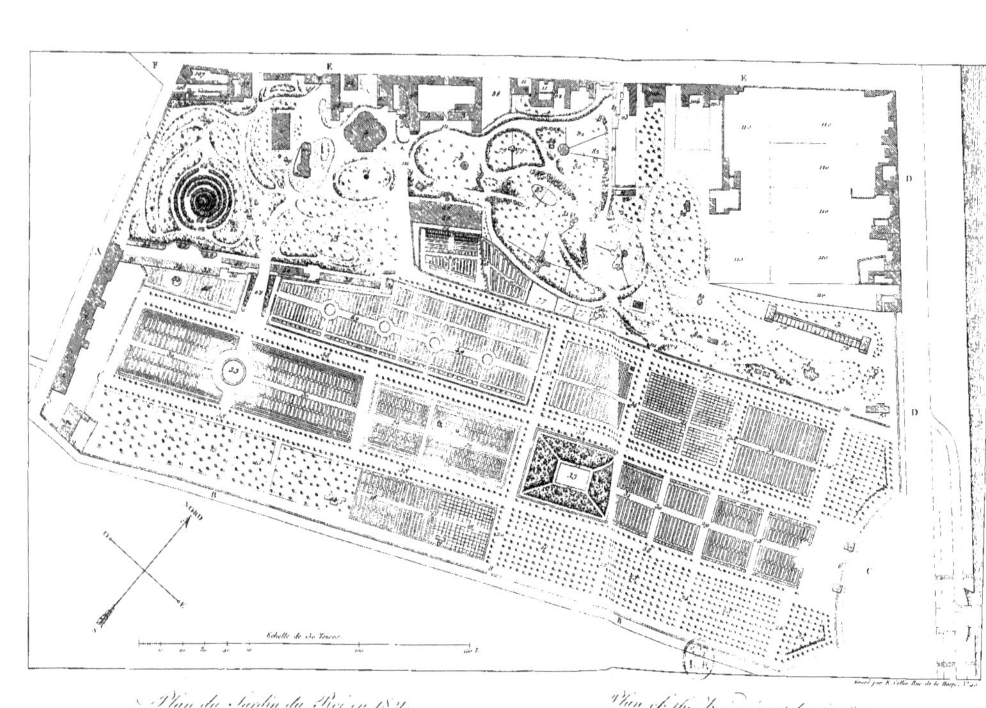

Plan du Jardin du Roi en 1821. — Plan of the King's garden in 1821.

remarkable species which was not known till it bloomed two years ago in the King's Garden. The trees transplanted to the opposite side of the garden, have been grafted in different ways, and abandoned to their natural growth. Among them are the double flowering cherry, *cerasus sylvestris flore pleno,* the sweet-scented crab-apple, *malus coronaria,* the Chinese apple, *malus spectabilis,* and a Chinese quince, described and figured by M. Thouin, in the 9th. vol. of the Annals of the Museum, when it bloomed for the first time in 1811. This enclosure is sheltered towards the quay by a tall hedge of the Chinese arborvitæ, *thuya orientalis,* and is bounded on the west by an avenue of Judas trees, *cercis siliquastrum,* which is the most beautiful of the garden, in the beginning of May, when the branches, still destitute of foliage, are covered with innumerable clusters of purple flowers.

Returning by the terrace to the gate, we see in front a narrow alley extending to the basin, and on the right and left parterres improved by different species of culture. The four first contain medicinal plants for the poor, arranged in beds properly labelled for the convenience of herbarists and students in pharmacy and medicine. Two of these squares are assigned to indigenous plants, and two to exotics. Among the latter are several species of rhubarb, *rheum*, and some vegetables lately introduced, to which great virtues are ascribed in their native countries, and of which our physicians are thus enabled to make trial. The two next divisions contain duplicates of the most beautiful vivacious plants of the botanic garden, which have here more room and grow more luxuriantly. They are not labelled, that the pupils may exercise themselves in determining the characters. In the two last squares are cultivated the most beautiful border flowers; care being taken to vary the species every year, and to renew them during the season, so that the bloom may continue from the middle of spring till the middle of autumn.

We now arrive at the square basin enclosed by an iron railing, by walking round which we may observe the various shrubs that adorn its sloping sides. From the beginning of spring to the end of summer it offers a splendid confusion of roses, snow balls, lilacs, venus sumachs, *rhus cotinus;* fontanesias (1), and bladder-nuts, *staphylea.* Several of the shrubs having begun to decay, they were all cut to the ground during the last winter, but they will be renewed with

(1) A pretty shrub described, and dedicated to M. Desfontaines by M. Delabillardiere, who brought the seeds from Syria in 1788.

increasing beauty a year or two hence. This basin, originally intended for the cultivation of aquatic plants, a purpose which it very imperfectly answers, receives its water by filtration from the Seine. It would be desirable to fill it up, or to open an easier communication with the river, as the stagnant water often becomes offensive in the summer (1).

We next cross an alley bordered on one side only by the *mespilus linearis*, and the *koelreuteria* (2), placed alternately and offering a striking contrast in their foliage, flowers, and general form—the one extending its branches horizontally, and the other collecting them into round masses. The next square is the nursery, which is also surrounded by an iron railing.

Here are raised the trees and shrubs for different parts of the garden ; and by means of slips, layers, and grafts, interesting species newly introduced, and not yet exposed to sale, are propagated and diffused among the cultivators who correspond with the Museum.

(1) When new hot-houses are constructed, they will probably form a continuation of the present, and by diminishing the botanic garden on the west, necessitate its prolongation towards the south. The great basin will then be filled up for the reception of the school of fruit-trees or of agriculture.

(2) An elegant tree brought from the north of Asia in 1789, and naturalized in the Museum, whence it has spread.

Among the shrubs, a beautiful horse-chesnut, *æsculus macrostachya,* brought from North America by A. Michaux, is seen spreading its branches and long bunches of flowers at the distance of only three feet from the ground. There are a number of stocks of Christ's thorn, *rhamnus paliurus* (1), as beautiful as those which grow in the south of France; the cissus, the periploca, etc. The culture necessarily varies every year, as some plants are removed, and others substituted in their place. In the centre of the nursery is a wooden shed covered with red canvas, containing thirty beehives, of the simplest and most complicated structure, of which several are composed of glass for the purpose of watching the labour of the bees. M. Lasseray, to whose care they are committed has made valuable observations on the method of collecting the honey without killing the bee, and has found out a mode of extracting it from the comb without destroying the cells.

On the south is a bed covered with bog-earth and shaded by the lime-trees of the great avenue, where are cultivated such plants as require a cool temperature and peculiar care; as several

(1) A pretty ornamental shrub which would form impenetrable hedges, as it has two axillary spines at the base of each petiole, the one straight, the other curved: Virgil mentions it—*spinis paliurus acutis.* It is commonly called *porte-chapeau* from the form of its fruit, which resembles a slouched hat.

species of heath, the rose bay, azalea, kalmia, whortleberry, itea, and andromeda, which it is not possible to rear as well in the botanic garden. We may also observe a fine swamp magnolia or white bay, *M. glauca,* a species in great request for its odoriferous flowers.

Beyond the nursery are two parterres enclosed with a trellis, and devoted to the multiplication and naturalization of such foreign vivacious plants as pass the winter without shelter in our climate. Among these are a row of the beautiful phlomis brought from the Levant by Tournefort; different species of fennel, *ferula,* ten feet in height; the fraxinella, *dictamnus,* and the aletris of Abyssinia, all which have been multiplied in the garden; several species of lychnidea, *phlox,* and two chelones, *C. campanulata,* and *C. barbata,* lately received from Mexico. There is also a collection of roses, less numerous than that of the Luxembourg, but comprising the most remarkable species. On the sloping borders are tufts of bulbous or tuberous plants, such as the piony, the crown imperial, lilies and martagons of different colours, the albuca, the amaryllis, etc. Each bed is bordered with flowers proper for edging, which are varied from year to year, and of which several have been introduced into other gardens, as the catchfly, *silene,* brought from Barbary by

M. Desfontaines, that now borders the parterres of the Luxembourg. In the fine season, beautiful trees from the orangery, such as the grewia, the corymbose cassia, the many flowered staff-tree, *celastrus*, the justicia, the African kiggelaria, *K. Africana*, the carob-tree, *ceratonia*, and the plane-leaved sterculia, *S. platanifolia*, are placed in the interval which separates these squares, and at the extremity towards the cabinet, magnificent rose bays, double flowering myrtles and palms.

Between the two squares is a circular basin for the cultivation of aquatic plants, some of which are exotic, as the drooping lizard's tail, *saururus cernuus*, and the American jussieua, *jussieua grandiflora*, both of North America. Upon the brink are scatterred pretty species of saxifrage, and other plants which require constant humidity.

This basin is in the form of a foot-glass, and has a subterranean passage round it, where cryptogamous plants which grow in obscurity might be placed to advantage: it produces a continual moisture beneath without suffering the water to escape.

Opposite to the squares just mentioned, on the right, is a garden with an iron railing, where the plants of the orangery are exposed in the summer. The eye is here arrested by the beautiful Adam's needle, *yucca*, the flax of New Zealand, *phormium tenax* (1), the Mississipi laurel, a large magnolia, the casuarina (2), the clethra, and many plants from the south of France and the Alps. In the rear of the garden is the orangery, whose walls are covered with climbing plants, especially the beautiful Virginian trumpet flower, *bignonia*, so remarkable for its large bunches of red flowers.

The orangery is divided into two parts, one of which only is arched. It has no stove: the windows are defended by mats in very cold weather, and in the spring are seen blooming with Alpine plants, which cannot support our climate though milder than their own, because

(1) The *phormium tenax* with which the inhabitants of New Zealand make their cloth, has some affinity with the aloe. It was first made known by the voyage of captain Cook. The Museum having received several slips at the return of captain Baudin, one was given to M. Faujas, who multiplied it in his garden near Montelimar, and spread it through the southern departments of France. It is a precious acquisition, as the ropes made of its leaves are twice as strong as those of hemp. See the Annals of the Museum, vol. 2 and 17.

(2) The casuarinas, which belong to the family of the *coniferæ*, are entirely destitute of leaves like the ephedra. Their slender, drooping, articulated branches give them a very peculiar appearance. They might be multiplied in the South of France, and their wood which is hard, agreeably veined and susceptible of a high polish, would be very useful in cabinet making: in New Holland it is preferred to every other for ship-building.

they are not as in their native soil covered during the winter with snow.

By the side of the orangery is a small enclosure sheltered on the north and west, containing hot-beds and frames for such delicate plants as are multiplied by slips.

On leaving the garden of the orangery, we find ourselves near the slope conducting to the two hills.

The labyrinth, so called from its numerous intricate paths, is of a conical shape. On the ascent we observe a cedar of Lebanon, which yields a profusion of seeds every year, and is the oldest stock in France. When Bernard de Jussieu visited England with du Fay in 1734, Collinson, a wealthy physician of the society of friends and a lover of botany, who had received a cone from mount Lebanon, presented him with two plants a few inches in height. They were cultivated with the utmost care: one of them placed in the old botanic garden, no longer exists; the other is the majestic tree which spreads its branches at the foot of the labyrinth, and whose offspring people the pleasure grounds of France. It would have attained a still loftier stature, if the summit had not been accidentally broken: it is well known that trees of this class increase from the end of the branches, and that when the shaft

G. Cathelineau del.

E. Aubert sculp.

Le Tombeau de Daubenton.   Daubenton's Tomb.

is broken they cease to grow in height. Below
the cedar of Lebanon, towards the south are two
stone pines, *pinus pinea*, of remarkable size,
which give an idea of their effect on the Ap-
penines, whose ridges are covered with them,
and in the gardens of Italy, where they are cultivated for ornáment. There are also several
larches, *pinus larix*, Weymouth pines, *P. strobus*,
hemlock spruce fir, *P. Canadensis*, balm of Gilead
trees, *P. balsamea*, a great number of yews,
taxus, silver firs, *pinus picea*, and cypresses,
and some very beautiful red cedars, *juniperus
Virginiana*.

Ascending by the path which winds several
times round the hill, we arrive at an elegant pavilion, encircled with bronze pillars and a bal-
lustrade. From this elevated spot the view extends over the garden, the greater part of Paris,
and the distant landscape in the direction of
Montmartre, Vincennes, and Sceaux. On the
eastern slope, between the pavilion and the
cedar of Lebanon, is a small enclosure of trellis,
in the centre of which a simple granite column
resting on a base of different minerals, marks
the grave of Daubenton : flowers often renewed
blossom over it, and a marble bust is to be
placed upon the column. In descending the hill
on the north side, we notice a beautiful Mont-

pellier maple, *acer Monspessulanum,* and below it the largest plane tree in Paris. Between the two on the verge of the slope, is a dairy-house, to which students who pass the morning in the garden repair to enjoy a rural repast, and inhale a cool and balsamic air beneath the shade. Below the dairy a path compassing the hill conducts to the terrace of the *rue du Jardin du Roi,* at the end of which is an entrance to the galleries of natural history.

Continuing the circuit, and still descending, we find ourselves opposite the smaller hill, which is of an oblong form, intersected with winding paths, and like the labyrinth planted with evergreen trees. Amongst these may be noticed several species of pine, particularly the Aleppo pine, *pinus Halepensis,* the cedar of Lebanon, *P. cedrus,* the black and red spruce firs, *P. nigra* and *P. abies,* evergreen oaks, phillyreas, the Japan medlar and a small clump of the blotch-leaved aucuba, *aucuba Japonica.* On the top of the hill is an esplanade with a picturesque view towards the river.

Among the herbage of these hills are often found exotic plants, sprung from seeds of the botanic garden; thus we have several years observed the cretan rye, the smyrnium of the Levant, *smyrnium perfoliatum,* etc.

Cathelineau del.

E. Aubert sculp.

L'Amphithéâtre.		The Amphitheatre.

At the foot of the smaller hill is a large oval
enclosure, in front of the amphitheatre, with the
seed-garden, the green-house, and the menagerie
on the right; and the dwellings of several professors, and beyond, the gate of the *rue de Seine*
on the left. This enclosure is destined for the
exposure during the fine weather, of the most
beautiful trees of New Holland, the Cape of Good
Hope, Asia Minor and the coast of Barbary, which
have passed the winter in the green-house. They
are placed in picturesque groupes, and the turf
is interspersed at small distances with ornamental
flowers. In the centre is a large stone table,
surrounded by the tallest trees, and covered with
plants remarkable for the beauty and brilliancy
of their flowers.

At the door of the amphitheatre are two beautiful Sicilian palms, *chamærops humilis*, 25 feet in
stature, which were sent to Louis the Fourteenth
by the Margrave of Bade-Dourlach, at the beginning of the last century, when only 12 feet high.
These vegetables never attain so great a height in
Italy or Spain. They are attached by cords descending from the summit to the angles of the
boxes, to prevent their being broken by the
wind. As they grow from the summit, and not
by lateral buds, they exhibit no layers in the
trunk, and as they every year put forth new

leaves and shed the old, the number of external rings indicates their age, like the concentric circles of ordinary trees (1). They grow also at the base of the trunk, so that the first ring is now more than a foot above the level of the soil—a peculiarity that has been observed in no other tree.

Near the amphitheatre is the entrance of the menagerie. The varied surface of the ground, the diversity of the plantations, and the singularity of the structures, give this part of the establishment the appearance of a landscape garden, though the interest excited by it is different. In English gardens, where the intention is solely to produce picturesque scenery and to vary the prospect, we are ready to ask if the effect repays the expence ; and grow weary of magnificence without utility.   But here every thing has its use : each of the numerous parks is appropriated to some species of animals, and is proportioned in extent to the exercise it requires.   The trees are disposed so as to afford it shade, and the building for its retreat recalls the habits of the animal and the country whence it came; while the

(1) This fact, however singular, is not an exception to the ordinary laws of vegetation.   The pivot being forced out by the inferior roots, which were stopped in their descent, forms at present the base or pedestal of the column.   Near the earth we observe a contraction of the trunk, as if it was nourished only by the central roots.

undulations of the ground enable the observer to examine it in every position. The basins are destined for aquatic birds, and a streamlet renews the water in which they bathe.

The spacious edifice in the centre of the menagerie contains the quadrupeds which require warmth in the winter, and suckling females that need particular attention. The pheasant-walk is sheltered towards the north, and has an open court in front. The dens of the beasts of prey, completed last year on an elegant but simple plan, are exposed to the south, and placed in the immediate vicinity of the river, for the conveniency of procuring water and of discharging it by a subterraneous drain. We shall describe these objects more particularly in revisiting them; we have barely hinted at their use, that their variety might not be attributed to vain caprice.

Taking alternately the alleys on the right and left from the entrance, we make the tour of the different parks, pass before the cages of the apes and birds of prey and the pheasant-walk, and reach the rotundo. Beyond this we visit the new buildings on the left, arrive at the lodge of the wild beasts, and return by winding through other enclosures, to the terrace leading to the green-house. Below this terrace are seen to advantage the garden of naturalization and that of the seed-beds. From the green-house we descend a small declivity, and by hollow paths regain the spot from which we started. Each of the enclosures is encircled by a trellis of different workmanship; and the alleys are shaded by trees equally remarkable for their variety and their beauty.

In this part of the establishment are assembled all the foreign trees and shrubs which pass the winter in the open air, and no where do we see so great a number of species arrived at their full dimensions. As they are nourished by a rich soil, and carefully cultivated, their growth is surprisingly rapid. They have all been planted since 1797, but some of them had already attained a considerable size, and only a few of their branches were retrenched, in order to preserve their natural appearance: they prove the inutility of lopping trees that are to be transplanted; a practice censured by M. Thouin both in his writings and lectures.

As these trees are labelled we shall mention only a few, deserving of notice for their rarity or perfect vegetation. The paper mulberry, *morus papyracea*, the first received in Europe, was sent to the King's Garden by Sir Joseph Banks, who brought it from Otaheite. After remaining a long time in the seed-garden, where it suffered

from the shade, it was removed to the menagerie. The black and grey walnut, *juglans nigra* (1) and *J. cinerea* which bear fruit every
year; different species of nettle-tree, *celtis;* the
ash and maple-trees of America, the mountain-
ash of Lapland, the white lime-trees, *tilia alba,*
the plum and cherry-trees of North-America, the
willow leaved pear-tree, *pyrus salicifolia,* the
sumachs and the acacias, are of extraordinary size:
the glutinous acacia, *robinia glutinosa,* brought
by the younger Michaux from North America, and
highly esteemed for its foliage and rose coloured
flowers, which are borne twice in the year, has
been extensively spread from seeds produced in
the garden. Most of the trees in the menagerie are what are termed *seed-bearers,* or stocks
destined for the multiplication of the species.

Some years ago a greater number of shrubs and
plants were found along the alleys, such as the

(1) This tree grows in the forests of Kentucky and Ohio. It is 70 feet
in height and 6 in diameter. Its wood is of a deep brown colour, compact, susceptible of a fine polish, and not liable to warp nor to be attacked by insects. it forms beautiful cabinet work, and is used in
ship-building. It would be advantageous to multiply this tree in
France, as it shoots quicker and grows more rapidly than the common
walnut, and is preferable in every thing but its fruit. M. Michaux
proposes to form nurseries of it, and to graft the stocks when 6 or 8 feet
high with the European walnut, thus uniting the advantages of the
two species. See North American Sylva, vol. 3, p. 164.

The grey walnut is also a very beautiful tree, but every way inferior to
the preceding.

spiræa, the honeysuckle, the clematis, etc. which have disappeared, as they were over-shadowed by the larger growth; but we still see round the edges of the parks the iris, St. John's wort, the hipericum and other species distinguished by the the beauty of their flowers. Behind the green-house is a pump kept in motion by dromedaries, which supplies water for the basins.

Quitting the menagerie and turning to the left, we pass before the green-house and approach the botanic garden. If we wish to reserve the other parts of the establishment for a more particular examination, we may return to the quay by the avenue of horse-chesnuts.

On the left, as we proceed, are seen the seed-garden and that for the naturalisation of exotic plants, and beyond, three deep paved courts with cells constructed in the sides for certain animals. Several bears formerly occupied the two first, and afforded much amusement to the public; but a person having perished in one of them, the animals were removed till an iron railing could be placed upon the wall. The third of these pits contains a number of wild boars, which have several times unpaved it. The parks of the menagerie occupy the rest of the space on that side, to the ancient lodge of the wild beasts upon the quay.

On the right succeed in order the botanic garden, an avenue of the sophora of Japan alternated with young stocks of the arbor vitæ, which last will be suppressed; the *school* of fruit-trees; an avenue of planes; the school of vegetables used in domestic economy and the arts; an avenue of Virginian catalpas; the school of agriculture; and lastly the avenue of judas-trees, and the plantation already mentioned.

Having thus taken a general view of the establishment, we shall enter upon the more particular description of its parts, beginning with the botanic garden, and ending with the cabinets, menagerie and library.

§ II. THE BOTANIC GARDEN.

———

THE Botanic Garden, which is encompassed with an iron railing, extends east and west from the ascent between the level grounds and the two hills, to the square planted with fruit-trees; and north and south, from the avenue of horse-chesnuts to that of lime-trees. It contains 3,064 square fathoms, divided by longitudinal and transverse alleys three feet wide, into sixteen compartments, forming one hundred and forty-four beds of unequal size, but generally 60 feet in length and 5 in breadth. In the main alley are four basins at equal distances, which furnish water for the garden. Two of its gates are always open, except when the gardeners are at their meals, though few persons are seen in it besides students (1).

This section of the garden, which is the foundation and most essential part of the whole, connects all the portions destined for culture by fixing the nomenclature of the vegetables and

(1) Children are kept at a distance on account of the damage they might do, and the danger to which they would be exposed from noxious plants.

determining their affinities; exhibits the actual state of our botanical riches; and ensures the propagation of the science, by enabling pupils to study the plants in all the stages of their growth, to compare them with each other and with the descriptions in books, and to observe their changes and physiological phenomena, according to the season, the hour, the intensity of the light, and the state of the atmosphere: in fine it affords the means of gathering seeds without fear of mistake, when the absence of the foliage renders it impossible to distinguish the species of the plant.

It is not in the season of the lectures only that the botanic garden is frequented by students; as soon as the snow-drop, *galanthus nivalis,* and the mizerion, *chamæ daphnæ,* announce the revival of vegetation, they resort to it, and every day some opening flower attracts their attention. Aided by the lessons of the professor they learn to analyze the organs and distinguish the characters of plants, and thus to recognise in the herbarium those which are not cultivated in the garden: neither dried specimens, nor figures however perfect, can supply the place of living nature.

There probably exist elsewhere collections of plants as numerous as that of the Museum, but there can no where be found an equal assemblage disposed in the order of their natural affinities, and consisting of the vegetables of every climate which it is possible to raise from the seed in Europe. By comparing the list with the *species* published since Linnæus, we are enabled to distinguish such as are well ascertained from such as are doubtful, and to seek abroad those that are still wanting.

This garden at first consisted of a collection of medicinal plants, mostly indigenous; it was enriched by Tournefort with a few exotics, but it extended only to the extremity of the hothouse in front of the small hill, and though the trees and shrubs had been removed, it was too small to contain all the species. At the solicitation of M. de Jussieu, Buffon trebled its extent, and encircled it with an iron railing, in 1774: at the same time the plants were disposed in their natural order. Not long after it was found still too confined, and was augmented one fourth in 1788: in 1802, M. Desfontaines assembled all the species lately arrived, those dispersed in the parterres, the green-house, and the seed-garden, and others in the possession of individuals, and planted it anew. It now contains six thousand five hundred species ; and as more are received than perish, the number is annually increasing.

All these plants cannot remain constantly in the botanic garden, as some are too delicate to support the air, and others too rare to be exposed without precaution; but they are all exhibited when the family to which they belong is treated of in the lecture, and they may be studied in the hot-houses, when in flower, by applying to the professor or headgardener. Many annual plants of short duration are sown at different periods, that they may be found in the garden after they have disappeared from the fields; and some which grow naturally in marshes, in forests, and on the sides of hills, and which it is found impossible to rear, as several of the *pediculares* and *orchideæ*, are renewed from the country as often as they perish.

All the plants are labelled: the large red tickets indicate the classes and families; the labels of the genera are placed above those of the first species; and those of the species, besides the Latin and French names, the country of the plant, and the signs of its duration and habits, indicate by coloured lines its use in medicine, domestic economy, the arts, or ornamental gardening, and the character of its juices (1). The catalogue by

(1) The *red* stripe signifies that the plant is used in medicine; the *green* that it is employed in domestic economy; the *blue* in the arts; the *yellow* in ornamental gardening; and the *black* that it is venemous.

M. Defontaines contains an exact list of the names and synonymes.

The plants seen in the middle of the beds without labels, are species lately received and not yet determined, but whose general appearance indicates their family and genus.

To examine the garden in its order we must begin at the western extremity of the beds parallel to the avenue of lime-trees. The plants are disposed in two rows at suitable distances. The first orders, or the *fungi, algae,* and *musci,* are suppressed, from the impossibility of cultivating them, and we begin with the ferns, *filices,* of which there are more than fifty species(1): next come the *naïades,* the *aroïdeæ,* the *cyperoïdeæ,* and the *gramineæ,* which commence towards the end of the first row, and form the second in returning along the other side of the beds: the number of species is about five hundred and forty, three hundred and ten of which are grasses. The greater part of these plants would suffer by intense heat, and are pleased in the shade of the avenue.

Towards the right, in the angle of the garden near the green-house of Dufay, begin the palms.

(1) Several of the *filices* which grow in damp and shady situations soon decay, and as it is difficult to collect the seeds, the species enumerated in the catalogue are not always found in their places.

We must now trace and retrace all the paths
which divide the beds, advancing from left to
right, to the bottom of the garden. The plants
are in double rows in the families which contain few trees; and in those which require more
room, and pass the winter in the air, as the
aceræ, the *rosaceæ*, the *amentaceæ* and the *coniferæ*, they occupy the middle of the bed. Wicker
cages open on one side are placed over those
which it is necessary to defend from the north
wind or from the sun, and bell-glasses over those
of the torrid zone which require a concentrated
heat. The aquatic plants, as the *naiades*, the
nympheæ, water-lilies, and some species of ranunculus, are kept in tubs of water; and those
which thrive best in rocky ground, as several
species of fern and houseleek, *sempervivum*, are
provided with an artificial soil. To the palms,
of which we have only twelve or fifteen species,
succeed the *junceæ*, the *liliaceæ* where are found
the aloes; the *irideæ*, the *orchideæ*, and the *hodro-
charideæ* among which is the *valisnerea*, an
aquatic plant remarkable for the manner of its
fecundation (1). Here terminates the series of
monocotyledon plants, of which there are eleven

(1) A plant deep-hid in Rhone's impetuous tide
    Ten lonely months beholds his waters glide;
    Th'eleventh, when genial love renews the year,
    The females rising o'er the stream appear,

hundred species in the garden, or one fifth of the whole; which is the proportion observed between the two great divisions, in the vegetation of the globe.

The dicotyledon families begin with the *aristolochiæ*, which form a separate class : after them are the *eleagneæ*, the *proteaceæ*, the *laurinæ* and the *polygoneæ* which comprise the rhubarbs, the *amaranthi*, the *labiatæ*, and the *solaneæ*, in which the genus night-shade, *solanum*, alone contains more than sixty species : there are also several species of thorn-apple, *datura*, and tobacco, *nicotiana*, lately received.

A little further, at about one third of the length of the garden, is a bed of bog-mould surrounded by a trellis and sheltered from the north and west, for the heath family, *ericeæ*.

We next arrive at the numerous class of compound plants, of which eight hundred and fifty

> The males, that fetter'd near the bottom lay,
> Burst their weak stems and cleave the eddying way,
> Crowd round each fair, exhale their amorous pain,
> And form her court above the moving plain;
> As if with pomp th' auspicious bridal god
> In festal triumph o'er the waters rode.
> But soon, accomplish'd Venus' pleasing law,
> Again the females from the light withdraw—
> Sink on their spires, and 'neath the wave secure
> Brood o'er the joy, and fruitful seeds mature.
>   . . .
> CASTEL, *Poëme des Plantes,* Chant I.

species are cultivated in the garden : they bloom chiefly in the autumn, and are most of them of the finest vegetation. The family of the *rubiaceæ* so diversified between the tropics, presents but a small number of species, such as the gardenia, the houstonia, the coffee-tree, etc. which thrive only in the hot-houses. Towards the middle of the garden after the *gerania, malvaceæ,* and *magnoliæ,* where the *caryophillæ* begin, is an arbour of climbing plants, covering the entrance of the subterranean passage to the seed-garden. Beyond this are the *myrti,* consisting of forty-five species, thirty of which have lately arrived from New Holland, and are esteemed for the decoration of gardens ; the *rosaceæ,* which comprise the greater part of our most beautiful flowers and delicious fruits : the *leguminosæ,* of which we have six hundred species, and among them several mimosas unknown till the last voyage of captain Baudin ; and the *terebinthaceæ,* a remarkable family composed of trees and shrubs resembling each other is the pungent odour of their leaves : here are seen the mahogany, *anacardium,* the pistacia, the mango-tree, *mangifera,* and the sumachs, of which two species are especially deserving of attention, the trailing sumach, *rhus toxicodendron,* and the upright poisonous ash, *rhus radicans,* whose juice is so caustic, that a single drop upon the skin causes an inflammation that soon extends over the whole body. Not far from these ligneous vegetables are seen recumbent on the ground or supported by a trellis, the *cucarbitaceæ* and *passifloræ,* two families distinct in properties and appearance from those that surround them. The *amentaceæ* and *coniferæ,* among which are our largest forest-trees, terminate the series. In the last bed are several plants, such as the side-saddle flower, *sarracenia,* the begonia and the coriaria, whose place in the natural order is not determined.

We shall not particularize the species which are not found elsewhere, as they are enumerated in the catalogue of M. Desfontaines, but shall mention only a superb eastern liquidambar, or sweet gum; a beautiful assortment of gleditschias, among which is that of the Caspian Sea; a fine virgilia (1), for the knowledge of which we are indebted to M. A. Michaux; and a *mimosa julibrissin* or silk-tree, remarkable for the delicacy of its foliage, and the colour and elegance of its tufted flowers.

Opposite the third basin, amongst the leguminosæ, is a fine Corsican fir, *pinus laricio,* which

(1) *Virgilia lutea* (yellow wood, Mich.) the buds of this tree, like those of the plane, are enclosed in the petiole, and do not appear till

was left standing when the garden was enlarged.
Near one of the gates on the avenue of limes, are
two rose-trees grafted upon the eglantine, whose
tops are covered with flowers of different species, at fifteen feet from the ground.

Varieties are not admitted into the botanic
garden unless they are constant, and in other respects remarkable, as they would occupy time
and space which are better bestowed upon
primitive types of species. In the fine season
the contents of the hot-house are exposed upon
a mound in front of the building (1).

Botany is too extensive a science to be learned
in a single year. Those whom a decided taste
attaches to this study, should proceed with method, and avoid embracing too great a variety of
objects, at once : during the first season they
should content themselves with verifying on
several species the characters of the genera
treated of by the professor, and the following
year, should resort to the garden at the opening of vegetation, and examine all the plants in
succession, as they bloom and mature their seeds.
Between the spring and autumn, persons who

the falling of the leaf. M. Michaux has named it the *virgilia lutea* on
account of the yellow colour of its wood.—See History of the Trees of
North America, vol. iii. page 266.

(1) It is of less extent this year than formerly, as one-third of it is
occupied by the new hot-house, constructed in 1821.

wish only to obtain a general idea of vegetable physiology and the affinities of plants, may behold each species surrounded by those which it most nearly resembles; and, with a few intervals, may pass by regular gradations from the *lilaceæ* to the *coniferæ*, from the tulip and the hyacinth to the fir-tree and the cedar of Lebanon. They will notice some families, such as the lilies, the labiated and the cruciform plants, so natural that a person the least accustomed to observation is struck by the analogy; others united by less sensible affinities, as the *rhamonïdeæ*, the *euphorbiaceæ*, etc.; some extremely numerous, as the compound and papilionaceous plants, and others which are limited to a few species, as the *aristolochiæ* and the *hypericeæ*; some whose characters run into each other, as the *cruciferæ* and the *capparideæ*; others again abruptly separated, as the *umbelliferæ*, the *rannunculaceæ*, the *euphorbiaceæ*, and the *cucurbitaceæ* : in fine, they will observe some genera so peculiar, that their place has been determined only by systematic considerations, such as the reseda, the *nasturtium tropæolum*, and the parnassia ; and others whose species cannot be distinguished when found alone, but which it is nevertheless important to separate, as the sixty species of *aster*, which vary by almost insensible shades, but bloom at different periods;

and many grasses, scarcely distinguishable by scientific characters, but differing in the rapidity of their growth, and of very unequal value in rural economy. Flowers constantly succeed each other throughout the year, from the daphne to the species of hellebore called Christmas-rose, and January is the only month in which none are found except in the orangery.

Many curious phenomena are observed in this vast assemblage of plants, such as the opening of flowers at a fixed hour —some unfolding at the dawn, as certain species of rock-rose, *cistus ;* some at mid-day, as several *ficoïdeæ,* and others at sunset, as the *mirabilis jalapa,* several of the *onagræ,* etc. Some species foretel the weather, as the small Cape marygold, *calendula pluvialis,* and the Siberian sowthistle, *sonchus sibericus,* the first of which opens and the second shuts, in the morning, when the day is to be fine. The sleep of plants, which presents itself under forms so different, not only in different families, as the *leguminosæ* and *malvaceæ,* but in neighbouring genera, as the mimosa and cassia, and even in different species of the same genus, as the milk-vetch, *astragalus,* can be properly observed only when a great number of species are assembled in the same exposure ; and these phenomena are the more interesting as they are connected with the secret of vegetable life. A fact still more curious is, that many flowers shed their perfume only after sunset, as the *geranium triste*, the *gladiolus tristis*, the *cestrum nocturnum*, some species of wall-flower, *cheiranthus*, etc.; and that this singular property is almost always announced by their colour, which is a mixture of a brownish red, yellow, and white (1).

Every year may be seen plants newly arrived, and destined one day to adorn our gardens. Shrubs with brilliant flowers, such as the *camelia japonica*, and plants used in the decoration of gardens in foreign countries, as the Indian chrysanthemum (2) might reach us by the ordinary

(1) See a memoir on the cruciform plants, by M. de Candolle, in the Memoirs of the Museum, vol. vii. page 184.

(2) This species of *anthemis* is much cultivated in the gardens of China, and its different varieties offer every shade of colour except blue. It was not known in Europe till 1789, when M. Blanchard, a merchant of Marseilles, brought it from China, and presented several stocks to the abbé Ramatuel, who sent them to the King's Garden in 1791. It was first placed in the orangery, afterwards in the open air, and as it multiplies rapidly it was soon spread in our gardens. It was introduced into England in 1795. M. Ramatuel published a description of it in 1792, in the Journal of Natural History, vol. ii. in which he proves it to be a distinct species from the *chrysanthemum indicum*. It is a valuable acquisition, because it blooms in the middle of autumn, when other flowers have disappeared, and resists the first attacks of frost. There is no ornamental plant which accommodates itself so easily to a variety of climates; it blooms in Belgium and in the Isle of France.

The abbé Ramatuel is the author of an excellent memoir on buds, of which M. Desfontaines has given an extract in the Journal de Physique, vol xlii. page 62. He died in Paris in 1795.

channels of commerce; but the smaller species would never be introduced, if their seeds were not gathered by naturalists and transmitted to botanic gardens. In this way we have obtained the reseda, sent from Egypt by Granger in 1736; the turnsole, from Peru by Joseph de Jussieu in 1740, and the *silene bipartita*, brought from Barbary by M. Desfontaines in 1786. The same may be said of the pretty Minorca sandwort, *arenaria balearica;* of the Mexican lopezia, which we received from Spain some years ago ; of the sweet-scented coltsfoot, *tussilago,* found on the Pyrenees, and described by Villars; of the *cacalia sagittata,* lately received from Java; and many other plants now common in our flower-markets.

The intervals left for new species, in replanting the botanic garden, have in many instances been filled up, especially by the accessions from New Holland; and probably ten years will not elapse without the necessity of again enlarging it.

§ III. THE SCHOOL OF FRUIT-TREES.

Th is enclosure, situated between the botanic
garden and that of economical plants, is 240 feet
in extent from east to west, and 166 from north
to south. It is divided into sixty beds 3 feet and a
half wide, separated by paths of 3 feet, and bordered by varieties of the strawberry; and contains more than eleven hundred species or varieties of fruit-trees, methodically arranged, and
placed from 3 to 9 feet apart, according to their
size. The first beds are occupied by trees or shrubs
yielding berries, as the currant, raspberry, grape,
and mulberry; the second division, by those bearing drupes or stone-fruit, as the cherry, plum,
apricot, etc.; the third, by those with ligneous
seeds, as the medlar, source, and date plum; the
fourth, by those whose seeds are a kernel covered
with a pellicle, as the apple and sorb apple, and
by those with juicy fruit, as the fig; and the fifth,
by those of which the kernel is contained in a
hard or cartilaginous shell, and is the only part
eaten: this class may be subdivided into two, in

one of which the shell is naked, as the fir-apple and hazel-nut, and in the other, enveloped in a husk, as the walnut and chesnut.

At the bottom of the plantation are peach-trees affording models of wall-fruit; and other trees singularly pruned.

This enclosure was planted in 1792, while M. de St. Pierre was intendant of the garden. Two individuals of each species were obtained from the celebrated nursery of the *Chartreux,* and from that of Vitry which furnished Duhamel with the subjects and names of his treatise; it consequently offers a type of that work, and a standard for the nomenclature.

The trees are generally grafted at the surface of the ground, and pruned in the shape of a distaff, which has been chosen, not because it is generally the best, but because it economises room, is the most convenient for observation, and gives birth to longer and more vigorous shoots; the object being to facilitate study and multiply the different species, and not to exemplify the method of rendering trees the most long-lived and productive.

Complete series of the species cultivated in this garden have been sent to Ghent, Strasburg, and Vienna, where nurseries have been formed upon the same plan and with the same nomenclature: fruit-trees have hitherto been designated under different names by different cultivators, and it is often difficult, even with the aid of books, to identify them.

Few remarkable species exist that are not found in the garden of the Museum. At first it contained only those described by Duhamel, but others have since been procured from foreign countries or distant provinces, and the original number has been nearly doubled.

The aspect of this plantation is amusing to the careless observer and instructive to the student. In the winter may be examined the characters which depend upon the colour of the wood and the form of the buds—a very necessary branch of knowledge to the cultivator, as plantations of fruit-trees are formed after the falling of the leaf.

The publication of the catalogue has been delayed in order to ascertain the synonymes used in other countries, but a list corresponding with the numbers on the trees is in the possession of the gardener, who readily names the species with which they are unacquainted to those who desire it. Grafts may be obtained by applying to the professor, and designating the species by the common name or by the number of the tree.

Among the species which are still rare, or

lately introduced, may be noticed a very fine individual of the two-flowered date plum, *dios-pyros kaki,* whose fruit, of the size of a small apple, is highly esteemed in Japan ; a beautiful Japan medlar, *mespilus Japonica,* which passes the winter in the open air, and yields a palatable fruit ; the pine of Monteray, sent from California during the voyage of La Peyrouse, the kernels of which are preferable to those of the stone pine ; the Japan quince, *cydonia Japonica,* and the pear-tree of Mount Sinai, both curious species ; some beautiful hazel-nuts from the Levant, etc. This part of the establishment is not open to the public, but admittance is never refused for instruction.

§ IV. THE SCHOOL OF PLANTS

USED IN DOMESTIC ECONOMY AND THE ARTS.

THIS garden, which is 220 feet long and 185 broad, is divided into twenty-three beds 6 feet wide, and subdivided into five hundred and fifty-two compartments, destined for as many species or varieties of plants, arranged according to their properties and uses, and distributed into three principal groups, as they serve for the food of man, the nourishment of animals, or the purposes of art.

Each group comprises several sections: the first is divided into cerealious plants, others with farinaceous seeds, pot-herbs, and plants yielding oil; and each section is subdivided according to part of the vegetable most esteemed: thus the culinary plants are distinguished into those of which we eat the root, as the potatoe and Jerusalem artichoke; the leaves, as the cabbage, spinage, etc.; the calix or the flower, as the artichoke and cauliflower; the fruit, as the pumpkin and melon; or the aromatic seeds, as fennel, coriander, etc.; and lastly, those which are eaten in salad, as lettuce, succory, etc. Of the vegetables used in the arts, those which furnish tissues, as flax and hemp, are distinguished from those which afford die-stuffs, as madder and woad. Those which serve for the nourishment of animals are divided into grasses, leguminous plants, and pasture herbs. A separate section comprises vegetables of a peculiar nature, as the tobacco, hop, teazle, etc.

In a botanical garden the plants must be detached for the convenience of studying them, but for the object under contemplation they should be disposed, as they are cultivated in fields or gardens, in masses; which should be of four times or at least twice the extent we have allowed them, where the space will permit.

To avoid putting the same plant into the same ground, the series is begun every two years at opposite extremities of the garden ; which preserves the order, and changes the place of each species. It is necessary also at intervals to renew the seed, as the varieties degenerate when long cultivated in the same spot, and often fecundate each other, thus producing mixt species that replace the primitive type.

Twenty thousand packets of seeds from this garden are annually distributed to cultivators, with inscriptions indicating the name and season of the plant, the soil best adapted to its growth, and its principal use. Among the species which have been thus propagated may be mentioned the flax of New Zealand, *phormium tenax,* and the Pensylvanian potatoe, *convolvulus batatas,* which are now spread through the south of France; the cresses of Para (1), the *tetragonia expansa* (2), the *cleytonia cubensis* (3), etc.

At the formation of this garden the ground was raised eight feet, to bring it to the level of the surrounding soil: the details of its cultivation may be found in the memoirs inserted by M. Thouin, in the 2d volume of the Annals, p. 162.

(1) *Spilanthus oleracca;* it excites salivation, and often dissipates the tooth-ache when rubbed upon the gums.

(2) A good esculent brought from New Zealand by Sir Jos. Bankes.

(3) A culinary plant, for which we are indebted to M. Bonpland.

§ V. THE SCHOOL OF AGRICULTURE.

Next to the garden of economical plants is the
School of Agriculture, which was formed by
M. Thouin in 1806, for the purpose of exemplify-
ing the processes of cultivation, and verifying the
principles of vegetable physiology. Though less
extensive than the object requires, it is not the
least interesting part of the establishment.

The enclosure is 210 feet in one direction, and
180 in the other, and is divided into forty-eight
beds, from 3 to 6 feet wide, according to their
destination to one or other of the following objects: the birth of vegetables, their preservation,
their multiplication by other modes than from
seed, and the exhibition of the uses in agriculture
and gardening of assemblages of living plants.

The examples of the third division are placed
immediately after the first, that the slips and
layers, which are liable to injury from the sun,
may be shaded by the catalpas of the avenue and
other trees.

As there is but one mode of giving birth to

vegetables, whilst there are many of preserving and multiplying them, the general plan is divided into ten sections succeeding each other in the following order:

1st. Seeds sown in the open ground, in hot-beds, under glass, in pots, in water, and on other vegetables; with the methods of ensuring their developement.

2d. Slips, and the means of facilitating the success of such as do not easily take root.

3d. Layers.

4th. Grafting; with heteroclite grafts (1).

5th. Plantations and their various treatment, according to the nature of the trees, their original climate, and the period at which the sap begins to circulate.

6th. Pruning, either to prolong the life of trees, to give them a particular form, or to change the quantity or quality of their fruit.

7th. The training of vines, whether detached or in arbours or espaliers.

8th. Hedges; divided into those of defence; simple, double, grafted, and forage hedges.

9th. Palisadoes of spring, summer, and winter.

(1) These experiments repeated with great care and in various ways prove that trees having no analogy cannot be grafted upon each other; and that all that has been written from Columella to the present day, on the means of procuring hybridous species and extraordinary fruits, is void of foundation.

10th. Ditches; their formation, and the manner of covering the declivities, and crowning them with hedges.

The choice of these examples exhibits the different processes of cultivation, and the experiments most proper to throw light on the phenomena of vegetation. But their utility cannot be appreciated without attentive examination, detailed comparison, and explanations of the manner in which they have been brought to perfection. Their distribution forms an assemblage of objects, pleasing by their contrasts and instructive by their analogies.

In the models of pruning we see side by side the forms of the distaff, the sphere, the bush, the vase, the fan, the espalier and the arc (1) ; and also the methods of hastening the maturity of fruit and preventing its untimely fall, by removing a ring of bark from the branch, or interrupting the course of the sap by ligatures.

Among the grafts are seen trees joined by their trunks, so that a single summit is nourished by four or five stocks: others with their branches curved and united, so as to form arbours covered

(1) This method, which consists in bending the branches into a semicircle to divert the course of the sap, has been abandoned ; its result is a surprising fecundity at first, but after the second year the curved branches become sterile, and the stock decays.

with fruit and garlands of flowers; trunks bearing as many species of fruit as they have limbs; stocks of a year old 15 or 18 inches high, grafted from an adult and laden with fruit; arbours formed by grafting together two rows of shrubs, which are covered with foliage from the surface of the ground, and form an impervious arch; and lastly, examples of a proposed form given to the trunks and branches of trees, in order to furnish solid natural curves for building.

The professor of agriculture delivers a part of his course in this garden; persons who wish to visit it, without attending the lectures, should read the memoirs in the Annals of the Museum, in which he has explained the various processes, and the result of a great number of experiments, some of which are not immediately useful in augmenting the product of vegetation, but which are all calculated to throw light on questions of vegetable physiology equally important in theory and practice.

§ VI. THE SEED-GARDEN.

This indispensable accessary to the botanic garden has existed only since 1786: the ground was procured the preceding year by Buffon, and confided to the direction of M. Thouin: to whose memoirs in the 4th and 6th volumes of the Annals of the Museum we refer the reader, who wishes to become acquainted with its details, and with the methods employed in rearing the vegetables of every climate.

This enclosure is 885 square fathoms in extent, and is 10 feet below the general level of the Garden. It is sheltered by a wall on the east, by the green-house on the north, and by the small hill of ever-green trees on the west: its plan and distribution may be seen from the avenue of horse-chesnuts or the terrace of the menagerie. The entrance is by a flight of steps at the end of the terrace of the green-house, which is 200 feet long, and which in the fine season, when filled with trees and shrubs that have passed the winter in the building, presents a splendid picture from the avenue of horse-chesnuts. Along the wall of this terrace is a line of eight hot-beds with glazed frames of iron for the seeds of warm climates, and a small hot-house in the middle (1); below, is another line with frames of stone for the bulbous and tuberous plants of the Cape of Good Hope and similar climates, which are covered with glass only in the winter; and at the extremity of the two, a second flight of steps to the terrace.

In front of the last division are two rows of simple beds 6 feet wide, 150 feet long and 2 feet and a half deep when newly made, but only 10 inches when settled, for the seeds of the temperate zone, containing 5,000 pots arranged in conformity with the numbers in the gardener's catalogue : a part of these beds are sometimes covered with glazed frames. In the opposite direction, extending from north to south, are thirty-seven other beds of unequal length and 5 feet in breadth, of which the first twenty-nine are destined for annual plants which arrive at matu-

(1) This hot-house belonged to the underlibrarian, M. Delaunay, a very learned amateur, and author of a well known book entitled *The Good Gardener's Almanack*. At his death it was purchased by the Museum, and it is now employed for raising the plants of New Holland and the Cape, which require a temperature above that of the green-house.

rity in a few months, and which are not sown till the advancing season has warmed the earth; and for those which have germinated in pots, and from which we are desirous of obtaining seeds. The eight following cold beds are destined for the transplanting, separation, and rooting, of young stocks, obtained by sowing the seeds of shrubs and vivacious plants from the frigid and temperate zones. The pots are buried in the earth, and sheltered for a few days from the sun; and the young plants, when they have taken to the soil, are transported to the green-house or the parterres.

Next are two sunk beds with a western exposure, for the pots of seeds that have not germinated the first season, but which may spring the following year.

East of these beds is a range of lofty arborvitæs, between which and the wall is a lodge, or work-shop, for the head gardener and his assistants, and a receptacle for their implements, tools, mats, bell-glasses, etc.

Returning on the side of the garden next the avenue we pass a line of sunk beds, defended from the noon-tide by the foliage of the horse-chesnut avenue, and covered with pots containing the greater part of the seeds from cold countries, and a few, which are extremely small, from temperate climates: at each end are vases of water for the seeds of aquatic plants.

Under the wall is a bed, divided in the middle by the subterranean passage to the botanic garden, which receives the rays of the rising and setting sun, and is perfectly sheltered at mid-day, where the shrubs and large vivacious plants of very cold regions are sown and nurtured: here are seen beautiful ferns from the north, veratrums, umbelliferous plants, daphnes, gentians, and alpine geraniums.

Along the terrace wall on the west is a mound composed of five stages, each 10 inches high and a foot wide, divided into compartments of 8 feet 10 inches in length, which is exposed to the north-east, and visited by the sun only till ten in the morning. Preparatory to its construction a fosse was scooped out and plastered over with impervious mortar; it was then filled with stiff mould, upon which were placed shelves of oak plank, supported by stakes and covered with bog-mould: in this manner a constant humidity is preserved, by preventing the water from losing itself in the sandy under-soil.

Here are reared with success the plants of the polar regions, and those which grow on mountains in the vicinity of the snow. They are almost all of low stature, and are covered in the spring with pretty flowers. Specimens of each are carried to the botanic garden, where they soon wither, while on the mound they retain their freshness till the ripening of the seed. We here see, in all their native elegance, the mossy mæhringia, the two flowered violet, androsaces, primeroses, and saxifrages of the Pyrenees, the alpine soldanella, the wormwood of the glaciers, *artemisia glacialis,* various species of willow-herb, *epilobium,* catchfly, and ranunculus, which grow in the edges of the snow, with several dwarf-willows and other shrubs of Greenland and Upper Canada. At the end of this stage we reascend the terrace.

In the middle of the garden are a basin and a well; unfortunately the water is so strongly impregnated with selenite that it forms a crust about the roots, which necessitates the use of river water for the delicate species : this inconvenience will be remedied by conducting the water of the canal of the Ourcq to the Museum. About the basin are troughs for aquatic plants.

The above concise description affords an idea of the various labours of this garden. It is hardly necessary to add, that it every year furnishes undescribed plants, the product of seeds sent without names by travellers. The superintendance is exclusively confided to M. John

Thouin the headgardener, who devotes the greater part of his time to it, particularly at the season of sowing the seeds; an operation which he entrusts to no other hand.

In the pot with each seed is placed a bit of lead with a number engraven on it, corresponding with that of a catalogue containing the name of the plant, of the country whence it came and the person who sent it, with the date of its reception, and the time when it was sown.

The seed-garden, as we have already remarked, is the nursery of the establishment; but for science it offers another species of interest in the opportunity of observing the germination of seeds, and the early growth of plants. It is well known that the evolution of the radicle, plumule and seed-lobes, affords important characters for classing vegetables, and determining their affinities. These characters, it is true, may be discovered by the dissection of seeds; but they are much more apparent in their incipient germination and the successive developement of the parts: the first phenomena of vital action are of the highest importance in vegetable physiology. It was here that M. Mirbel made the beautiful observations on the characters and developement of seeds, which he first communicated in the Annals of the Museum, and which he has since republished with additions in his Elements of Botany and Vegetable Physiology.

This garden is not open to the public; the headgardener and the professors only have keys. The necessity of this precaution is obvious, but persons visiting it for scientific observation obtain admittance from M. John Thouin, who sends a gardener to attend them, or accompanies them himself.

In every part of the establishment is found, what perhaps is not to be met with elsewhere in Europe, the utmost readiness in communicating whatever can gratify the curiosity of the public, aid the progress of the student, or contribute to the advancement of science.

A djoining the Seed-garden and on the same level is that for the naturalization of foreign plants, which is 132 feet long, and of the same breadth as the preceding at the entrance, but narrower towards the other extremity. On the eastern side, which is sheltered from the north and west by the walls and by the arbor-vitæ hedge, are exposed, during the summer, the greater part of the trees and shrubs from New Holland, which have passed the winter in the green-house, the metrosideros, the melaleuca, the leptospermum, the eucalyptus, the banksia, the embothrium, etc. Under the walls of the three remaining sides, which are 15 feet in height, are placed different trees and shrubs according to the exposure ; thus on the south we see the pistacia-tree, the ziziphus, the pomegranate, the *ephedra altissima* (1), brought from Barbary by M. Desfontaines, a

(1) A leafless shrub, with slender, pendent, ever-green branches, which climbs upon other trees, and covers them with thick tufts.

lagerstrœmia from the East-Indies (1), a very beautiful oak with sweet acorns from Spain, the caper-tree, *capparis,* etc. On the north are placed the shrubs and plants of cold countries, such as the spiræa of Siberia, with several species of orchis and fern.

This garden is divided transversely by two adjoining walks of arbor-vitæ, under which are raised in pots, such vegetables as grow in the thickest forests, and require an impervious shade. Before these alleys is a well and basin, and near them a paper mulberry-tree, sprung from the stock presented by sir Jos. Bankes, which has been transplanted to the menagerie. The remainder of the space is divided into beds for the culture of the rarest and most interesting vivacious plants which grow in the open air, and of others newly introduced or imperfectly known, which are removed to the parterres as their species are determined.

The aspect of this garden from the avenue of horse-chesnuts, or the terrace of the menagerie, is rendered extremely picturesque by the beauty and variety of foreign plants, growing in a rich soil and developing the full luxuriance of their flowers and foliage.

(1) *Lagerstrœmia indica,* L. a shrub cultivated for the beauty of its flowers in China and the East-Indies.

§ VIII. THE GREEN-HOUSE.

THE green-house was begun in 1795, after the design of M. Molinos, and finished in 1800. It is 200 feet long, 24 wide, and 27 high to the centre of the arch, and has seventeen windows, 9 feet and a half wide and nearly 12 in height, besides the circular part which does not open, placed on a wall of free stone 19 inches from the ground, and separated by stone pillars: the great door on the west is 10 feet wide and 24 in height. Two stoves furnished with pipes are lighted when the thermometer without is 4° of Réaumur, or 9° of Fahrenheit below the point of congelation; but as the building faces the south, the least ray of sunshine produces the necessary warmth by day, while mats and inside shutters fence out the cold by night. The window sashes, which would be in danger of drooping from their excessive weight, if supported by ordinary hinges, turn on pivots, placed at one quarter of their breadth, with such ease that they are opened by the pressure of a finger.

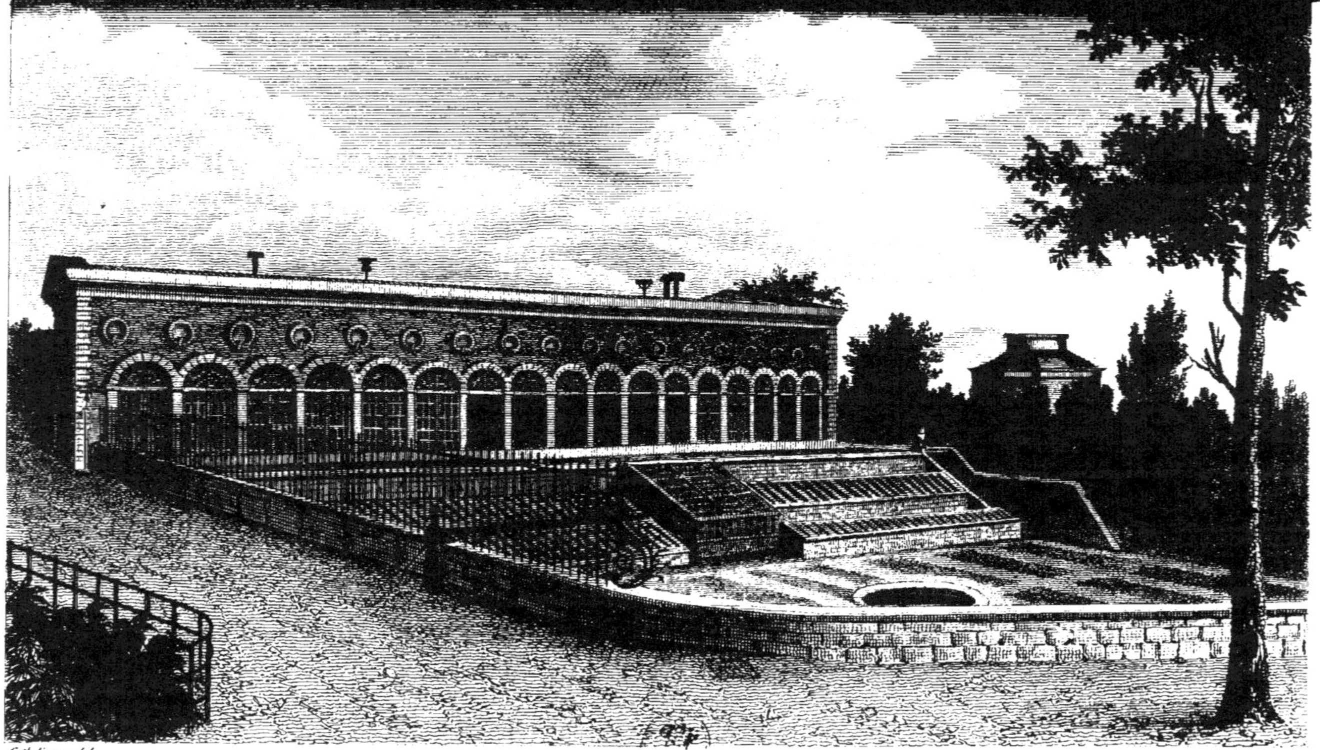

La Serre tempérée.   The Green-house.

The green-house is destined for the trees and
shrubs of Asia Minor, Greece, Florida and other
countries of the northern hemisphere, resembling
the south of Spain in temperature, and for those
which inhabit regions as cold as France, as Vandieman's Land and New Zealand, but which notwithstanding would perish in our climate. This
phenomenon, the cause of which has been already
hinted at, is easily explained. Much has been
said of the possibility of accustoming vegetables
of warm countries to a colder climate; but this
resource is confined by nature within very narrow limits: a tree that would be destroyed by
6° of cold in its native country, will never support 10° with us. The orange-tree, olive and
fig-tree, have been cultivated for ages in France,
yet they cannot resist a severe winter, and in
1820 almost all the orange-trees and olives perished in Provence.

The trees which at first yield to our climate,
which we may hope at length to cultivate with
success, are those which bloom in the winter.
The eucalyptuses, the banksias, and the casuarinas
of Vandieman's Land, fructify in January, February, March and April, the summer and autumn
of that latitude, and on the opposite side of the
equator they vegetate at the same period and
perish by the frost; but in the green-house fer-

tile seeds are obtained, which, sown at the proper period, conform to our seasons. The first attempt does not always succeed, but by continuing the practice the end is ultimately attained. Thus the marvel of Peru, *mirabilis,* now so common in our gardens, where it diffuses its odour at the setting of the sun, was first placed in the hot-house by M. Lemonnier, who received it from Mexico in 1760; the next generation it flourished in the orangery; and since that time it has passed the winter in the garden. The superb dahlias of Mexico, the roots of which were sent to the King's Garden by M. Cavanilles in 1802, were placed at the end of the winter in a hot-house of the temperature of 70°, where they advanced slowly, and bloomed at the end of autumn ; they have since been multiplied from the seed, and we now possess beautiful varieties that require no particular attention : M. Thouin announced this result, but did not promise that it would be so speedy (1). The melaleucas and metrosideroses raised from the seed, bloom later also than those brought from New Holland.

The plants are lodged in the green-house in October, and carried out at the latter end of April or the beginning of May : during the sum-

(1) See Memoir on the Dahlias in the Annals of the Museum, vol. iii. p. 420.

mer the building is entirely empty. About the month of March it is in its greatest perfection, as most of the trees are then in blossom. It is impossible to conceive the elegance and beauty of the mimosas of New Holland, some of which are covered with long spikes and others with tufts of flowers variously coloured(1). With these are mingled the *sophora microphylla, S. tetraptera* (2) and *cassia flavescens;* the sparmannia (3), the pittosporum (4), the camphire-trees and laurels of Madeira, banksias, melaleucas, hakeas, visneas, etc., and the two palms already mentioned. In front of these trees, which spread their

(1) There are in the green-house sixteen species of arborescent mimosa from New Holland, of which the *M. lophanta* and *M. botricephata* are the finest, at least, in foliage; the simple-leaved species as the *M. floribunda, M. armata* and *M. sophora,* are very beautiful when in flower : it is to be hoped that they will one day be cultivated in our southern departments. Some of the mimosas are employed in New Holland in cabinet-work and ship-building.

(2) Both originally from New Zealand, and first introduced into England by sir Joseph Bankes.

(3) A beautiful shrub, originally from the Cape of Good Hope, introduced into the King's Garden at the beginning of the century. It is covered with bunches of flowers from September to May. If by multiplying it from the seed we can retard its bloom, as there is every reason to hope, it will become one of the finest ornaments of our gardens.

(4) *Pittosporum undulatum,* Vent. H. Cels; and *P. Tobira,* Hort. Kew. The first is a shrub brought from Teneriffe by Riedlé, whose flowers have the smell of jessamine. The juice of the stalk, oozing through the bark, concretes in the form of a resinous powder. The second is a tree recently known in Europe, which is cultivated in China for its odoriferous flowers.

branches unconfined at 20 feet from the ground and sometimes reach the vaulted roof, are ranged different ligneous vegetables, as the oriental coronilla, the indigo-tree, a very beautiful collection of geraniums, etc., and on the shelves before the windows, alpine plants which bloom in February. In the beginning of May, the most curious trees, such as the meterosideros, the melaleuca, the eucalyptus, the leptospermum, the banksia and the mimosas (1), are placed in the oval inclosure before the amphitheatre, the duplicates, and stocks of inferior size, in the seed-garden, and the other shrubs and vivacious plants, on the terrace in front of the green-house. On the northern side of the building the ground-floor is occupied by workshops, and that above by the lodgings of the gardeners, and a laboratory for the preparation and packing of seeds.

Though this green-house is spacious and convenient it might have been more advantageously constructed; it should have been lighted from

(1) The eucalyptus, banksia, mimosa and casuarina, are the largest trees of New Holland, where they are used in the ship-building. M. de la Billardière and M. Perron observed in Van Diemen's Land eucalyptuses from 160 to 180 feet in height, and 25 feet in circumference. The savages hollow these trunks by fire and inhabit them, which does not arrest their vegetation. See Labillardière's Voyage in search of La Perouse, vol. 1. p. 131 ; Peron's Voyage of Discovery, vol. 1. p. 252 ; Freycinet, *idem.* p. 40.

above as well as in front. On several of the trees the branches opposite the windows are covered with flowers, while above and towards the wall they are barren. The young plants also lose their shape in turning towards the light— an effect which is scarcely perceptible in the orange-tree, but very sensible in the eucalyptus, mimosa, etc.

§ IX. THE HOT-HOUSES.

—

THERE are five hot-houses in the Museum, each of which has its peculiar destination. The largest is situated at the upper end of the botanic garden, and is sheltered on the north by the small hill. The entrance is at the head of the horse-chesnut avenue, through a small court, where the most curious plants are exposed in warm weather: on the left is a court still smaller, containing hot-beds and frames for such as require an elevated temperature and particular care, especially when young; and a cabinet for the use of the head gardener, where he renews the pots.

This hot-house consists of three distinct parts; one above, to which an exterior flight of steps conducts through a porch with double glass-doors; a second, 5 feet lower, in advance of the first, and separated from it by glazed sashes, which is entered through the gardener's cabinet; and a third in front of the preceding, but of less extent, which communicates with them by an

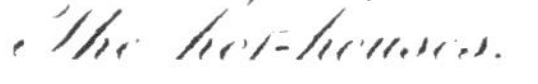
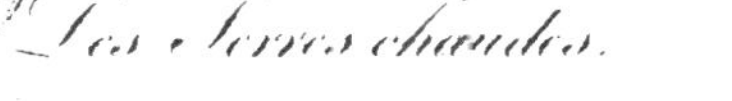

Cathilineau del.

Eugène Aubert sc.

Les Serres chaudes.   The hot-houses.

Durand imp.

interior staircase, and opens into the botanic garden. The roof is glazed, and is covered with mats in very cold and sometimes in very hot weather, and at the approach of storms.

The upper part, which was built by Buffon in 1788, on the spot formerly occupied by the seed-beds, is 125 feet and a half long, 12 feet 4 inches broad, and 15 feet in height, with four stoves, and a bed of spent bark in the middle for the pots. It was at first intended for a hundred fruit-trees of the tropicks, which were to be planted in the soil, and made to fructify for the purpose of disseminating them in the south of France; but so many plants from those regions have since been received, that the original project has been abandoned. Among its contents are many very rare and beautiful vegetables: we shall mention only two large stocks of the *pandanus odoratissimus* (1), green-spined screwpine, whose trunk, swollen at the top and spirally furrowed by the impression of the former leaves, is surrounded near the base by sprouts which take root and support it like buttresses on every side; the ravenala (2), of the musa or banana family, which is

(1) The male flowers are in request for their odour : a small bunch of them suffices to perfume an apartment, for which purpose they are sold in Egypt.

(2) The ravenala equals the palms in height; its naked trunk is crowned with leaves from 6 to 10 feet in length, and 2 in breadth, dis-

extremely useful to the inhabitants of Madagascar, who cover their houses with the leaves and make flour of the seeds, after extracting an oil from the pellicle; the strelitzia, a plant of the same family lately introduced from the Cape, of which the flower, partly scarlet and partly of a beautiful blue, is singular in its form; the *caryota urens,* a very rare species of Indian palm, with pinnate leaves composed of notched, triangular leaflets; the *littæa,* a shrub of the narcissus family recently procured from Italy by M. Bosc, taken at first for a *yucca,* on account of its narrow pendant leaves, fringed with white threads like those of the *yucca filamentosa;* the pimento, or allspice, a species of myrtle from Jamaica, which owes its name to the smell and taste of its leaves, branches, and fruit; the *psydium,* or guava; the *eugenia jambos,* whose fruit sheds a rich perfume; the *brucea ferruginea* (1), brought by Bruce from Abyssinia; the olive-tree with notched leaves, a superb tree of India and Madagascar with esculent fruit, of which M. du Petit-Thouars

posed in the form of a fan. The inhabitants of Madagascar call it the traveller's tree, because the sheaths of the petioles form reservoirs, which are always filled with fresh and limpid water.

(1) M. de Lamark, who first described it, gave it the name of the *brucea anti-dysenterica,* because in its native country the leaves are regarded as a specific for the dysentery. Bruce employed them with success.

makes a distinct genus under the name of *noron-
hia*; the mahogany-tree; the *cecropia* and *cocoloba* (1), from the West Indies, remarkable
for the size of their leaves, which are buckler-
shaped, thin and silvery underneath, upon the
first, and on the second thick, coriaceous and
of a gloomy green; the litchy of China, which
yields a very palatable fruit of a beautiful red
colour; the *citharexylum,* or fiddlewood, so
called from its eminent fitness for musical instruments; the *sterculia fetida,* a large tree of
the West Indies, with digitated leaves, whose
flowers are of an insupportable odour, though
an oil is extracted from the seeds; the sa-
pium of the West Indies, as poisonous as the
manchineal; the *gardenia thunbergia,* whose
flowers are more beautiful and odoriferous than
those of the Cape jessamine; the violet sugar-
cane of Otaheite, a species or variety, of quicker
growth than that of India; begonias, of singular
appearance from the form and colour of their
leaves; and lastly, superb indian fig-trees, among
which should be particularly noticed the *ficus
elastica,* elastic gum fig-tree, the milk of which
forms the indian rubber, and the *ficus macro-*

(1) The *cocoloba uvifera* is called *grape-tree* in the Antilles, because
its fruit which is of a very agreeable taste is borne in clusters resem
bling those of the vine though larger.

phylla, large leaved fig-tree, which the gardener, Riedlé found while at Timor, and recommended on his death bed to the companions of his voyage. This hot-house is constantly kept at the temperature of 59° of Fahrenheit.

The second part, called Baudin's hot-house, constructed in 1798 for the plants brought home by Riedlé, who accompanied that officer to the West Indies, is appropriated to the shrubs and smaller vegetables of the tropicks. It is somewhat larger than the former, being 140 feet long, 9 and a half wide, and 11 in height; and is heated by three stoves. Slips are here reared under sashes, and the most curious herbaceous plants are cultivated, with infant shrubs which are afterwards removed to the division above. The plants are ranged on the sides instead of occupying the middle. Along the windows in front is a shelf for the *liliaceæ*, the *irideæ* and the *orchideæ*, which require a strong light; and below, a bed of bark 15 inches deep for slips, and for pots containing the smaller plants of the torrid zone: on the other side is a similar shelf, and stages for plants and shrubs of larger growth.

We may here remark the xylophilla, a singular shrub, whose numerous flowers are situated in the indentations of the leaves; the *crinum* and the *pancratium,* which diffuse the most delicious odour when in bloom; some superb species of amaryllis, as the *amaryllis belladonna,* belladonna lily, and the *A. sarniensis,* Guernsey lily; different *orchideæ,* among which is the vanilla; several *aroïdeæ,* such as the colocassia, the *arum pictans* and the *pothos crassinervia* ; some pretty species of sensitive plant, and the *hedysarum girans,* or moving plant, from the banks of the Ganges, so called because two of the three leaflets which compose the leaves are constantly in motion ; the banisteria, and the Barbadoes trumpet flower, *bignonia unguiscati,* with various species of passion flower which climb on the partitions and hang in festoons from the roof. The *passiflora quadrangularis,* square stalked passion flower, and *P. princeps,* are remarkable for the size and beauty of their flowers, which are single and variegated on the first, and collected in long red clusters on the second (1). Among the trees and shrubs which are too young to bloom, and which are kept in this hot-house till their increasing size requires ampler room, are a beautiful

---

(1) The *passiflora quadrangularis* is of such astonishing vegetation that it shoots fifty feet in the year ; its flowers quickly fade, but they bloom successively during several months : it last year produced an esculent fruit of the size of a small lemon.

The *passiflora princeps* has been known to us only three years. It was presented to the garden by M. Cels, who received it from England,

individual of the tamarind; the *hura crepitans,* or sand box-tree, the fruit of which is common in cabinets; the *tamnus elephantipes,* elephant-foot black bryony, brought from Caffraria by M. Lalande, singular for the enormous channelled and glandulous tubercle which gives rise to the stems; the *carolinea princeps,* digitated carolinea, brought from Schœnbrunn by M. Bonpland, and the *carolinea insignis,* great flowered carolinea, raised from seeds gathered in the Brazils by M. Auguste St. Hilaire, two superb trees remarkable for their large digitated leaves, and for their flowers which are ten inches in diameter and furnished with innumerable stamina; an *adansonia baobab,* brought from Cayenne by M. Perrottet, which is seven feet high and of vigorous growth, but which we have no expectation of seeing in bloom; at Senegal this tree is from 25 to 3o feet in diameter, and by the calculation of Adanson is capable of vegetating at least four thousand years: and lastly, an *araucaria,* or Chili pine, also reared from seeds furnished by M. de St. Hilaire, which is already three feet high, and which will soon be sent into the orangery: hopes are entertained of multiplying it in the south of France, where, if it attains the same dimensions as in its native country, it will be of eminent utility for the masts of vessels.

The thermometer in this hot-house is kept at 65°, and several of the plants remain in it during the summer for fear of injury from the night-air. Certain *liliaceæ* and *cacti,* which require to be kept extremely dry, are placed upon the stoves. At the extremity is a small room, similar to that at the entrance, where the painter of the Museum delineates the plants without exposing them to the air.

This repository is visited with peculiar interest by botanists, on account of the new plants that every year spring from seeds collected near the equator. A great number of those sent from India by M. Leschenault, and from the Brazils by M. de St. Hilaire, germinated last year; and several of those brought home by MM. Freycinet and Delalande have already succeeded. These plants are not mentioned in the printed catalogue, as the species can be determined only by the flower, but they are regularly numbered and registered, with the date and other circumstances of their reception,

From November to April the hot-houses are filled, and in February and March they present the greatest variety of plants in bloom. Towards the end of May almost all the pots are taken out; one individual of each species is carried to the botanic garden, and the rest are ranged in the small courts at each end of the building, or before the wall, where they can be seen and studied to advantage.

The care of these hot-houses for the last twenty-two years has been confided to M. Riché, a man perfectly skilled in this branch of culture, who shews them to strangers with permission of the professors. As he is acquainted with the amateurs and nursery-men who cultivate rare plants, he is informed of the reception of new species, and procures them by exchange.

The remaining part of this hot-house was constructed in the summer of 1821, for the fine collection just received from India and Cayenne, and named Philibert's hot-house, after the captain on board whose vessel M. Perrottet, gardener of the Museum, took passage with the plants, of which he had collected the greater part himself, and which he brought in the finest condition to the Museum (1). It is 75 feet long, 12 broad and 10 high, and is heated by two furnaces: it is worthy of remark, that the division immediately behind is improved since its construction.

(1) Several of the trees cultivated in Cayenne, especially those affording spices, were furnished by M. Poiteau, author of several memoirs inserted in our Annals, and successor of M. Martin in the management of the royal plantations at French Guyana.

The collection of M. Perrottet consisted of eighty-five cases, containing five hundred and thirty-four individuals from six inches to six feet in height, of one hundred and fifty-eight different species; the greater part of which were wanting to the Museum, and many of them, to all the gardens of Europe. The most interesting were the true bread fruit-tree, *artocarpus incisa*, and the variety produced by cultivation, whose fruit without seeds forms the ordinary nourishment of the inhabitants of the South Sea islands (1); the Jaca-tree, *artocarpus integrifolia*, of which the fruit, shaped like a melon and often two feet in length, is eaten in the Moluccas, though inferior to the bread-fruit and full of large seeds; the betle pepper, *piper betel*, from which, with a mixture of the nut of the cabbage-tree and a little lime, the Indians make the preparation which they continually chew; the cabbage-tree, *areca faufel*, a species of palm, whose fruit forms an article of commerce in India; the *cyclantus bifolius*, considered by M. Poiteau as a new genus of palm, together with several other species of this family known only by the descriptions of Aublet; several cocoa nut-trees;

(1) The plants of the bread-tree without seeds brought from Cayenne by M. Perrottet, are the shoots of a stock sent to that colony from the King's Garden in 1797, and multiplied by M. Martin, director of the nurseries.

a Madagascar raphia, *raphia pedunculata* (Pal. de Beauv.), which furnishes the sago, and whose leaves are employed for tissues; a sort of calamus with winged leaves, and a foot-stalk like a filament, garnished with thorns an inch long disposed in the manner of leaflets; a nutmeg-tree, *myristica aromatica;* the *virola sebifera* of Cayenne, a species related to the preceding, whose fruit yields a substance employed in making candles; several species of the cinnamon-tree, *laurus cinnamomea,* one of which sent from Ceylon to Manilla, where it was obtained by M. Perrottet, is much superior in savour to that cultivated at Cayenne; the bitter quassia, *Q. amara,* an Indian tree well known in medicine; the *couroupita guyanensis,* a large tree covered nearly the whole year with beautiful, odoriferous flowers, and fruit in shape resembling a cannon-ball; the *carapa guyanensis,* which is 80 feet in height, and whose seeds yield an oil; the chocolate-nut-tree, *theobroma;* the long-leaved omphalea, *omphalea diandria,* a shrub of the spurge family, whose branches climb above the highest trees and decline again to the ground, and whose fruit contains an esculent kernel; the American genip-tree, *genipa americana,* from South America, whose flowers shed an agreeable odour, and whose fruit yields a deep violet dye; the laurel-leaved barringtonia, *B. speciosa,* from the East Indies, a tree of the myrtle family, remarkable for its shape, size, beautiful flowers, and fruit, which is known in cabinets by the name of square cap ; the *morinda umbellata,* an Indian-tree, the roots of which yield a yellow dye; the dyer's oleander, *nerium tinctorium,* a tree also of India, analogous to the rose-bay, whose leaves furnish a fecula resembling indigo and used for the same purpose ; the *cavanillea philippensis* (Lam. Ill.), whose fruit is eaten in the Philippine islands; a shrub not yet determined, but believed to be a *cookia,* all the parts of which are impregnated with an odour resembling that of the Chinese aniseed-tree, whose seeds are employed in making cordials; with many other plants hitherto unknown in our gardens.

These hot-houses, though far preferable in structure and position to those we shall next visit, do not correspond with the grandeur and beauty of the establishment. That of Buffon has frequent need of repairs, and as it is only 15 feet in height, it is necessary to lop the branches and retrench the summit of the trees, which sometimes prevents their fructifying. The other two are excellent in their place, but a receptacle is still wanting for trees whose increasing developement requires a more spacious habitation. For this purpose the upper hot-house should be at least 25 feet in height, and so arranged that the trunks might be placed on a level with the soil; by this means only can we preserve, multiply and render useful, the vegetables procured at great expence from regions about the equator: the Garden of Paris should not be destitute of accommodations which are found in that of Schœnbrunn. The buildings requisite for so vast an establishment cannot be erected at once, but this object will be steadily kept in view, and government, in requesting a plan for new hot-houses, has given room to hope for its speedy execution.

Issuing from the hot-houses just described at the western extremity, we cross a court and arrive by a narrow passage along the side of the small hill at one of those built by Dufay, the entrance to which is on the ascent from the lower to the upper part of the garden. This is called the shrub hot-house, and is 75 feet long, 9 wide, and 16 in height: a temperature of 50° is kept up during the winter by means of two furnaces. It is destined for the larger tropical shrubs, which are planted in boxes, and disposed on stages one above another. We here find large stocks of the ironwood, *sideroxylon atrovirens;* the small leaved schotia, *schotia speciosa* (1); the *erythryna corallodendron*, or coral-tree of the West Indies, so called from the brilliant red colour of its flowers and seeds; and the *globa nutans*, a plant lately introduced into Europe, and much esteemed for its beautiful clustering flowers.

The second hot-house, bearing the name of Dufay, is of the same size as the preceding, and has also two stoves, but the temperature does not exceed 45° 3o′. It is destined for succulent plants, mostly of African origin, such as the aloes, the cactuses, the house-leeks, *semperviva*, the agaves, the spurges, *euphorbiæ*, etc.: here are found a Canary spurge, *euphorbia canariensis*, whose straggling, leafless branches are supported by cords, and which is never removed from the hot-house; the lofty *aloe ferox*, great hedgehog aloë, and the *cactus monstrosus*, whose green, uneven stem resembles a stalactite. We possess the species which feeds the cochineal, but the insect sent with it has perished, and probably it no longer exists in Europe. The plants of this hot-house, except one of each species for the botanic garden, are placed upon the terrace in the summer.

(1) A shrub of the Cape of Good Hope, of a beautiful appearance when covered with its clusters of scarlet flowers, which grow upon the wood like those of the judas-tree.

We next enter a small, narrow hot-house with only one stove, which is the winter residence of the numerous genus fig-marygold, *mesembryanthemum*, and other analogous plants from the Cape of Good Hope: it is sufficient that the heat should not descend lower than 38° 45′ of Fahrenheit, or 3° of Reaumur.

Beyond, to the right, in the side of the hill, is the oldest hot-house of the garden, which was built in 1714, in the time of Vaillant, and called the coffee-tree hot-house, because in it was reared the coffee-plant sent from the botanical garden at Leyden to Louis XIV, the seeds of which have peopled the West Indies. It is 34 feet long, 14 broad, and 15 in height; and is heated by one stove to the temperature of 59°. The roof is glazed for only half of its breadth. Delicate plants of the East Indies and of the torrid zone are here cultivated in a bed of bark, which occupies the centre. This hot-house, though small, offers the most picturesque appearance from the singular form and varied foliage of the trees and shrubs. We here see a very beautiful broad-leaved cycas, *cycas circinalis,* a singular tree brought from the Isle of France by Joseph Martin, the pith of which yields a species of sago, eaten by the inhabitants of Madagascar; the plumeria; the lime-tree-leaved hibiscus, *hi-*

biscus tiliaceus ; the *spaendoncea tamarindifolia,* brought from Abyssinia by Bruce; the date palm; an *aletris fragrans,* whose pyramidal flowers shed a delicious odour in the night, and whose stem, remarkable among the *liliaceæ,* is 12 feet in height; a beautiful pandanus ; a *dracæna marginata* from Madagascar; and two arborescent grasses (1) that cover the wall in the rear. Almost all these trees pass the whole year in the hot-house. Upon the shelves in front is a numerous collection of stapelias, the singularity of whose flowers and stems, would cause them to be more generally cultivated, were it not for their disagreeable odour.

The last hot-house, situated immediately behind the labyrinth, is composed of three parts ; the two first of which, built in 1717 under the intendance of Fagon, bear the name of the Cactus hot-house, from their being divided by a large Peruvian cactus, surmounted by a glazed frame : the third, was finished in 1792 under the administration of the last intendant of the garden, and is called St. Pierre's hot-house: the entrance is opposite to the cactus, whose covering forms a separate compartment, heated to the temperature of 43°.

(1) One of them has not bloomed, the other is the *panicum latifolium,* of whose hollow stock the American savages make the stem of their *salumet,* or pipe of peace.

This cactus, whose roots occupy but little space, is never watered, and imbibes its nourishment from the air by the suction of the bark. It is already forty feet in height, a stature which it could never attain in its native soil, as its articulated branches would be broken by the wind. It bears a profusion of flowers every year, which fade in twenty-four hours, but succeed each other during a month: they nearly resemble those of the *cactus grandiflorus* of the Antilles, but are inferior in brilliancy and odour (1).

Beyond the cactus to the right are the tropical shrubs cultivated in separate vases or boxes, occupying the middle of the hot-house; and other plants, disposed on shelves before the windows. We may remark in this part beautiful specimens of the *dracæna draco*, or dragon-tree of the Canaries. The division on the left is also filled with shrubs, among which are some very fine stocks of the *dracæna reflexa* from the Isle of France.

These two divisions are arched, and are each

(1) This cactus was sent to M. Fagon, by the professor of botany at Leyden, in 1700; when planted in the King's Garden it was four inches high and two in diameter. In 1713 the top was seared to arrest its growth, but this operation did not prevent its shooting laterally. In 1717, when M. A. de Jussieu gave a description and figure of it in the Memoirs of the Academy of Sciences, it was twenty-four feet high and seven inches in diameter; subsequently a glass-case was placed over it, which is raised as the stem aspires.

L'Amphithéâtre vu de côté.   Side View of the Amphitheatre.

4o feet long, 9 wide and 11 in height: the heat from the two stoves is conveyed in tubes behind the wall. The third part is covered with glass, like the hot-house of Buffon, and is 36 feet long, 10 broad, and 12 in height. It has only one stove, but is warmer than the preceding, from the pots being placed in a bed of bark. Here are seen the cycas of India and of Japan, *cicas circinalis* and *C. revoluta* (1); the crinum, pancratium, dianella, and pitcairnia; a theophrasta, a very rare tree remarkable for the vase-like tuft formed by the long verticillated leaves at the extremity of the trunk; the *chamærops histrix;* the *sabal adansonii*, a sort of swamp dwarf palm; the *rhapis flabelliformis*, or the fan palm of China : the red-edged aloe from the Isle of Bourbon ; and the latania of China. On the right of the entrance is a *passiflora alata*, whose branches extend 5o feet along the roof, and are covered for eight months in the year with flowers similar in appearance, though inferior in size, to those of the *passiflora quadrangularis.*

In this hot-house a new species of cactus bloomed and fructified three years since, whose flowers, of a changeable colour frosted with gold,

(1) The Japanese eat the fruit, and extract a very nourishing sago from the trunk. They value this tree so highly that they expressly forbid its exportation.

are in form and brilliancy among the most beautiful with which we are acquainted. M. Desfontaines, who has given a description and drawing of it in the Annals of the Museum, has appositely named it *cactus speciosissimus*.

In the summer the plants are ranged along the wall, and on the terrace between this hot-house and that of the *ficoideæ*.

§ X. THE BOTANICAL GALLERY.

—

The botanical galleries occupy the first floor of
the building called the administration. On the
staircase leading to them is the trunk of a palm-
tree from Cayenne, perfectly cylindrical, 12 feet
high and 10 inches in diameter, enclosed in a
network or grating, formed by the flattened
stems of a large *liana,* a species of fig, naturally
grafted together. They have produced no impression upon its surface, but they would have
become imbedded in the trunk of a tree growing
by concentrical circles. There is also a section
of a smaller palm on which the *liana* is much
thicker.

Beside the door of the gallery is a trunk of the
chamærops covered with large scales, which are
the base of the footstalks, that persist after the
falling of the leaf.

On entering we turn to the right, and pass
through three galleries communicating with
each other by an arch in the centre. The first
contains samples of wood; the second, the her-

bariums; and the third, the various species of fruit. On the left of the entrance is another gallery divided into two parts, of which that beyond the partition is reserved as a cabinet for study.

The saloon appropriated to specimens of wood is furnished with glazed cabinets. The three first, nearest the window on the right, contain the specimens more particularly required in a course of botany, consisting of different samples of epidermis, bark, roots, stems, thorns, sap-vessels, pith, grafts, ridges, wounds, excrescences, and sections of trunks, exhibiting the organisation of the wood. There are a root of the *polypodium barometz,* or Scythian lamb (1); some beautiful specimens of the bark of the lacewood, detached or adhering to the stock; trunks of the *bauhinia anguina,* Roxb. whose sinuous form is met with in no other plant (2); trunks of the sandbox-tree, *hura crepitans,* and of the

(1) The root, or rather the base of the trunk, of this fern rises horizontally above the soil, and as it is clothed with a thick coat of silky down, it has the appearance of a lamb; hence have arisen the numerous fables concerning this tree, which is said to feed on the surrounding plants. The specimens seen in the cabinet are from the north of China. The plant has not yet appeared in our gardens.

(2) The stocks are flat stems from 1 to 6 inches in breadth, and from 3 lines to 1 inch in thickness, according to their age, curved into a succession of arcs, so that each portion forms the letter S. The tendrils spring from the convexity of the arcs, and the plant rises to the top of the highest trees. It is the *naga-mu-valli* of Van Rheede, Hort. Malabar. vol. viii. tab. 5o and 51.

zanthoxylum with large prickles; and stems of
the cactuses 8 inches in diameter, covered with
long bundles of thorns.   We may also remark
a section of a tree with the impression of
letters traced in 1750 still visible on the bark
and in the interior of the wood, though no vestige of it appears in the intermediate layers; and
the horn of a stag projecting from a trunk in
which it had been arrested while the tree was
young.

The three following compartments contain the
species of North American wood, assembled by
M. Michaux, author of a valuable history of the
forest-trees of that country.   This collection, the
most complete in its kind in existence, is the more
interesting, as all the species may be cultivated
in France, and as many of them are eminently
proper for cabinet-making and joinery : the wood
of the black walnut, the red maple and the
sugar-maple, is as beautiful and susceptible of
as high a polish as the trees of India.   All the
samples are from middle sized stocks, and the
bark has been preserved wherever it offered
interesting or peculiar characters.

Beyond the arch the first compartments are
reserved for monocotyledon trees, and contain
trunks and sections of the palm, the baquois,
the yucca, the arborescent ferns, the bamboo,

rotang, papyrus, etc. designed to shew the difference between their organization and that of trees with two seminal leaves. There is also a trunk of the *xanthorrea resinosa,* which affords a resin resembling *pitch* and proper for the same uses (1); its wood of a red colour and clothed with a very thick bark, is remarkable for the appearance of medullary rays, though upon attentive examination it is found to be organized like other trees of this class.

The third cabinet contains specimens of indigenous wood, in plates 6 inches by 8, together with transverse sections; the fourth, a collection of wood from Cayenne, proper for cabinet-making, labelled with the names used in commerce (2). The four opposite compartments, between the window and the passage to the second gallery, are filled with a miscellaneous collection of the most beautiful species of foreign wood used in cabinet-making, which formed a part of the former cabinet: most of these samples are remarkable for their colour, and several of them from China are inscribed with the Chinese names.

Of the six remaining compartments, the two

(1) See Freycinet's Voyage of Discovery, p. 41.

(2) This collection was presented to the Museum by M. Jacob, cabinet-maker.

first contain longitudinal and transverse sections of the wood of Porto Rico and St. Thomas, brought by the gardener Riedlé on his return from the voyage of captain Baudin : the specimens were damaged on the voyage, but the species are known.

The adjoining compartment contains one hundred and seven samples of the wood of Cayenne, sent by M. Duclerc, 5 inches by 3, polished on one side and rough on the other, named and disposed in series, according to their uses for carpenter's work, joinery, cabinet-making, etc. The lower part of this compartment, with the two following, contains beautiful specimens of the wood of the Isles of France and Bourbon, and various species from different countries, remarkable for their texture, their rarity, or their uses. In the last compartment is a numerous miscellaneous collection brought from Portugal in 1808, and presented to the Museum by the brother of marshal Serrurier, which would be more valuable if the samples, which are only 4 inches by 3, were larger.

The foreign specimens in these cabinets do not all bear the name of the species, which cannot always be determined by inspection, and we are often ignorant of the systematic name answering to that in trade : travellers who transmit objects of this nature, should if possible add a branch of the tree in flower to shew the genus and species.

The gallery of the herbariums, which follows, is fitted up with wooden cases 10 inches high, 11 wide, and from 17 to 18 deep—three hundred and forty-four on the right of the entrance, and two hundred and sixty-six on the left—secured from the dust by curtains raised and let down at pleasure. Those on the right contain the general herbarium, composed of about twenty-five thousand species, of most of which there are several specimens. Beside the ascertained species are placed the varieties, and such as are not positively determined, with duplicates exhibiting both the flower and the fruit, and samples gathered in different countries.

The basis of this herbarium is that of Vaillant, which contained the plants labelled by himself, with a memorandum of the place where they were gathered, and the synonymes of the authors known in his time ; and also specimens sent by different botanists, whose writing was known, so that the synonymes became certain on the publication of their works. M. Desfontaines has affixed the best ascertained and most common modern names on a separate label to each of these plants, and they have all been compared

with those in the herbariums of MM. de Lamark
and de Jussieu.

The botanists, who have published monographies of genera or families, on consulting the
herbal, were invited by M. Desfontaines to annex
the names adopted in their works to the plants
which had not been described: thus many specimens are original types corresponding to the
descriptions of M. Decandolle, in the two first
volumes of his *Regnum Vegetabile;* of M. Duval
in the monography of the *solanum;* of MM. Bonpland and Kunth, in that of the *melastoma;* and
of MM. Mertens, Aghart, and Lamouroux, in that
of the *fuci,* etc.

The herbarium of Vaillant, which was arranged
after the method of Tournefort and shut up in
boxes, was of little use, when, in 1797, M. Desfontaines undertook to reduce it to the natural
order, and to insert the other collections of the
Museum, viz. that made by Commerson in his
voyage round the world, and during his residence in the Isles of France and Bourbon; that
brought by Dombey from Peru and Chili; the
plants of the South Sea Islands, given to Buffon
by Forster; an herbarium sent by Macé from
India; one from Cayenne, by Martin; one from
Madagascar, by Chapelier; one from St. Thomas
and Portorico, by the botanists who accompanied

captain Baudin ; one from St. Domingo, presented by M. Poiteau ; one from Java, given by M. Leschenault; and one from Persia, Egypt, and the Levant, composed by MM. Olivier and Bruguière, during their residence in those countries.

The numerous species from New Holland have been added to the general herbarium ; which has been still further enriched by the presents of travellers and botanists, and principally of Mr. Robert Brown. As it is essential that it should not be too voluminous, the duplicates have been reserved for particular collections; thus we have herbals of New Holland, of Cayenne, of the Antilles, of the Cape of Good Hope, of India, and of the Mauritius ; which, besides the convenience of studying the plants of a given country, are particularly useful in the composition of monographies. The analysis of the plants in the great herbarium is strictly forbidden, but with the leave of the professor the duplicates may be examined in detail, and they are also given in exchange.

We have other particular collections from which the duplicates only have been taken for the general herbarium, and which form the type of printed works ; such as that of the elder Michaux, containing the species described in his *Flora Boreali-Americana;* that of the trees of

North America, by his son; that of the plants of
Taurus and Caucasus described by M. Marshall;
that of the plants of France presented by M. de
Candolle, which is the type of his French Flora;
and lastly, that presented to the Museum by M. de
Humboldt, containing all the species collected in
his travels, and published by him in concert with
M. Bonpland, in the Equinoxial Plants, and with
M. Kunth, in the *Nova Genera and Species,* etc.
The herbarium of Tournefort has been scrupu-
lously preserved, from respect to the father of
French botany: almost all the plants collected in
his voyage to the Levant, which are only indi-
cated by a phrase in the *Corollarium institutionum
Rei herbariæ,* are found in it, labelled by himself,
or by Gundelsheimer.

The species contained in these three collections and wanting in the general herbarium, are
at least six thousand, making the whole number
more than thirty thousand. It is admitted by
botanists, who have visited the principal cabinets of natural history in Europe, that there no-
where exists so numerous a collection; and it is
daily increasing by the contributions of travellers
and by exchanges (1).

(1) In 1821 were received eight hundred specimens, collected on the
borders of the Black Sea by M. d'Urville, who has lately published
a catalogue with the synonymes of the species already known, and a
description of the rest: one thousand two hundred specimens of plant-

The space between the windows, at one end of this saloon, is occupied by a beautiful collection of *fungi* in wax, presented to the Museum by the Emperor of Austria, and at the other, by models of foreign fruit admirably imitated in wax or plaster. Leaves of the great fan palm, *corypha umbraculifera*, fifteen feet in diameter, are attached to the ceiling.

The last gallery, which is that of fruits, has twenty glazed cabinets, eight on each side and four at the ends, twelve of which contain fruits dried, or preserved in spirits, arranged according to de Jussieu's method (1), and the others, pro-

that grow in the eastern provinces of Russia, presented by M. Fischer, director of the botanic garden of Gorenki near Moscow, carefully labelled with the names they bear in his descriptions and in the printed catalogue. A herbal of three hundred plants from the shores of the gulph of St. Lawrence, collected and preserved by M. La Pylaie ; one from Martinique, sent by M. A. Plée ; one from the Philippines and Guyana, brought by M. Perrottet ; one from Caffraria by M. Delalande ; a number of beautiful plants from the Brazils, by father Leandro di Sacramento ; in fine, the plants collected by M. Gaudichaud in his voyage with captain Freycinet, which has been given to the Museum, with all the other collections of natural history, by the minister of marine. The most interesting of these plants will be described and figured by M. Gaudichaud, in the relation of the voyage about to be published by M. Freycinet. We expect this year collections not less considerable from India, Senegal, Guyana, and the Brazils.

(1) The fruits of a given family are found in the same case, but the species and genera are not arranged in the order of their affinity, as it was necessary to place the objects so that they might take up the least room and be seen to the best advantage.

ductions of the vegetable kingdom employed in medicine or the arts.

The two first on the right, towards the window, are appropriated to the fruit of the palms and other monocotyledon vegetables. Here are seen beautiful cocoa-nuts of the Maldives (1), one with four lobes and another enveloped in its husk, with male catkins of the same tree of the size of a man's arm; the fruit of the *doum* or palm-tree of Thebaïs; of different species of cocoa-tree; of the beautiful screwpine of India, *pandanus;* of the *lontarus,* the *xanthorrea resinosa,* the cycas, and the nelumbo; a large specimen of the *maripa,* etc. The two following compartments contain the fruits of the laurel family, among which may be noticed the nutmeg with its mace; and other species not less curious, employed in domestic economy. Next comes the fruit of the different species of banksia, which was purchased in England after serving for the beautiful figures of White's Voyage to New South

(1) This fruit was anciently so called from its being found floating on the sea near the Maldives, the sovereign of which claimed it as his property. It was fashioned into vases of great price, and the wildest conjectures were formed as to its origin. In 1768 Commerson discovered the tree that produces it, in the *Sechelle* or Mahe Islands, and named it *Lodoicea.* M. de la Billardière has given a description of it in the Annals of the Museum, vol. ix. made on the spot by M. Lillet, to which he has added his own observations: he retained the name *Laodoicea,* adding the epithet *Sechelarum.*

Wales; and lastly, some very remarkable and rare specimens of the *bignoniæ, apocyneæ,* and *sapoteæ,* as cerberas, strychnoses, the *omphalocarpum* of M. Palissot de Beauvois, the *nerium tinctorium* of India, etc.

The eight compartments opposite the windows, on the right and left of the entrance, complete the series of fruit. In the four first are the families which follow the *sapotæ* to the *leguminosæ* inclusive; several specimens of the *rubiaceæ, sapindi, aurantiæ* (1), *malvaceæ, tiliaceæ* and *myrti,* are worthy of remark for their form and size, especially the Indian elephant apple, *feronia elephantum,* Roxb.; the silk-cotton-tree, *bombax;* the downy-leaved ochroma, *ochroma lagopus;* the boabab, or monkey's bread; five-lobed capsules, divided by membranes, with their seeds enveloped in a cottony down, which is reserved by the Indians for the cushions placed under the statues of their gods (2); the fruit of the coronet-flowered sterculia, *sterculia balanghas,*

(1) M. Poiteau has presented to the Museum the specimens described in his work on orange-trees: they occupy the bottom of the first cabinet.

(2) This fruit was sent from Pondicherry by M. Leschenault; the indians name it *konnamarum*; it belongs to a new genus found in India and South America, which M. Kunth has described under the name of *cochlospermum.*

with the five capsules open and garnished
with seeds; the species of cocoa described by
M. Bonpland; the *carolinea insignis*; the seeds
of the *bertholetia*, which were formerly known
by the name of Brasil chesnuts, though we
had no notion of the tree which produced
them, till it was described by MM. de Humboldt
and Bonpland (1); the fruit of different species
of *lecythis*, in the form of a vase with a
moveable cover, whence the largest are called
monkey's porridge-pots; of the *couratari* of
Guyana; of the *butonica,* in form like a square
cap, etc.

The seeds of the leguminous plants, which
occupy the lower part of the third press and the
whole of the fourth, are very numerous; among
them are those of the *eperua falcata;* the mountain ebony, *bauhinia racemosa;* the wart-podded
locust-tree, *hymenœa verrucosa;* of the *moringa
zeylanica,* or ben-nut, which differ from the rest
of the family, in having three valves instead of
two; some pods of the *mimosa scandens,* and
mimosa entada, which are four or five feet long
and three feet wide, though the flower is extremely small; lastly, various brilliant coloured

(1) M. Bonpland had not seen the flowers, which M. Poiteau has
observed, and brought home dried and in spirits; they resemble those
of the Lecythis, and prove the affinity of the two genera.

seeds employed for necklaces and other or-naments.

The two compartments on the left of the en-trance, contain the fruit of the nettle and gourd families, *urticæ* and *cucurbitaceæ*, and of the *terebinthaceæ*, or resinous trees.

There is also a collection of the walnuts of North America, by M. Michaux, exhibited by dif-ferent sections; one of gourds presented by M. Duchêne, who has published a treatise on that family (1); cashew nuts with the fleshy husk in which they are embedded; the apple of the manchineel, *hippomane mancinella*; the fruit of the sandbox-tree, *hura crepitans,* which is bound with wire to prevent its exploding; the *ambora,* whose fruit forms the transition between the fig and the mulberry; the jaca-tree; and all the va-rieties of the bread-fruit, *artocapus*, either dried or preserved in spirits.

The next case contains a collection of cones of the pines and larches, most of which were ga-thered in North America by M. Michaux, and some beautiful specimens of the Chili pine, *araucaria,* brought over by Dombey. The last on that side comprises the fruit of the genera not yet classed, or, as they are termed by botanists, *in-*

(1) The drawings for this work have been procured for the library of the Museum.

certæ sedis, with others presented by travellers, of which the species are not known.

In the collection of fruits only those have been admitted which were too large to be preserved in the herbarium. M. Thouin possesses a very numerous collection of the smaller seeds, which will be added to that of the gallery, and arranged in a cabinet of drawers, with a separate compartment for each species.

This collection is labelled with reference to the figures of Gœrtner, for all the specimens described by that learned carpologist; and in the vase or pasteboard-box, which contains them, are often found notes of travellers, concerning their uses (1).

The eight remaining compartments are reserved for the ancient drugs, with those that have been since added. In the first are the roots, in the second, the barks, and in the third, the leaves and flowers used in materia medica and domestic economy, in France, and in foreign countries : the fourth is filled with miscellaneous

(1) The collection of seeds has increased more rapidly than the herbarium, since the observations of Gœrtner have attached more importance to the subject. M. Freycinet has brought home a great number ; very rare ones have been sent from India, Guyana and Martinique, and others have been presented by amateurs of natural history. The names of the donors are carefully inscribed upon the objects.

objects of curiosity, as the beautiful stems of certain grasses (1); spaths of the palm-fruit; a bunch of the sagus; misletoe on a branch of oak; unwoven cloth fabricated in the South Sea Islands, etc. Two of the compartments between the windows contain the gums and resins employed in medicine and the arts; the third, different vegetable preparations, as indigo, Indian rubber, perfumes, tapers made of the common candleberry myrtle, *myrica cerifera,* and a phial of the poison with which the natives of Java taint their shafts; the last is destined for vegetable substances, of which the nature is not well understood.

The greater part of these objects, many of which are from China, are contained in bottles, and those of the ancient collection preserve the labels of Vaillant and Geoffroy, to which the modern classical names have been added. On one of the tables in the middle of the gallery is a vase, formed from the trunk of a palm, nineteen inches high and eighteen in diameter, brought from Porto Rico and presented by M. Maugé, at the return of captain Baudin.

Of the saloon already mentioned, on the left of the entrance, the first part is occupied on one side by cases containing the herbarium of

(1) *Arundo sagittata.* Persoon. *Gyncrium,* Humboldt and Bonpland.

Tournefort. By the window are two fruit-. stalks of the sago palm, *sagus farinacea* (1), six feet in height, covered with fruit. Here are deposited also trunks of palm-trees and ferns, with curious specimens of wood that could not be conveniently placed elsewhere ; for example, a section of an elm felled in 1784, with the layer of 1709, almost destroyed by the frost.

The second part, or the study, is furnished with glazed cabinets filled with flowers and fruit preserved in spirits, and a collection of *gnaphalia* and *xeranthema* or everlastings, from the Cape, contained in large glass jars.

The galleries of botany are not open to the public, and like those of comparative anatomy, they offer little to interest spectators unacquainted with natural history ; herbariums in cases, bits of wood covered with bark, or sections exhibiting the texture, and fruit, dried or preserved in spirits, are objects little calculated to strike the eye, though extremely useful for study : but they are at all times accessible to amateurs and to persons engaged in scientific researches, or desirous of comparing their herbariums with that of the Museum. The monographies and

(1) One was sent from India a few years ago, and the other was given by M. Fulchiron.

.descriptive works of botany, published for some
years past, bear testimony to the unreserved liberality with which the treasures of the establishment are exposed, and every explanation given
that can aid the progress of science.

Les Galéries d'Histoire Naturelle.   The Galleries of Natural History.

Durand imp.

CHAPTER II.

CABINET OF NATURAL HISTORY.

---

§ I. CURSORY VIEW OF THE WHOLE.

In the historical notice which forms the first part of this work, we have related how the cabinet of natural history became remarkable from the improvements completed or begun by Buffon, and afterwards continued; observing that it was necessary to increase it one third in 1808, although the anatomical and botanical collections had been separated from it, and galleries expressly constructed for them. It is therefore useless to recur to those details, and we will confine ourselves to its present state and the distribution of the objects it contains.

The building which bears the name of Cabinet or Gallery of natural history, and of which one room is devoted to the library, is 390 feet (or 60 *toises*) long. It is exposed to the east on the side of the garden, from which it is separated by a court and an iron railing. The front, which has thirty-three windows on the first floor and the same number on the second, is divided into three equal parts. The middle part, which has a small projecting wing on each side, was formerly the dwelling of the superintendant and the cabinet. The southern part, which contains the library, was almost all built in the time of Buffon; and that division which extends from the second wing to the hill was added in 1808. The gate and staircase opposite the great avenue were then suppressed. The present entrance from the street into the garden opens into that part of the court which is in front of the house where Buffon lived. The door of the cabinet and the staircase have been placed at the other end in the angle next to the guardhouse and orangery. In the first wing there is another staircase which leads to the library and galleries on the days they are not open to the public (1).

The ground floor is composed of the porter's lodge to the south, and of several rooms with doors and windows of iron grating which open into the court. The largest of them contains models of agricultural tools, and is the lecture-

(1) The Cabinet of natural history is open to the public every tuesday and friday, from three o'clock until six in the summer, and from three until dark in the winter. Admission is given on mondays, wednesdays and saturdays, to those who have students' cards, or who present a ticket signed by one of the professors.

room of M. Thouin; the others serve as store-
rooms for such objects as cannot be placed in
the galleries; they are lower as they approach
the hill from the elevation of the soil in that
direction; so that the ceiling, which is 12 feet
from the ground on the south, is only 5 feet on
the north. Large trunks of petrified wood are
placed between the gratings.

In the middle of the second floor of the building is a very beautiful clock, of which we see
the mechanism, as it occupies the space of a
window and is between two glasses. The windows of the second floor are merely for ornament, as it is lighted from the top.

The interior of the cabinet is composed of six
saloons on the first floor without including the
library at the end, and five on the second. The
first floor is devoted to geology, mineralogy, and
the collections of reptiles and fishes: the second
is occupied by the quadrupeds, birds, insects,
shells, etc. Some of the semicircular sashes,
which give light from the roof, are raised and
lowered at pleasure for the admission of air.
Curtains are placed over the cases when not
open to the public. This second floor, the middle
of which is a long gallery, has a door leading to
the terrace by the side of the street.

We will now enter the great staircase by the

guardhouse, and go through all the rooms, beginning with the geological collection. Those who enter by the small staircase, and wish to follow the order which we point out, will turn to the right, and go through the mineralogical rooms to reach the further end.

§ II. GEOLOGICAL COLLECTION.

On the landing place of the stairs by the side of
the door is a very large jointed basaltic column
from La Tour, in the department of Puy-de-
Dôme, given by the late M. Desmarest of the
academy of sciences. This column is surmounted
by a beautiful pyramid of rock crystal 2 feet
6 inches in diameter at the base: it was found in
Le Valais. Next to it are two other jointed
basaltic columns from the giant's causeway in
Antrim in Ireland, and other irregular columns
from Saint Sandoux, in Puy-de-Dôme. These objects announce our approach to the geological
collection which occupies the three first saloons
of the first floor.

The entrance hall contains the remains of vegetables and invertebrated animals which are found
in a great number of strata. These remains, which
almost all belong to lost species, are classed geo-
logically, that is according to the date of the formations in which they are found. The greater
number are accompanied by a portion of the rock
which contained them.

In this hall we also see several series of rocks, designed to illustrate the geology of different parts of the french territory. They are placed here for a time only, as they will be arranged in the third hall as soon as the Museum is possessed of a sufficient number of fossil vegetables or invertebrated animals to fill the cases which are intended for them.

The fossil vegetables are placed in the cases to the left and opposite to the entrance: they are arranged according to the order of the formations to which they belong, and the age of which they characterize. Although their assemblage in this point of view is recent, we already remark some interesting specimens, among which we must be content to notice :

1st. A series of the larger herbaceous plants, exclusively found in beds of sandstone and coarse slate accompanying coal.

2d. A large trunk of dicotyledonous wood, which has been changed into silex after having been bored full of holes by ines: it comes from Maestricht.

3d. A large plate of quartzose sandstone covered with various impressions of leaves : it was found near Le Mans, by M. Ménard de la Groye, who presented it to the Museum.

4th. An enormous trunk of palm-tree, easily

recognised by the scales or remains of the leaf-stalks with which it is covered: it was found at Vailly, near Soissons, by M. Menat, who gave it to the Museum.

5th. Two beautiful impressions of leaves of the chamærops palm, given by M. A. Brongniart, taken out of the quarries in the neighbourhood of Aix, near Marseilles.

Lastly. A numerous series of impressions of leaves on the newer limestone, found at Monte Bolca, on the confines of the Veronese and Vicentine in Italy.

The invertebrated fossil animals are in the cases to the right of the entrance: they are divided into three sections: the zoophytes or radiated animals, the articulated animals and the mollusca. Each section is subdivided according to the age of the formations from which the specimens were taken: that is, we see in each section the species which characterize either the transition or intermediary, the secondary, or tertiary formations. This distribution, although merely geological, is so much the more advantageous here, as many specimens belonging to the first and third sections, and remarkable for their insulated position and beautiful preservation, have been inserted in the great collection of zoophytes and living mollusca on the second floor (1), and in which we must study the genera and species, when we wish to become acquainted with them independently of geological considerations.

We will now notice the most remarkable objects of each section.

Amongst the zoophytes:

1st. A beautiful stem of encrinite, from the secondary limestone in the neighbourhood of Brunswick.

2d. Several polypis and echinites belonging to the chalk formation in the neighbourhood of Maestricht; drawings and descriptions of which are given by the late M. Faujas Saint Fond, in his work on the quarries of that city.

Amongst the articulated animals:

1st. Several fine specimens of trilobites (genus ogygia of M. A. Brongniart) from the slate quarries of Angers.

2d. A complete specimen of a trilobite (genus calymène of M. A. Brongniart) from Dudley in England, presented by M. A. de Humboldt.

3d. Several specimens of trilobites (genus calymène) found in the transition slate of La Hunaudière, in the department of the lower Loire,

(1) M. De la Marck has thought it necessary to bring the fossil and living species together, in the collection which serves as the type of his complete history of invertebrated animals.

by M. Regley, assistant naturalist, and given by him.

4th. A very distinct limulus, a precious specimen, long since described by Walch and Knorr. It was found in the secondary limestone of Pappenheim, and a good representation of it may be seen in the work of MM. Brongniart and Desmarest.

5th. A large palinurus contained in the tabular limestone of Monte Bolca.

Amongst the mollusca:

1st. Several hippurites and orthoceratites of a large size, from the ancient limestone beds of lake Erie, in the United States, and from the northern side of the Pyrenees.

2d. Radiolites, partly siliceous, and dicerates collected in the more recent limestone beds of the island of Aix, in the department of the Lower Charente, by M. Dorbigny, correspondent of the Museum.

3d. Nautilithes and ammonites, the shells of which have preserved their pearly lustre. They are from the sandy clay formation above the chalk. The greater number were given by M. Crow, and come from Sheppy in England.

4th. The cast of a gigantic ammonite of whose locality we are ignorant, but which probably comes from the lower part of the chalk formation

Lastly. In the other cases, which are at the bottom of the hall to the right, we find the following series of rocks, which are only placed there temporarily, as we have before said, and which are intended for the completion of the geographical collection already begun in the third room. These are:

1st. The principal rocks of the tertiary formation, which constitute the soil of the environs of Paris. It is well known that this formation has been the subject of a complete monography, published by MM. Cuvier and Brongniart.

2d. Various rocks from the neighbourhood of Nantes, Rennes and Paimpol, given by M. Dubuisson of Nantes and M. Regley. Amongst the latter we remark the ancient lavæ discovered at Treguier in 1821, by M. Regley.

3d. A fine series of rocks from the environs of Cherbourg, Caen and Havre, collected and given by M. Constant Prevost.

4th. A series of rocks from the neighbourhood of Aix and Marseilles, collected by M. Fontanier, travelling naturalist to the Museum. Amongst them we see a large specimen of compact limestone more recent, containing scoriæ, which was brought from the volcanic mountain of Beaulieu by M. Ménard de la Groye.

5th. A considerable number of specimens of

argentiferous lead from the neighbourhood of Vienna, in the department of Isère : the Museum is indebted for them to the viscount Héricart de Thury.

6th. Some lavæ from the department of Ardèche, given by the late M. Faujas de Saint Fond; amongst which we distinguish a column of basalt, containing in its center a fragment of granite altered by the heat, and various fragments of secondary limestone from Villeneuve de Berg, changed by its contact with a basaltic vein.

Lastly. A numerous suite of lava, tufæ and other rocks, which constitute the departments of Cantal and Puy-de-Dôme. The cabinet is indebted for them to viscount Héricart de Thury and to M. Lucas, keeper of the galleries of the Museum.

The second hall contains the rich and numerous series of fossil vertebrated animals, and a general and methodical collection of the different formations which compose the mineral crust of the earth. This last collection is arranged in two large chests, 20 feet in length, with drawers on both sides, placed in the middle of the room. We will first examine the fossils of this room, to complete the description we began in the entrance hall. We shall thus pass in review the whole of the organic remains of the former world, beginning with the most simple, and ending with the most complicated (1).

(1) Almost all the animals which we see here existed before the last revolution which changed the surface of the globe. The greater number of them are lost species; several cannot even be traced to any of the genera we now know; they belong to different periods, and the more distant the period the less they resemble those which now exist.

The presence of fossil organized bodies, conjoined with the superposition of heterogeneous layers, demonstrate the relative age of formations, and give positive notions, independent of all system, on the theory of the earth, and the changes which successively took place on its surface. There are no organized bodies in primitive formations; madrepores, shells, some crustaceæ and fishes, appear first in the transition formations; other shells, fishes, and some reptiles belong to a second period of formation; new families of shells, fishes, and reptiles occur in the third period, they are accompanied by reptiles, birds and mammalia of lost genera. The species of mammalia resembling those which are living, as elephants, rhinoceros, and bears are only found in alluvial formations caused by the last deluge.

The prodigious number of remains of the same species of animals, their arrangement in beds, and the preservation of certain shells, lead us to believe that several ages have elapsed between the several revolutions.

We do not find human bones in any of the formations we have mentioned, which does not prove that man did not exist upon the earth at the time of the last catastrophe, which gave the continents their actual form, but that he did not inhabit the places which were swallowed up by the waters, or that his foresight and industry furnished him with the means of retreating elsewhere.

It is at the end of the last series, and only in alluvial formations, that some species appear analogous to our domestic animals, such as the ox, horse, and rabbit.

It is remarkable that the fossil vegetables, whose species can be determined, appear to belong to genera and even families which are no longer found in our climates. Thus we see trunks of palm-trees in the neighbourhood of Paris, and trunks of arborescent ferns in the ancient coalpits of northern countries.

The fossil vertebrated animals are divided into four great sections : fishes, reptiles, birds and mammalia.

The first section is subdivided according to the age of the formations to which their localities belong. It occupies all the cases to the left as we enter the hall. We must begin our survey by the cases at the bottom. The objects are as follows :

1st. Several fossil fishes from the transition slate of Plettenburg, in the neighbourhood of Glaris in Switzerland : they are the most ancient vestiges of this class, which have yet been discovered in the bosom of the earth.

2d. A considerable number of fishes, whose skeletons are filled with sulphuret of mercury. They were taken from the sandstone which contains the quicksilver mines of the Palatinate. They were discovered and collected by M. Beurard.

3d. Some ichthyolites from the coalmines of Saarbrück. Their gangue is formed of the compact argiliferous carbonate of iron, which is worked so advantageously in Saarbrück, and especially in England, for the use of large furnaces.

4th. Some good impressions in the marno-bituminous cupriferous slate of Mansfeldt and Thuringia.

5th. Some large fish discovered in France in that species of secondary limestone called gryphite stone.   We also observe a species of elops, which comes from Grandmont, near Beaune, in the department of Côte d'Or, and a species of carp sent from Elbe, in the department of Aveyron, by M. Pacat.

6th. Some sharks' teeth and palates of rays from the chalk formations.

7th. Various impressions from Aichstaedt in Franconia, from mount Lebanon in Syria, and the tertiary formations in the neighbourhood of Paris.

8th. Some impressions on the foliated bitumen called dusodile.   They were presented by M. de Humboldt, and were collected near Rott, three leagues from Bonn, on the right bank of the Rhine.

9th. Lastly, the magnificent suite of fossil fish from Monte Bolca, collected by the count Gazola, of whom the government bought it for the Museum in 1798.   It is composed of more than four hundred specimens belonging to a great variety of species.   Drawings and descriptions of them are given in the great work entitled *Ittiolitologia Veronese*, printed at Verona in 1796, large folio.

The fossil bones of quadrupeds, birds, and reptiles occupy the twelve glazed cases opposite the windows.   The space not allowing of a geological distribution, they are arranged according to the order adopted by M. Cuvier in his great work on fossil remains, which contains a description and figure of almost every specimen in this collection.

Beginning with the first case at the side of the door leading into the cabinet of mineralogy, we see the teeth of fossil elephants, named mammouths by the Russians, dug up in different parts of the globe, and principally in France. Perhaps the most remarkable for their size are those found in digging the canal of Ourcq, near Paris, and which were given by M. Girard. We ought also to observe those from North America, sent by M. Jefferson, and those from Mexico, presented by M. de Humboldt. The next case contains tusks, portions of jaws, and long bones of fossil elephants from different countries. The most astonishing specimen amongst them is the part of a tusk which was found near Rome, by the duke de la Rochefoucauld and M. Desmarest, which at first sight we are tempted to take for the trunk of a tree. Some hair with a portion of the skin of the elephant that was found on the ice at the mouth of the river Lena, by M. Adams, is preserved here as a very interesting specimen of that animal, which at the time it was discovered had still its flesh and skin on.

In continuing the examination of each case from left to right, always beginning from the top shelf, we again see fossil remains of elephants, and amongst them the lower part of a femur, lately found in the Bog or Hypanis, and given by the chevalier Raynaud: the dimensions of this bone shew, that the animal to whom it belonged must have been 14 feet high.

The drawings in the first cases of fossil fish, which represent the head of an elephant from Siberia with its lower jaw, deserve to be mentioned here. They are a present from the imperial academy of St. Petersburgh, who had them made for M. Cuvier's work on fossil remains.

We then see the remains of another great animal called mammouth by the Americans, and to which M. Cuvier has given the name of *mastodon*. Most of the pieces were sent to the Museum by M. Thomas Jefferson, then president of the United States; above the case which contains them we see a glazed box, in which are bones of large *mastodons*, a fine tusk, for which the Museum is also indebted to the generosity of M. Jefferson, and a femur from the Ohio, brought by M. de Longueil in 1740.

We next observe teeth of different species of *mastodons*, smaller than the preceding, which have been found both in the old and new continent. Those of America are all from the southern part of that continent, and were collected by Dombey and M. de Humboldt; the others were found in different parts of Europe, but principally in France. Amongst the latter are those discovered at Simorre in the department of Gers, formerly employed in making what was called occidental turquoise.

Then follows a collection of fossil bones of the hippopotamus, almost all from the upper Vale of Arno, and brought by M. Cuvier in 1810 and 1813. The two thigh bones are very remarkable for their preservation; the remains of several species of hippopotamus, much smaller than the preceding, are not less so: they are in the two glazed boxes next to those of which we have just spoken. One of these species was found between Dax and Tartas in the department of the Landes; a second comes from a calcareous stratum near Blaye in the department of the Gironde, it was given by M. Jouannet; a third from Saint-Michel de Chaisine in the department of Maine and Loire, given by M. Dubuisson, keeper of the cabinet at Nantes.

Next follow the teeth and bones of horses, the greater number found in France, accompanying those of elephants and rhinoceroses.

After these bones are placed those of the rhinoceros, several of which have also been dug up in France: such as a considerable portion of a skull found in the environs of Figeac, given by M. Delpon, and different bones from the neighbourhood of Abbeville, for which the Museum is indebted to MM. Fraullé and Baillon: here is also an entire head of a blackish brown colour, given by the reverend M. Buckland, professor of geology at Oxford. But what is still more worthy of remark is a glazed box containing rhinoceros bones of a very small size, found at a very considerable depth; they were given by the baron de Tours, mayor of Moissac. Two drawings of heads of rhinoceroses from Siberia, also sent by the academy of St. Petersburgh are placed above the first cases of fish, and by the side of those of the elephant.

We next come to the bones of a genus nearly related to the tapir, to which M. Cuvier has given the name of *lophiodon*. They are contained in several glazed boxes, each of which holds the different species coming from a particular spot.

On examining these boxes we find that this new genus comprehends a number of species found hitherto only in France. M. Bollinat presented those dug up in the neighbourhood of Argenton in the department of the Indre, and M. Hammer, professor of natural history at Strasburgh, those from Buchsweiler in the department of Bas-Rhin. The femur and ribs which are still incrusted with calcareous stone, and which are contained in a glazed box placed above the cases, were presented by M. Boirot Desserviers, physician at the mineral waters of Neris; they come from Gannat in Bourbonnais: they probably belong to a very large lophiodon; the same may be said of another femur found in Auvergne, and given by M. Lacoste.

After these lophiodons we observe three glazed boxes full of teeth and other bones of gigantic tapirs, also found in France. Some come from Chevilly, near Orleans, and were given by M. Rousseau of Etampes; others were collected at Carla-le-Comte in the department of Arriége, by M. Lourde-Seillans, and were sent by M. de Mortarieu, prefect of that department. They are followed by the remains of the antracotherium, recently discovered in the coalmines of Cadibona, a new genus of the pachydermata, a family which seems to have lost more than all the others together: they were sent by M. Laffin and M. Borson of Turin.

We here commence the very curious series of fossil bones found in the quarries of gypsum in the environs of Paris, and which M. Cuvier has discovered to belong to lost genera, differing from all those now known in a living state. This series is contained in several cases and in glazed boxes which are placed above them.

We first observe heads and fragments of heads and jaws of various species of palæotherium and anoplotherium, next those of the genera adapis and cheropotamus, with the feet bones of these genera, their long bones, and those of their body. We must look above the cases for considerable portions of the skeletons of the *anoplotherium commune,* so remarkable for the size of its tail ; the *palæotherium magnum* and *minus,* still incrusted with the gypsum which forms their gangue (1).

At the end of the genera of which we have just spoken, and which all belong to the pachy-

(1) The system which M. Cuvier has introduced in comparative anatomy has enabled him to determine to what genus even an insulated bone belongs, although the animal should have no living analogue. When he established the genus anoplotherium, it was from the scattered bones of different individuals that he determined the general form and distinguishing characters; a short time after, the almost entire skeleton was discovered, which we see above the cases, and it was found perfectly conformable to the description which he had given of it.

Fossil vegetables do not offer the same means of determining their place, because the form of the leaves of an unknown plant cannot enable us to guess at the characters of its fructification; we can only tell whether it should be classed with the monocotyledons or the dicotyledons. It is somewhat singular, that the fossil plants found in the oldest formations all appear to be monocotyledons.

dermata, are placed the bones of carnivora, the didelphis and the rodentia; and that the fossils from the plaster quarries of Paris may not be separated, we here see the remains of birds, tortoises, crocodiles and fishes, which are found there mingled with those of mammalia. Lastly, we find the palæotheria, foreign to the soil of Paris, the greater number of which are from the neighbourhood of Orleans.

After these bones from the quarries near Paris, we remark the remains of ruminantia and rodentia, found in alluvial formations or in the bony breccia on the shores of the Mediterranean. Amongst the former, several are from the neighbourhood of Abbeville, and were collected by M. Traullé and M. Baillon, correspondent of the Museum: others are from America.   Amongst the latter we must remark a model in plaster of part of the skull of an aurochs, which must have been of a prodigious size. This model was sent to the Museum by M. Peale. We should here speak of two heads of a gigantic species of elk from the peat bogs of Ireland, which we see over each door of this room, one of which was presented by the trustees of the British museum, the other by colonel Thornton; and of the head of a large ox, placed above the last mentioned cases, which was found in the peat bogs of Arpajon, and given by M. *le comte* Dumanoir.

It is a curious fact, that the breccia of Gibraltar, Cette, Nice, Corsica, Pisa, Naples, Romagnano in the Vicentine, Dalmatia, and the island of Cerigo, all contain the same bones and present the same appearance, which leads us to suppose that they were formed at the same time and in the same manner, although at great distances from each other.

A little further are the fossil bones of the carnivora, which are found lying on the ground, and sometimes only covered with a very thin bed of stalactite, in the caverns of Germany, Hungary and England. The greater number of these bones belong to bears of a larger species than those now existing. The other carnivora, which are less numerous, belong to the genera of the cat, hyæna, and wolf.

Near these boxes are two canine teeth of a tiger; one found at Paris in digging a well, and given by M. de Bourienne, the other at Abbeville, and sent by M. Baillon; which shew, that at the period when elephants and rhinoceroses inhabited our countries, the larger species of carnivora, which prevented their too rapid propagation, also lived with them, as the tigers and lions now accompany these enormous animals in Asia and Africa.

Then come plaster casts of different bones of

the extremities of the great animal of the sloth genus, which M. Jefferson first discovered, and named megalonyx: they are found in America, in the caverns of Virginia, similar to those in Germany. M. Peale of Philadelphia presented these casts to the Museum.

These are followed by the bones of lamentins and seals, almost all found in the department of Maine and Loire, and presented by M. Renou, professor of natural history. Some come from the Isle of Aix, and were given by M. Fleuriau de Bellevue.

Further on we see bones of cetaceæ, collected in different places. The most remarkable are: 1st. The head of a small species of whale, which appears to differ much from those now existing; it was found near the sea shore at Sos in the department of the Bouches du Rhone. The Museum is indebted for it to M. Raimond Gorse, surveyor of roads and bridges. 2d. Another small head of a whale, different from the preceding, dug up in the trenches made for the basin at Antwerp, and sent to the Museum by M. *le comte* Dejean, then senator. 3d. An enormous radius of a whale or physeter, found in digging the canal of Caen, and sent by M. Roussel, professor of natural history. Several vertebræ have been collected from the basin at Antwerp, and given by M. le Chanteur, inspector of the mines, and by M. Ducos: that which is remarkable for the greatest diameter was recently found at Paris in digging the foundation of a house: it was presented by M. de Ferussac.

Here end the mamalia. The series of the other vertebrated animals commences with bones of birds found in the quarries of limestone of Chaptuzat and Gannat, some given by M. *le comte* de Chabrol, prefect of the Seine, and others by M. Boirot, physician.

A little further we find the remains of tortoises, discovered in the limestone quarries of Maestricht; a part were given by the late M. Faujas, professor of geology to the Museum, and the rest were procured at the sale of his collection.

Then come the vertebræ of different species of crocodiles, which are found in the rocks called the *Vaches-Noires,* on the borders of the Channel: they belong to those with a narrow muzzle, and are nearly the same as the crocodile of the Ganges.

Immediately after are the vertebræ and various bones of the great animal of Maestricht, classed by M. Cuvier amongst the monitors : in a glazed box above the case which contains them is placed the head of this animal, one of the finest remains of former creations that has been hitherto discovered. Drawings of most of these bones from Maestricht are given in the work of M. Faujas on the mountain of Saint Peter.

Next to those reptiles which may be called in the present existing genera, or which are at least very little removed from them, are : 1st. The remains of those singular animals called in England proteo-saurus, or ichtyo-saurus, which appear to have filled the same place amongst the reptiles, which the cetaceæ occupy amongst the mammalia ; that is to say, they were essentially swimming reptiles, and formed only to inhabit the waters, as their extremities are flattened like the hands or fins of the dolphins. The greater number of these fragments were procured at M. Bullock's sale in London, and some were given to the Museum by professor Buckland. Lastly, in two small glazed boxes are plaster models of the flying reptile called by M. Cuvier *pterodactylus,* which has as yet only been found in the quarries of Aichstaedt, and which holds amongst reptiles the same rank as bats amongst mammalia. They were sent from Munich by M. Soemmering.

Such is the rapid view of the collection of fossil remains of vertebrated animals. It is easy to imagine, that a collection of this kind may receive unlimited augmentations, and that the space it now occupies will be soon insufficient ;

but as the present order will be preserved, and all the new species carefully labelled, it will be easy, whatever are its additions and the space it may occupy, to adapt the present description to its future state.

This important examination being finished, we will not dwell so long upon the collection of formations, which is contained in the two sets of drawers in the middle of the room.

This collection is designed to represent the structure of the solid crust of the globe according to the actual state of our knowledge. It was only begun in 1821, and therefore is far from being complete. As space is wanted to display it properly, it has been necessary to shut it up in drawers, and a small number of large specimens, which may be considered as a sort of index to the contents of the drawers, which cannot be seen at all times, are placed in glass-cases above them. When the two sets of drawers are filled they will contain about ten thousand specimens from 3 to 4 inches square: it would more than furnish the shelves of a room as large as the one we are now considering. The drawers therefore answer the purpose of a much larger space.

The formations, or, in other terms, the different strata which, by their superposition, constitute the solid crust of the earth, are placed

according to their relative ages.   Each of them
is represented by its principal and subordinate
rocks, to which are added the most abundant or
remarkable organized fossils contained in the
formation.   Besides these, after each rock with
its organic remains, are placed all the vegetables,
articulated animals and mollusca, which are too
small to be properly seen in the cases of the
first room.   Lastly, that nothing may be wanting
to complete the history of each formation, specimens of the metalliferous or sterile veins, most
frequently found, are added to each rock which
contains them.   By going round the glass cases
we may form a very good idea of the above
collection, which has been thus summarily explained.   We must begin at the end near the
entrance.   We first observe specimens of primitive formations, they are rocks entirely formed
of crystallized mineral particles, generally hard,
and never containing organic remains ; precious
stones and minerals are contained in them.

Immediately after these are specimens of transition formations, a knowledge of which is no
less important, as they are almost as rich in metals
and polishable masses as the preceding ; they
shew besides the volcanic productions of the first
ages of the world, and the vestiges of the first
organized beings.

Then follow the numerous series of rocks which compose the secondary formations, properly so called. It presents sediments and transported matter more easily recognized than those of transition formation, although almost as perfectly consolidated and cemented. It is poor in metals, but abounds in fossil remains without any trace of mammiferous animals. It contains coal-mines and the largest mines of sulphur and rock salt. The volcanic productions of the middle age of the globe are here seen dispersed under different forms.

The rocks of tertiary formations immediately fellow. The sediments and transported matter which compose them are generally but imperfectly cemented, and of a very variable consistence; many are still loose; with the exception of iron ore, improperly called alluvial, we no longer find amongst them metals which can be worked. We meet with marls, clays, fuller's earth and potter's clay, and common building stones, materials easily worked because of their softness; but there are no more of those masses, remarkable for the variety and brilliancy of the polish which they are capable of receiving. The only rocks, which are an exception, belong to the volcanic productions of this period.

The specimens of modern formations terminate the collection. We remark first, the rounded silex, sand, mud and fossil remains of the great diluvian formation. 2d. The different alluvial productions which are daily created by the sea, rivers, torrents and springs. 3d. The organic detritus, which are mingled or alternated with these productions, such as peat, and conglome-rated shells or madrepores belonging to animals whose species are now living. 4th. The various substances proceeding from volcanoes either now burning, or extinguished since the last diluvian catastrophe gave the continents their present outline. We observe with some in-terest among the lavas of this last epoch a slab of scoria from mount Vesuvius, bearing the name of Dolomieu. It was torn away with long pincers from the sides of the running lava in 1805, and moulded while still hot. We also find various specimens of the saline or sul-phurous incrustations which line the interior of craters; they complete the collection of forma-tions, which we now quit to pass into the third room.

It bears the name of the rock-room, and prin-cipally contains a systematic collection of rocks, classed according to their composition and tex-ture. There are also the first elements of a geo-graphical collection, as well as a collection of geological and mineralogical specimens, which have been cut and polished. We will begin with the geographical part.

This was begun only in 1821; it is intended to offer, if we may be allowed the expression, a text of the data on which the arrangement of formations in the preceding room has been founded. It is composed of a series of rocks, which are generally from places very distant from each other, and which represent the peculiar constitution of these different points on the surface of the globe. When the series is more numerous and complete, they may be considered as so many witnesses, proving the exactness of the place which shall have been assigned to each formation in the structure of the solid crust of the earth. They are dispersed around the room, and in the order of the space which was unoccupied at the time of receiving them. Beginning by the cases to the right on entering, the lower parts of which they occupy with only a few exceptions, we successively find:

1st. The rocks of Greenland, collected and presented by M. Giesecke, geological professor at Dublin.

2d. The rocks of the islands of Saint Pierre and Miquelon, near the bank of Newfoundland, sent by M. de la Pilaye.

3d. A considerable series of the rocks of the United States, sent by M. Milbert, correspondent of the Museum.

4th. Some specimens collected in the Antilles by various persons; particularly in Guadaloupe, by M. L'Herminier, director of the garden of naturalization in that colony.

5th. Some beautiful specimens from the environs of Rio Janeiro, in the Brazils, procured by the expedition of captain Freycinet. To these has been added, a superb elastic plate of quartz-rock, from the province of Minas-Geraës, which the Museum owes to the generosity of M. Bouch.

6th. Several series more or less detailed, but all important, collected in the islands near Cape Horn, and in those of the Pacific Ocean, by the naturalists of captain Freycinet's expedition.

7th. A considerable number of rocks, brought from New Holland by M. Freycinet; to which are added, those found by Peron in the same country.

8th. Specimens from Manilla and Sumatra, some given by M. Perrottet, and the others procured by MM. Duvaucel and Diard, travelling naturalists to the Museum.

9th. A small series collected in the mountains of Rajemahal, on the borders of the Ganges, by M. de Saint Yves, correspondent of the Museum.

10th. A more numerous series, illustrating the

composition of the mountains of Nazam and Hy-
derbad, in the Madras government, presented by
M. Milne Ricketts, member of the supreme council of Bengal.

11th. Some lavas from the Isle of Bourbon,
given by different travellers, but chiefly by
M. Des Étangs.

12th. Specimens from Caffraria and the Cape
of Good Hope, brought by M. de Lalande, assistant naturalist to the Museum.

13th. Lavas from Teneriffe, given by different
collectors.

14th. An interesting suite of rocks, from Scot-
land and the North of England, given by Dr. Boué.

15th. Some others from the Shetland isles, collected by M. Wilson, and presented by M. Lucas.

16th. A remarkable series of ancient volcanic
productions from the Ferroë isles, and of the
substances which they contain; it was given by
Prince Christian of Denmark, during his Royal
Highness' visit to Paris in 1822.

17th. Several specimens from the Netherlands,
collected by MM. de la Jonkaire and de Basterot.

18th. A suite of ancient lavas from Kaiserstuhl,
in Brisgaw, collected by M. Eckel of Strasburgh.

19th. An interesting series of calcareous formations from Vienna, collected and presented by
M. Constant Prevost.

20th. Some specimens from Hungary, which made part of a rich collection of minerals presented to the Museum by the Emperor of Austria: interspersed with these are a hundred beautiful specimens from the same country, the recent gift of professor Zipser of Neusohl.

21st. Various rocks, formerly collected by Olivier on the shores of the Bosphorus and the Grecian Archipelago.

22d. Several numerous and instructive series brought from Sicily, the Lipari and Pumice isles, from Naples, the environs of Rome, Padua, the Vicentine and Veronese, by M. Lucas. To these are added a number of large specimens from the same countries, given by Dolomieu and Spallanzani.

23d. Several polished specimens from Piedmont, given by count Daru.

24th. A magnificent series of primitive and transition rocks from Corsica. All the specimens are polished, and rival each other in richness and variety of colours.

We will not quit this *geographical* collection of rocks without remarking, how desirable it is to give it a greater extent. We hope to attain this object, when we recollect the zeal and liberality of the numerous individuals who now cultivate the study of geology even in the most distant countries.

The *systematic* collection, at which we are now arrived, is placed in the upper part of the cases which are to the right of the room on entering. It is divided into two sections: one is formed of specimens for study, which are displayed in a horizontal line of shelves extending throughout this part of the hall, and which divides the height of the cases into two nearly equal parts: these specimens are almost all of the same size. The second section contains either large specimens, which could not be mixt with those for study, because of their disproportionate size, or pieces cut and polished, which shew the different uses made of them in the arts: they occupy the whole of the upper part of the cases.

The collection is classed according to the characters which each species of rocks derives from its composition, its texture, and its origin, without considering the situation which it occupied in the formation of the solid crust of the earth: its object is to furnish the means of recognizing rocks even when they are out of their original position. The outline of this methodical collection was laid down by the late M. Haüy. It comprehends five classes, which may be successively examined, beginning at the extremity of the room, and alternately directing the attention to the stages and upper shelves of each case.

In the first class we find the earthy and saline substances: as, 1st. The pegmatite, which, under the vulgar name of *pétunzé*, furnishes the glazing matter for chinaware, and the detritus of which produces the earth called *kaolin,* of which that valuable sort of ware is made. 2d. Common porphyry. 3d. The syenite, which forms the rocks of the cataracts of the Nile, and from which the Egyptian obelisks were hewn. 4th. The common granite, which is the lowest bed of the crust of the earth within our reach. 5th. The phthanite, of which touchstones are made. 6th. The potstone, which in several countries is formed into cooking and other domestic utensils.

In the second class, that of combustible substances, we see various species of coal, and a species of fossil wood used by jewellers under the name of jet.

In the third class, containing *metallic* substances, we find specimens of the principal metalliferous rocks, from which are extracted copper, iron, lead, tin and zinc.

The fourth class contains the rocks of *igneous* origin, according to many geologists; of *aqueous,* according to others. These rocks offer a problem, for the solution of which, we may compare their characters with those specimens which are in the preceding and following classes.

The last class is that of rocks incontestably volcanic; in which we see, not only vitreous and scorified substances, which clothe the upper and lower surfaces of the current of lava, but also the materials which constitute the internal masses of these currents; substances which, in coagulating more slowly than those of the surface, produce aggregates of a stony aspect, and so similar to crystallized rocks of ancient formation, that unless apprised of the fact, we should be in danger of considering them as such.

We will terminate the examination of this room by noticing the works of art, which are for the present placed in the five cases to the left on entering.

On the first shelf, the uppermost, we see four large vases of the Vesuvian lava, a large and beautiful cup of limpid rock crystal, a large slab of greenish serpentine, and a mirror of black obsidian similar to those used by the Peruvians before their conquest by the Spaniards.

On the second shelf are several cups of agate, chalcedony, and jasper of different colours, another of rock crystal, and one of violet-coloured fluate of lime, two of greenish jade, a vase of the same matter, and a small one of lapis-lazuli.

On the third shelf we find a numerous suite of small slabs of jasper, agate and chalcedony, a row of small columns of amethyst, some little cups of chalcedony, chrysoprase and amethyst, with several cut precious stones, such as diamonds, oriental rubies or red corundum, oriental sapphire or blue corundum, chrysolite, etc.

On the fourth shelf we observe, amongst a second series of polished slabs analogous to the preceding, variously coloured specimens of rock-crystals, some facetted, others *en cabochon,* that is, simply polished. In the front row are several specimens of artificial precious stones, from the manufactory of M. Douault-Vieland, jeweller, who presented them tot he Museum.

Objects varying greatly in form and substance are exposed on the fifth and sixth shelves; amongst which may be mentioned, a beautiful box of yellow amber, several large slabs of Florentine marble, different tomahawks of savages, a cup of red jasper, and a large spoon of greenish jade, which is considered as a rare and precious object.

§ III. COLLECTION OF MINERALS.

—

The most striking character of minerals, when compared with the productions of the two other kingdoms, is a want of organization, and the absence of that internal motion which, in animals and vegetables, contributes to the developement and preservation of the individual. Considering minerals in their perfect state, they are of a simple structure, and consist of a symmetrical arrangement of particles similar to each other; whence results a regularly formed exterior, analogous to that of geometrical solids.

The diversity of forms, of which the same substance is susceptible, establishes a new contrast between minerals and organized beings. In vegetables, for example, the different individuals of one species bear the mark of a common model, on which they seem to have been formed; the primitive type always existing amidst slight and accidental variations. The same mineral, on the contrary, frequently presents itself under a multitude of different forms, all equally regular, and sometimes having no outward resemblance to each other. But this singular metamorphosis, submitted to simple laws, and the effects of which may be calculated, confines itself to modifications of the external appearance, without altering the mechanism of the internal structure, which is constant and uniform in every variety.

Such is the action of these laws, to which nature is subjected, that, when undisturbed, she inclines to produce the most simple forms, and those best characterized by regularity and symmetry. This operation is called *crystallization*, and the bodies so formed, *crystals*. But it frequently happens, that local circumstances and disturbing causes interfere, to interrupt or derange the ordinary progress: hence incomplete and irregular forms are produced, either misshaped structures, into which the primitive form insensibly degenerates, confused aggregations of foliations or needles, fibres or grains, or lastly, masses entirely compact.

Every mineralogical collection should possess a series of species methodically distributed, and each species should have its varieties placed according to their degree of perfection. The collection in the Museum presents a picture of the mineral kingdom, divided into four great classes, according to the method of M. Haüy.

The first class is that of earthy substances containing an acid, the salts of former systems; the second class comprehends earthy substances, or stones; the third presents an assemblage of different inflammable substances; and the fourth comprises the metals.

The cases which contain this collection are numbered and divided into shelves. At the height of the eye are placed the specimens specially intended for the use of students; they are arranged on smaller shelves, connected together like steps, forming little stages. The specimens follow uninterruptedly in the same case, beginning from the lowest shelf; so that in passing from one case to another, the first specimen of the one comes immediately after the last specimen of the upper shelf of the preceding. Above and below the stages, on the shelves, are pieces remarkable for their size; they belong to those species on the stages to which they are connected. They shew these same species in their various associations with other mineral substances, and recall the different localities in which they have been found. As these specimens are of various sizes, and are not all of equal interest, it was impossible to arrange them in the same systematic order; the larger pieces have therefore been placed in the lower part of the cases, whilst those which are rare and precious are placed most in sight, and those least worthy of attention are on the upper shelves.

The mineralogical collection occupies the two rooms immediately following those devoted to geology. To examine them methodically we cros sthe first room, and begin with the first case, at the extremity of the second room, in the corner, by the window.

This first case contains several instruments for ascertaining the characters of minerals; amongst others the goniometer, by which are measured the mutual inclinations of the faces of crystals. In the space between the windows are several shelves, on which are arranged models in wood; some of which serve to explain the structure of crystals according to the theory of M. Haüy, and others represent the principal varieties of the regular forms of crystallized bodies. Below these models are polished slabs of red porphyry, etc.; others which are of a rounded form, and which present tolerably regular patterns of that description of argillo-ferruginous limestone, generally known under the name of *ludus helmontii.* Small tablets are made out of it for covering brackets.

We will return to the first room, which contains the two first classes of minerals, the salts and earthy minerals; and beginning with the

case on our right, we shall follow their numerical order.

The first series is that of calcareous substances,
or those which contain lime in a greater or less
quantity. The carbonate of lime begins the
range: the numerous modifications of this substance, the most abundant on the globe, almost
entirely occupy the five first cases. We may first
observe on the stage a suite of crystallized varieties for study, each of which has a characteristic
name. Some places, although vacant, are labelled
to mark those known varieties which are still
wanting to complete the series. Below the
stage, in the first case, is a very fine crystal of
Icelandic calcareous spar, which exhibits to peculiar advantage the property of double refraction,
for which this mineral is remarkable. This specimen, of an uncommon size and clearness, is
nearly seven inches thick. We also see some very
interesting groups of crystals on the shelves;
amongst them the primitive carbonate of lime
from Ratieborztiz, in Bohemia; also metastatic
crystals from Derbyshire, given by M. Heuland,
a German mineralogist, to whom the Museum is
indebted for a great number of specimens, all
remarkable for their size and freshness. Here
too we see the equiaxe variety from Andreasberg,
in Harz. The two following cases present a

continuation of the geometrical forms of carbonate of lime. In the second, above the stage, we remark a beautiful crystal of the variety called *continue*; and in the third, below the same stage, a superb specimen of the *soustractive* variety, given by M. Heuland.

In the fourth case is a series of irregular bodies; amongst which is the fibrous carbonate of lime, presenting the undulating reflections of watered silk, and is hence called satin-spar. It is used in jewellery.

The lamellated variety, more generally known by the name of Parian marble, employed by ancient sculptors for the representations of illustrious personages.

The saccharoïdal carbonate of lime, so called from its resemblance to sugar; it is the marble now employed by statuaries, and comes from the famous quarry of Carrara, in Italy.

The coarse carbonate of lime, or building stone of Paris.

The variety called *liais*, or very hard freestone, only employed for ornamental building and sculpture.

The earthy carbonate of lime, or the substance commonly called chalk, from which is formed the whitening used in house-painting and glaziers' work.

Lastly, the *pulverulent,* more generally known under the name of *fossil-flour.*

Above the stage is the lithographic stone, discovered a few years ago at Chateauroux, in the department of Indre; and which has all the qualities of that of Bavaria, which was first employed for that purpose. A specimen of the latter, from Ingolstadt, is at the bottom of the case.

The series of carbonate of lime (fourth and fifth cases) is terminated by specimens of those varieties of it, which have been denominated concretions; they result from the filtration of a liquid, charged with calcareous particles, through the roofs of subterranean cavities. In proportion as the drops, which remain suspended to these arches, dry, the stony particles unite in a tube, which elongates from successive deposits like icicles: this kind of concretion is called *stalactite.* A portion of the liquid falling from the roof to the ground forms other deposits, generally mammillated, named *stalagmites.* These deposits sometimes increase so as to unite, and consequently form enormous columns. We see such in the grotto of Auxelle, in the department of Doubs; whence the beautiful stalactite, which is placed between the windows in this room, was taken. But of all the grottos of this kind, the most celebrated is that of Antiparos, in the Archipelago.

When Tournefort visited it, he imagined that stones vegetated in the manner of plants. He brought home the beautiful concretion placed at the bottom of the fourth case. The calcareous alabaster, of which vases and statues were formerly made, and of which we have here several specimens, is the result of successive deposits of carbonate of lime : the layers of which form undulations more or less distinct, and shew, when polished, zones of different colours.

When water, having calcareous particles in solution, remains in a cavity of small extent, these particles incrust the sides of the cavity, and line it with crystals : this is called a *geode*. We see on the stage of the fourth case an example, the interior of which is lined with crystals belonging to the metastatic variety.

The liquid sometimes deposits particles, which are held suspended, on the surface of different organic bodies, and clothes them with a stony envelope; under which are preserved the principal features which characterize the body. It is this variety, named *incrusting concretion,* which covers the branches and nests, which are at the bottom of the fifth case. Next to these is an incrustation of a milk-white colour, presenting the likeness of Galileo, from the baths of San Philippo, in Tuscany. Immediately over the stage we observe the *inverse quartziferous* variety, commonly called *crystallized sandstone of Fontainebleau;* but which in reality is only carbonate of lime mixt with sandy particles. Its crystals frequently form groups of considerable size, such as those we see on the shelves of the same case.

The bituminiferous carbonate of lime constitutes the black marbles of Dinant and Namur, and is employed for paving churches.

In the sixth case we have another species, called *arragonite*, of which there is at the bottom a considerable block, given by M. Lacoste de Plaisance, professor at Clermont-Ferrand. The most remarkable of its varieties is the coralloïd, formerly known under the improper denomination of *flos ferri,* which frequently rivals snow in whiteness: it is a stalactite, the twisted branches of which are entwined together. Here are very fine groups of it from Eisenerz, in Styria.

The third species is the *phosphate of lime.* In the seventh case we observe the earthy variety from Estramadura, where it is employed in building. Its dust, when thrown upon red-hot charcoal, emits a beautiful phosphoric light.

In the eighth case is the *fluate of lime;* a substance known in the arts under the name of *fluorspar,* remarkable for the diversity of the colours

which ornament its crystals.   Amongst the regular forms it most commonly assumes, we observe that of the cube exactly traced in a great
number of specimens. The most beautiful groups
in the collection are from Derbyshire and Nor-
thumberland, and were presented by M. Heuland.
On the upper shelves is a specimen of the *concré-
tionnée* variety, which is formed of bands and
zones like calcareous alabaster : in England it is
cut into slabs and cups of different forms.   The
acid contained in this substance is also used for
engraving on glass, on account of the corrosive
property it possesses.

The *sulfate of lime,* which we see in the
ninth case, is the mineral commonly called *gypsum,* or *plaster-stone.*   In fact, plaster is only a
mixture of this substance with carbonate of lime :
this mixture sometimes exists in nature, as at
Montmartre, which is almost entirely composed
of gypsum.   The *lenticular* variety, so called on
account of its rounded form, is usually found
there : on the lower shelves may be seen many
beautiful specimens of it.   The thin and transparent plates, which are detached from these
crystals, are vulgarly called *specular stones,* and
asses' looking-glasses.   The ancients used them
instead of window-glass: according to Pliny.
the temple of fortune, which was built of this-

stone, had no windows; and was only illuminated by the light that passed through the walls. Besides these, we observe in the cabinet the white laminar variety from Sicily, remarkable for its beautiful pearly lustre; the fibrous variety, of a shiny and silky whiteness, used in jewellery; the niviform, which has the appearance of a snowball; lastly, the compact variety, or gypsous alabastar, employed in making statues and ornamental vases.

In the tenth case is the *anhydrous sulphate of lime*, which only presents one remakable variety, the lamellar, of a sky-blue colour, and called Wurtemberg marble. There is a beautiful polished slab of it, which was presented to the Museum by the King of Wurtemberg.

The *nitrate of lime* is the substance daily formed on the sides of damp walls, and obtained by washing old plaster for the making of saltpetre.

The *arseniate of lime* follows immediately; its name indicates the presence of arsenic acid in it; it is only known to mineralogists. It has been called pharmacolite, or poisoned stone. It owes its pink hue to the presence of cobalt.

The *sulfate of barytes*, which occupies the eleventh and twelfth cases, is, next to the carbonate of lime, the species most abundant in regular forms. The stage for study presents a considerable number of specimens, most of which are from the departments of Puy-de-Dôme and Cantal. On one of the upper shelves is a beautiful specimen of the primitive variety from Czarles, in Transylvania. This species is remarkable for its great specific gravity, and is the heavy spar of former mineralogists. The radiated variety has been long known under the name of Bológna stone ; because the phosphorous, which bears the name of that city, is obtained by calcining this stone.

The *carbonate of barytes*, found principally in England, is in the thirteenth case. It is a poison for all animals, and is named in that country *ratsbane*. Regular forms of this substance are very rare. On the stage for study is a beautiful group of crystals, terminated by pyramids with six sides.

The *sulfate* and the *carbonate of strontian*, in the fourteenth case, are only interesting to mineralogists. In the bottom of the case is a beautiful group of crystals of the former substance, of a very great size ; it weighs twenty pounds, and was presented by Dolomieu. Some superb specimens of limpid crystals and concrete masses are distributed on the shelves. They were collected in Sicily by M. Lucas, whose last excursion in that country has enriched the Museum with a multitude of rare and precious objects. The compact variety, in grey masses, is common at Montmartre. The sulphate of magnesia is employed in medicine as a purgative; it has been known by the names of bitter salts, Epsom salts, Sedlitz salts, etc. The borate of magnesia is of no use in the arts; but its crystals are remarkable for acquiring electric properties, when submitted to the action of fire.

In the fifteenth case is the *aluminous fluate of silex,* or the substance called *topaz:* it furnishes several precious stones for jewellery; but, however, it must not be confounded with that called by lapidaries *oriental topaz,* which belongs to a species of the second class. This want of agreement between the scientific nomenclature, and that in use among artists, proceeds from the disposition of the latter to unite all substances, resembling each other in colour, under one denomination; although this similarity of appearance frequently only disguises essential differences in composition: besides which, the same species is frequently subject to a series of the most diversified tints. We have already observed such a variation of colour in the modifications of fluorspar; the topaz furnishes us with another example no less remarkable. In fact, we see on the shelf, for study, several specimens of a yellow colour; and, next to them, a blue crystal of the *quindecioctonal* variety, sixteen lignes thick, and weighing nearly 4 oz. 2 drams : on the shelf immediately underneath, a reddish yellow crystal, longitudinally striated; another red, vulgarly called Brazil ruby ; and a white topaz from the same country, cut like a brilliant, and given to the Museum by professor Geoffroy Saint-Hilaire. This last variety is that called by the Portuguese the *water-drop* and *mina nova*. Above and below, on the same stage, are various groups of pale yellow crystals, remarkable for their size.

The greater number of stones, sold under the name of Brazil rubies, and those which are called by dealers burnt topazes, are merely topazes of a reddish yellow, which have been exposed to the action of fire to give them a fine rose-coloured tint.

The nitrate of potash, which we see in the same case, is the salt called *nitre*, or *saltpetre*, employed in the manufacturing of gunpowder; which is a mixture of about six parts of nitre, one of charcoal, and one of saltpetre. Aquafortis is extracted from this substance, which is the reason why it is also called nitric acid.

After the nitrate of potash follow the species which have soda for their basis; and the most important of which is the *muriate of soda*, or *common salt,* so useful in domestic economy. The crystals are generally of a cubic form, as we may observe in several specimens on the upper shelves of the fifteenth case. Next to them is a beautiful indigo blue crystal from Ischel, in Upper Austria. The suite of varieties is continued in the sixteenth case. Below the stage is the red salt of Cardona, in Catalonia; still lower is a beautiful specimen of limpid muriate of soda, from the celebrated mine of Wieliczka in Poland, one of the most important saltmines known. It produces 120,000 cwt. of salt annually, is 900 feet deep, and extends nearly six miles in every direction.

A considerable quantity of muriate of soda is held in solution in sea water and certain lakes; being extracted by evaporation, it is then called marine salt; and it differs only from rocksalt, because the latter is crystallized by nature.

The muriate of soda, by dissolving, yields muriatic acid, so usefully employed in dying and bleaching.

The borate of soda, which comes next, is the substance commonly called *borax,* or *tinkal,* and comes from the East Indies; it is purified before it is used in the arts, and serves for soldering metals and gilding jewellery.

The *carbonate of soda* was formerly known

under the name of *natron,* and was obtained
by the evaporation of the waters of certain rivers
or lakes, especially in Egypt, where it is in great
abundance. In Europe it is found in an efflo-
rescent form on the surface of the ground, or on
the sides of old walls; it is also found in great
quantities in the ashes of several vegetables.
It is employed in manufacturing glass, in the
composition of hard soap, and is also used in
medicine.

The muriate of ammonia, which we see in the
seventeenth case, is better known by the name
of *sal ammoniac.* It is found amongst volcanic
productions. There is a specimen in this case of
a concrete form, from Vesuvius. It is now extracted from putrefied animal matter, and is employed in the arts for plating and soldering
metals.

Under the name of *alkaline sulfate of alumine,*
we here find *alum,* that substance so useful to
dyers, who make use of it to fix and strengthen
colours. It is only found in nature in small
quantities in the form of filaments, to which the
name of *plumose alum* has been given. We remark a fine specimen of this fibrous variety,
which the celebrated Tournefort brought from
the island of Milo, in 1703. The alum, which
is sold in commerce, is obtained by washing sub-

stances impregnated with it, or is composed of such as contain its principles. It is thus that the beautiful semi-transparent group, in the bottom of this case, was formed, which came from the manufactory of M. Curaudau. The greater number of the varieties on the stage is also produced by artificial crystallization.

The following species, the *alkaline fluate of alumine*, also called *cryolite*, is of no use in the arts; but is interesting to the mineralogist on account of its rarity. The shelves, next to the stage for study, present a very precious series of the different varieties of cryolite, all remarkable for their size, and were brought from Greenland by M. Giesecke, professor of mineralogy at Dublin. This learned professor employed nearly eight years in exploring, with indefatigable zeal, a country still new to the naturalist, and collected an abundant harvest of various productions, with which he has been pleased to enrich the collection of our Museum.

(Eighteenth and nineteenth cases). We are now arrived at the second class of minerals, that of stones, or earthy substances. These productions are generally of brilliant and beautiful colours; amongst them are classed those rare and desirable stones which art transforms into objects of dress and ornament. In the first place,

under the name of quartz, we see one of the
most abundant species in nature, and the modifications of which are the most numerous and
diversified.   We will examine this substance in
its purest state, the whole of whose varieties
bear the name of *hyaline* quartz. The first is transparent and crystallized, and is known by the
name of rock-crystal.   There are some beautiful
specimens on the shelves of this and the adjoining cases.   The most general form of these crystals is a solid of six sides, terminated with two
pyramids with six faces.   The Museum possesses
a fragment of such a crystal, the enormous size
of which prevented its being placed in this gallery: it was brought from Valais, and weighs
more than 800 pounds.   It has been placed on
the staircase, leading into the first geological
room.   A group of similar crystals of a remarkable size is placed in this room, near the windows; the crystals are nearly a foot long, and
altogether weigh 525 pounds: they come from
Fischbach, in Valais.   The colourless hyaline
quartz, or rock-crystal, is employed in making
lustres, vases of different forms, etc.

This substance is not always limpid, as in the
crystals of which we have just now spoken: it
is often coloured by other matter, without entirely losing its transparency, and then it is named

according to its different tints. After the colour-less hyaline quartz, on the stage for study, we observe the violet rock-crystal, commonly called *amethyst*; the rose-coloured or Bohemian ruby; the blue, the yellow, or Indian topaz; the yellow-brown or smoked topaz; the dark green, the *hœmatoïd* of a dull red, or *compostella hyacinth*.

The iridescent rock-crystal, which we see in the nineteenth case, owes its name to the colours of the rainbow, which are reflected by the air lodged in a flaw.

As we proceed, we remark some precious specimens of all the preceding varieties on the different shelves; amongst others, some masses of rock-crystals from Madagascar, some beau-tiful samples of rose-coloured lamellar quartz from Siberia, a crystal of a blackish brown hyaline quartz, from the same country, nearly 12 inches long.   In the bottom of the eighteenth case is a magnificent *geode* of a violet hyaline quartz, given by M. Brard, formerly assistant naturalist to the Museum.   The variety named *aero-hydre*, which is in the following case, con-tains a drop of water, which but partly fills a tubular cavity, so that the air bubble, which oc-cupies the void space, ascends and descends ac-cording to the motion given to the stone, as in the water-level.

With the twentieth case commences the series of specimens which present quartz or rock-crystal in its various modifications. They are generally called *agates*; the greater number is concrete bodies, some only offering the primitive form of quartz. Amongst them we may distinguish *chalcedony*, which is of a milky white, with a cloudy transparency; *cornelian*, of a cherry red; *sapphirine*, of a delicate blue ; *sardonyx*, of an orange colour ; *prase* and *plasma*, which present different shades of green; other varieties present coloured zones, sometimes parallel and sometimes circular, they are then called ribband and onyx agates. Artists cut them, as well as the preceding, into boxes, vases, seals, and other ornaments. In the lower part of the case is a simple onyx *agate* in chalcedony, from the environs of Oberstein. Two varieties, far less inviting from their appearance, are valuable for their use, as the gunflint and the millstone.

After the agates, come the quartz-resinite, which shines like rosin ; and the jaspers, which are opaque, the fractures of which are always dull. Amongst the former we observe the *hydrophane*, which becomes transparent when plunged into water; the *girasol*, which reflects red and gold yellow colours; the common *opal*, which is so much sought after for the variety and beauty of its colours. The varieties of jasper, receive like those of agate, their name from the colour they principally display. One of the most rare is the sanguine jasper, which is of a more or less dark green, spotted with deep red. In the same case we see several specimens of *pseudomorphous xyloïde quartz.* These bodies, commonly called *petrified wood,* were originally trunks or roots of trees, the substance of which has been replaced by quartz. The substitution takes place by degrees, so that the stony particles are successively lodged in the small cavities, formerly occupied by the vegetable matter, in proportion as the latter abandons them, whence the appearance of vegetable tissue is preserved. The greater number are from common wood, and marked with concentric zones, which answer to those we see in the transverse section of a tree. Others, and particularly the magnificent trunk which we see in the lower part of one of the cases, were originally palm-trees; the totally different organization of which is to be traced in the whitish or yellowish ground, sprinkled with little black spots.

The substances, which follow the quartz, in the twenty-second case, are those which furnish the rarest precious stones, next to the diamond; and are the most sought after for their brilliancy

and hardness. The first is the *zircon*, a fine series
of its crystals is to be seen on the stage for
study. The precious stones, furnished by this
mineral, are the *jargoon of Ceylon*, which is of a
greenish yellow, or a marigold hue, and the *hyacinth* scarlet mixed with brown.

The next case presents the corundum, which
of all minerals is the most rich in precious stones.
Those called *ruby, topaz,* and oriental *sapphire,*
by jewellers, are only varieties of this substance :
one of which is red, another yellow, and a third
of an indigo blue. In the collection for study
we find prismatic and pyramidal crystals of
these different colours, and afterwards specimens, polished, but not cut, of similar crystals ;
one of them is divided into two parts, one red,
the other yellow, so that it is at once a ruby and
a topaz ; which circumstance is the best proof,
that the colour of these stones is merely accidental, and cannot affect their nature. By the side
of the transparent variety, of which we have formerly spoken, we see others which are more or
less opaque, and extremely lamellated ; for which
reason they have been named, by M. Haüy, *har-
mophane;* their former name was adamantine
spar. The last variety, less remarkable in appearance, is the granulated corundum, vulgarly called
emery, the fracture of which is compact and

dull. It is used in polishing different substances, such as metals and looking-glasses. After the corundum comes the *cymophane,* which is very brilliant, and of a greenish yellow; with lapidaries it bears the name of chrysoberyl and oriental chrysolite. The *spinelle,* which follows, furnishes two varieties of ruby, found in commerce under the names of the *spinelle* and *ruby balais,* and which only differ in the deepness of their colour.

The emerald is also much sought after for the purpose of ornaments. The precious stones belonging to this species are the emerald of Peru, and the beryl or *aigue-marine.* We, see on the stage for study, a fine crystal of emerald from Santa Fé, which presents the primitive variety. The Peru emerald is the most esteemed, it is of a pure deep green. The beryl is of a greenish blue, or of a honey yellow. Below the stage are some long cylindrical crystals belonging to this variety, and which come from Siberia. We find in France, opaque emeralds of a considerable size, but which are of no value : such is the enormous crystal, placed in the lower part of the case, found at Barat, near Limoges.

After the emerald, we see the *cordierite;* amongst its varieties is the water sapphire of the jewellers. The name of *cordierite* is a compliment paid to the learned professor of geology in this Museum, to whom we owe the first exact description of this substance.

The *Euclase,* which we see in the same case, was brought from America by Dombey, travelling naturalist to the Museum. This mineral, which is only remarkable for its rarity, is of an agreeable green colour, and can easily be polished; but, for the great facility with which it splits, it is never worked into objects for ornament.

The garnet, which is distinguished by the size and exactness of its dodecahedral crystals, affords several precious stones; such as, the *Syrian garnet,* which is red mixed with violet; the *Bohemian garnet,* of a vinous red mixed with orange; and the *vermeille,* which is of a deep scarlet.

We will pass rapidly over the different substances which terminate the second class (twenty-fifth case); because they are mostly unknown to those not acquainted with the science, and are scarcely of any use in the arts.

The felspar, which we see in the twenty-sixth case, however, deserves attention for the beautiful varieties which it affords; amongst which is the Labrador stone, or opaline felspar, whose iridescent reflections may be compared to the wings of the most beautiful butterflies; the *moonstone*, or mother of pearl felspar, the ground of which is of a sky-blue; the *avanturine*, of a carnation or green colour, spotted with gold coloured and white dots; and lastly, the *amazon stone*, or green felspar, the surface of which in certain lights looks like satin. There is a beautiful specimen of this variety on the first shelf, above the stage for study. The *kaolin*, or *decomposed* felspar, is of a great use in making porcelain.

Several varieties of *tourmaline*, in the twenty-seventh case, are employed by jewellers; but it is chiefly remarkable for its electrical properties, when heated. Beyond the stage is a superb specimen of the red acicular variety, called *siberite*.

The *amphibole* and *pyroxene*, whose numerous modifications fill this case, are chiefly interesting to mineralogists for the important share they have in the structure of the terrestrial globe. On one of the upper shelves is a beautiful group of pyroxene crystals of the quadrioctonal variety, given by M. Muthuon, engineer of the mines of France. Below, in the twenty-eighth case, is a magnificent specimen of silky amphibole from St. Gothard.

The lazulite, which we see in the twenty-ninth case, is better known under the name of *lapis-lazuli*, or simply, *lapis*. That which is of a purple blue is most sought for by artists, who work it into slabs, and extract from it the ultramarine blue colour, which produces such a beautiful effect upon stuffs, and is so unchangeable. Several of the minerals, which follow, are not yet brought into use, and are not worthy of much attention. Amongst them, however, are some specimens of a large size, such as the red *stilbite* and whitish *analcime*, given to the Museum by M. Lucas, jun'.

Mica, of which we see some large slabs in the bottom of the thirtieth case, has been called Muscovy glass, because it is employed in Russia instead of window-glass. By the side of the mica is *asbestos*, the filamentous variety of which was known to the ancients under the name of *amianthe*, or incombustible flax, which they spun, and made cloth and napkins of it. They threw them into the fire when dirty; by which means they were made whiter than if they had been steeped in lye. They wrapt the bodies of their dead in this stuff, when they wished to preserve their ashes. We see, on one of the shelves, a speci-men of this sort of cloth, which was made in Italy.

The talc, which immediately follows, presents also several varieties, which are interesting for their uses; such as the *lapis ollaris*, or potstone, out of which vases are made on the turn-lathe; the Verona earth or green earth, which is em-ployed with oil for painting landscapes; and the lamellar or Venetian talc, the powder of which renders the skin smooth and shining. It is coloured with the plant named *carthamus,* and then sold as rouge. Before we quit this room, we should observe between the windows a superb vase of the *brecciated* porphyry of the *Vosges,* and two very large groups of prismatic crystals of colourless quartz. The following room contains the inflammable substances and the metals. Returning to the order of the numbers on the cases, we find in the thirty-second those substances which are combustible, but not metallic. The principal is native sulphur, or sulphur free from all combinations. There are some superb groups of translucid crystals, and a beautiful series of varieties, given by M. Lucas, and which he procured in his last journey in Sicily and Italy.

The diamond, which follows in the thirty-third case, is placed amongst the combustibles, next to the *anthracite,* or native mineral carbon, because it burns without leaving any residue; and the most exact experiments have proved, that the diamond is only mineral carbon in its purest state. Below the stage for study is a series of diamonds, rough and cut; the regular forms of the first are the effect of crystallization. In the same case are placed the different varieties of solid and liquid bitumens; amongst the most curious is that which comes from England, under the name of elastic bitumen.

Black coal is one of the most precious minerals for its usefulness. The French territory abounds with mines of it, which are worked with advantage. *Jet* is of a more brilliant black than coal; it is polished and used for different purposes, but chiefly as ornaments for dress in mourning. The yellow amber has been much used in making ornamental furniture; it is now cut in the same way as precious stones. Pieces, similar to those placed here, are particularly sought by the curious, for the insects that were enveloped by the amber when in its liquid state, without injuring their form.

With the thirty-fifth case commences the class of metallic substances. Their utility in the arts and their properties are too well known, to render it necessary to enter into long descriptions, to enhance the value of those principal sources of our wealth. We will limit ourselves to a rapid notice of them, merely remarking those most worthy of our attention.

Platina, which first presents itself, is the least fusible of all the metals; it takes a perfect and tolerably brilliant polish; as yet it has only been found in the form of little grains, such as are now before us. Watch chains and snuff boxes are made of it, as also mirrors for telescopes, crucibles, and small instruments for the use of mineralogists.

Gold has hitherto been found only in the native state. In the Brazils it is met with regularly crystallized, as appears from the beautiful specimens with which M. Geoffroy enriched the Museum. It more often exists in a ramified state on the surface of apparently common stones, the ground of which is a white or yellowish quartz. It is disseminated in grains amongst the sand of several French rivers, such as the Rhone, the Arriège, etc. In the thirty-sixth case we may observe an enormous piece of massive gold, from Peru, it weighs five *hectogrammes,* or sixteen ounces and a quarter, French weight.

Native silver, like gold, exists under the form of twisted filaments on the surface of certain stones; when in contact with air it loses the lustre it has immediately after it is taken from the mine; its surface is soon tarnished and covered with a blackish coating. We see in the bottom of this case a fine specimen of native silver, brought from Mexico by Dombey.

In the thirty-seventh case are the different combinations of silver with sulphur and antimony, and the carbonic and muriatic acids. Above the stage is a considerable mass of sulphuret of silver from Bohemia, and, below the same stage, a beautiful group of crystals of antimoniated sulphuret of silver.

Mercury rarely presents itself in its native state or as quicksilver, such as we see it in the thirty-eighth case. It is more commonly found in the state of cinnabar, or in combination with sulphur. The thirty-ninth case contains an interesting series of specimens of sulphuret of mercury from the famous mines of Almaden in Spain, and Idria in the Frioul.

(Fortieth case.) Lead is one of those metals which present themselves in the most varied combinations of form and colour. We should particularly remark the beautiful groups of cubic crystals, which belong to sulphuret of lead or galena, which were given to the Museum by M. Heuland. In the following case, under a glass, is a rare variety of carbonate of lead in acicular crystals of a dazzling white. In the forty-second case are several specimens of lamelliform molybdate of lead, from Bleiberg in Carinthia, and green phosphate of lead, from Brisgaw.

After having glanced at the specimens of the nickel ore, a metal which is of no use in the arts, let us pause before the two cases in which are the different varieties of copper. We should notice first, several specimens of ramose native copper, from the Ural mountains in Siberia; the beautiful copper pyrites from Bannat; and above all, those magnificent concretions of green carbonate of copper, commonly called *malachite*, which is polished and made into tables, chimney-pieces, and other valuable objects.

On the shelf immediately below the stage, in the forty-seventh case, where the iron ores begin, we see a numerous collection of stones, the origin of which will long be an inexplicable mystery. They are the *aerolites*, or stones which have fallen from the atmosphere, on the formation of which the learned have hazarded so many theories. It is now well authenticated, that from time to time, and in different countries of the globe, stones fall from the atmosphere, which not only differ from all others known, but which bear a very remarkable resemblance both in aspect and composition to each other.

Amongst the numerous specimens from the iron mines, which furnish the six following cases, we may remark on the lower shelves some large masses of compact oxydulous iron : these are improperly called loadstones, and furnish the natural magnets which are sold in commerce. The beautiful varieties of *oligiste,* or specular iron ore, deserve attention on account of their beautiful iridescent reflexions, and the lively tints which decorate their surfaces. They come from the famous mines of Elba, which were worked even in the time of the Romans. In the fifty-first case we see several specimens of sulphuret of iron, better known by the name of *martial pyrites :* settings for stones and jewels were formerly made of it, which were called *marcasites.*

The fifty-fourth case contains various specimens of oxyde of tin, from the mine lately discovered near Limoges. Zinc, which we see in the following case, is scarcely employed for any thing but alloy. There are in France several mines of oxyde of zinc, which is more commonly called calamine stone, or *lapis calaminaris*, because of its mixture with earthy substances. The filings of zinc are of great use in fireworks, and produce the brilliant stars and other beautiful effects. Bismuth, which is in the fifty-sixth case, also serves as alloy, and communicates its fusibility to the metals with which it is mixed, while it increases their hardness. Cobalt is used for colouring glass blue, and for painting on enamel. The azure blue, known by the name of *smalts,* is prepared from this metal.

(Fifty-seventh case.) Arsenic is well known as one of the most powerful metallic poisons. It is sold in its native state under the name of

powder for killing flies. In the state of sulphuret
it presents two varieties of colour, very useful
for painting : one is the realgar, of an orange-
red colour; and the other orpiment, of a beautiful citron-yellow.

Manganese is of great use in the manufacturing
of flint-glass and mirrors. Antimony, which immediately follows, is used in medicine, especially
as an emetic. It serves also in the casting of
printers' types, and for the composition of metallic mirrors. In the fifty-eighth case we see
some fine specimens of sulphuret of antimony in
needles of a reddish brown colour, and in very
silky filaments.

Uranium, molybdena, titanium, tungsten, tel-
lurium and chrome, the modifications of which
furnish the remainder of this and the next case,
are metals but little known, and almost without
use, if we except the last of them, which was
discovered by M. Vauquelin, and is successfully
employed in painting porcelain and staining
glass.

Here terminates the collection of minerals, properly so called; one of the most precious in existence, on account of the great number of choice
specimens which it possesses, and the order
in which it is distributed. It will be proper
to remind our readers, that the specimens of this

collection have been named and classed according
to the method of the late M. Haüy, and that the
collection for study is also indebted to that
learned professor for the greater number of its
specimens (1).

(1) We have only pointed out the principal riches of this collection.
Those who may wish for further details, will find them in the second
volume of the *Tableau Méthodique des Espèces Minérales,* published by
M. Lucas, jun[r], in 1813.   They will find there a full description of all
the specimens of the collection, up to that period.

§ IV. COLLECTION OF MAMMALIA.

Ascending to the upper story of the cabinet, by the grand staircase to the right, we enter the rooms which contain the zoological collections. The three first and that at the furthest end contain the mammalia, arranged according to the system of M. Cuvier. The intermediate gallery is occupied by the birds and animals without vertebræ. The number of mammalia now amounts to about one thousand five hundred individuals, belonging to more than five hundred species.

The first room contains the family of monkies. Between the two windows is a case containing five species of the genus oran-otan (1). The first, placed on the upper shelf, is the chimpanzee, (*simia troglodytes*, Lin.) a native of Congo and Guinea. This animal was brought to Paris, alive, to M. de Buffon. It was remarkable for its gentleness, and adroitness in walking upright, waiting at table, sitting down, and in taking its food like a human being.

(1) These words *oran-otan* signify, in the Malay tongue, *a reasonable being*, and were given to these monkies, because they resemble man.

The second species is the oran-otan (*simia satyrus*), a native of the most southern parts of Asia. This individual, which also lived in Paris, was very slow in all its movements, and the disproportionate length of its arms deprived it of all grace. It was silent and melancholy, but of a gentle disposition; and it possessed considerable intelligence, though not so much as what travellers have reported, not more than that of the dog.

On the second shelf is the gibbon (*simia leucisca*). The largest of the three individuals was sent from Java, by M. Diard. We know nothing of the habits of this species, nor of those which follow, and which were sent from Sumatra, by M. Duvaucel. One of these on the third shelf is the black gibbon, (*simia lar*): five individuals of this species, having differently coloured fur, shew as many varieties.

In the bottom of the case is the ape with united toes (*simia syndactyla*, Raffles), a new species. Besides the remarkable character of the toes being united as far as the last joint but one, the absence of hair on the neck, and the swelling of the throat, lead us to suppose, that the economy of this ape must be as curious as it is new in the history of animals.

Beginning our inspection of the cases by those

to the left of the entrance of this room, we first see the numerous family of apes, natives of the warmest regions of the ancient continent: they are extremely lively and active. The patas (*simia rubra*), from Senegal, is on the upper shelf; next to it the mangabey (*simia fuliginosa*), which Buffon believed to be from Madagascar, but which comes from Senegal. On the same shelf are the different sorts of apes, commonly called green apes. They have all been alive in the menagerie, and have only been well described and known to naturalists since the work of M. Frederic Cuvier. On the second shelf is the malbrouck (*simia faunus*), and its different varieties, all from Bengal. On the third, the varied monkey (*simia mona*), the spotted monkey (*simia diana*), the mustache (*simia cephus*), the vaulting monkey (*simia petaurista*), the white nosed monkey (*simia nictitans*). All these apes are very gentle, and come from Guinea. To the right of the third shelf is the douc (*simia nemæus*), a large and beautiful species from Cochin China. It was long believed that this species wanted the posterior callosities; but a young individual, lately obtained for the Museum, proves that it has this character in common with all its kind. On the same shelf is the kahau (*simia nasica*), remarkable for the excessive elongation of its nose, which grows with age. These animals are very common in Borneo, where they live in troops on trees: they cry *kahau.* The species on the fourth shelf are new; they are allied to the *simiæ entellus* and *maura* of Geoffroy; and were procured by MM. Diard and Duvaucel. The macaucos are at the bottom of this case: one of the most remarkable, for the long mane which surrounds its face, is the *ouanderou* of Buffon (*simia silenus*). After the macaucos we observe the magot (*simia innuus*), from Barbary; it is naturalized in Spain, on the rock of Gibraltar, and is one of the most common apes, and the easiest to instruct. One of these in the Museum is of a remarkable size.

On the side opposite the windows are the apes with long faces, called *cynocephali,* or dog-headed. The largest and most formidable of them is the hairy baboon (*simia porcaria*): these animals inhabit in troops the woody mountains near the Cape of Good Hope. M. Delalande's voyage has afforded the two sexes, and several young individuals of this species. On the lower shelves are the different ages of a species, nearly of the same colour as this, and the habits of which are equally brutal and ferocious; they belong to the *simia sphinx,* or baboon of Guinea. Below these we see the mandrill (*s. maimon* and *s. mormon,* L.), an inhabitant of the same country. This ferocious ape, so dreaded by the negroes, is rendered one of the most hideous and extraordinary of all animals by the red and purple-blue colours of its naked parts: its size is nearly that of a man.

On the right, in the corner of this case, we see the black ape without a tail, from the Soloo isles; it was given to the Museum by M. Dussumier, who has enriched our zoological collections with a great number of rare and curious species.

Opposite the door are two cases; on the upper shelf of the former are howling apes (*stentor*, Geoff.). These inhabit the equatorial countries of America, and derive their name from the tremendous cries with which they make the forests resound.

On the second shelf, more than fifteen individuals of the sai and sajou (*simia appella* and *s. capucina*) shew the numerous varieties which age and probably locality produce in these small species.

On the third shelf, the horned and the white faced sajou are remarked for their singular appearance, and the *simia sciurea* and its little congeners, for their round faces and elegant forms. In the upper part of the second case are several species of the genus *ateles*, established by M. Geoffroy. These animals, with long and slender limbs and prehensile tails, can take the most varied attitudes.

Below are the *sakis,* or night apes, to which a tufted tail, and long hair covering their body, give a peculiar appearance. One of the most remarkable is the capuchin of Orinoco, so named because of its long beard. M. de Humboldt, who first described it, after having observed it living in America, gave it the name of *chyropotes,* from two greek words, signifying *hand* and *drink;* because, when about to drink, this animal takes the liquid into the hollow of its hand, and pours it into its mouth, taking great care not to wet its beard (1).

On the third shelf of this case are placed the numerous species of *ouistiti,* very small monkies, of a pleasing form, differently coloured, very easy to bring up, and much sought for in Europe, on account of their genteelness.

Lastly, in the bottom of the two cases we see the *lemurs,* nearly allied to the apes in habits and movements, but differing from them by having their muzzle as long as that of the fox. They all come from Madagascar and the neighbouring islands: the most remarkable species are the mau-cauco (*lemur catta*), the lemur macauco and the

(1) The individual now before us, is that which M. de Humboldt observed, during his travels in Orinoco.

red lemur (*lemur ruber*, Peron). They breed in our menageries. Next to them is the *Indri*, first made known by Sonnerat, and which the inhabitants of Madagascar train like a dog for hunting. It differs from the others of its tribe by having two incisors less in the under jaw, and no tail.

This case is terminated by the loris of Bengal, the galago of Senegal, which has large ears, and by the tarsiers (*lenur spectrum*, Pall.) from Amboyna. All these animals are very slow in their movements, and lead a nocturnal life. The tarsi of these two last are elongated, which make their hind feet appear of a disproportionate length.

Passing into the second room, in the cases right and left of the door, we see the different genera of *bats*, so remarkable for the form of their noses and ears, the length of their toes, and their membranous wings. Several species roll themselves up into a ball, and remain torpid during the whole winter (1). The largest, which belong to the genus *pteropus*, are placed on the cornice. At the top of the case are the *phyllostomœ*, whose lips and tongues are furnished with warts,

(1) M. Geoffroy Saint-Hilaire, in the 8th and 15th vol. of the Annals of the Museum, first published a complete work on this family, which he divided into sixteen genera, and made known several new species. The genera which he established differ not only in their anatomical characters, but also in their habits and exterior forms. There are more than eighty species in the Museum.

which assist them in sucking the blood of large quadrupeds, even without awaking them from sleep. The most formidable species is the vampire (*vespertilio spectrum*, Lin.), which is very noxious in several parts of South America by killing cattle.

On the lower shelves of the case to the left are the hedgehog, the tenrecus, and different species of moles.

The first of the six cases, which cover the left wall, contains the bears. Those on the mountains of Europe retire into caves, where they pass the winter in a state of torpidity. The largest species, and the most celebrated by the exaggerated accounts of its voracity, is the sea or polar bear, whose fur is white. It lives on the borders of the frozen ocean, and pursues the seals and other marine animals, which it seizes swimming, when they rise to the surface to respire. It remains buried under the snow from October till March, and the female brings forth during this period: it dreads heat more than any other quadruped. The individual in this case lived in the menagerie, and had eighty pails of water thrown over him daily. By the side of this northern bear is a species from India, which feeds on honey; it was lately brought by M. Leschenault from the mountains of Gates, had not been well observed until now, and was classed among the sloths. After the bears are the racoons, which only differ from the bear in size and length of tail (1).

On the first shelves of the second case are the long-nosed coatis, the badgers, whose bristles are employed in making soft brushes, and the civet of the Cape (*viverra mellivora*), which has only been well known since M. Delalande brought it from that country. On the fourth shelf is the northern glutton, or the *rossomaque* of the Russians, which hunts at night, and whose fur is much esteemed. The lower shelves are occupied by weazels and martins, of which there are nineteen species. The most celebrated for its rich fur is the sable, which inhabits the frozen mountains of the north of Europe and Asia. Long and hazardous journeys are performed in the winter to hunt these animals; and it was in searching for their fur that the eastern countries of Siberia were discovered. The individual in the Museum was presented to Buffon by the Empress of Russia (2).

At the top of the third case are the European

(1) These animals inhabit America, where they are commonly called *washers*, from their habit of steeping their food in water.

(2) All the animals of this class have a more or less offensive smell; whence the names of *polecat* and *seamew* given to both genera.

and American otters. The most remarkable is the sea-otter (*mustela lustris*); the black fur of which has the eclat of velvet. The English and Russians seek this animal in the northern parts of the Pacific ocean, and sell its skin in China and Japan. In the same and in the following case are different varieties of dogs (1), and the two species of European wolves.

The fifth case contains thirteen species of foxes. The black fox of North America, the blue fox, and the *isatis*, are most sought for their fur.

On the first shelf of the sixth case are the hyænas: one species of which (*hyæna picta*) is but lately known to naturalists; and travellers formerly mentioned it by the name of the Hottentot's hunting dog.

Below the hyænas are the seals, amphibious animals, whose different species have been vulgarly called sea-calf, sea-lion, sea-elephant, etc. Peron has formed his genus, *otarius*, of that with projecting ears, and which is called the sea-bear (2). On the cornice of the same case is the

(1) The most beautiful of them all was presented to the Museum by Baron Laugier.

(2) The seals are mild and sagacious animals, and attach themselves to man. *Vide* Peron's Voyage, vol. ii, page 32; and **M. F. Cuvier's** Memoir in the Annals of the Museum, vol. xvii.

arctic walrus, vulgarly called sea-cow (1). In the projecting case, which terminates this side of the room are the civet and genet cats. The largest is the civet, bred in Abyssinia, because it furnishes the perfume of which it bears the name (2). The only species found in France is the common genet, the fur of which is an article of trade in the Pyrennees. At the bottom of the case is an animal nearly allied to the civet, sent us alive from Pondicherry, by M. Leschenault, and of which M. F. Cuvier has made a genus under the name of *paradoxurus*. It is called in India the palm-tree martin, because it generally lives on these trees.

In order to follow the classification that has been adopted, we must immediately pass to the third room; the first case of which contains the genus *mangouste*, of which there are ten species: one of them is the ichneumon, so celebrated for the high degree of estimation in which the Egyptians formerly held it. We owe it to M. Geoffroy Saint-Hilaire.

The other cases on the same side of this room

(1) This animal sometimes acquires 20 feet in length; it is caught in the frozen seas; its flesh yields much oil, and coach braces are made of its skin. The ivory of its tusks is employed in works of art, although it is far inferior to that of the elephant.

(2) This odoriferous matter is secreted by two glands, which are situated in a pouch between the anus and the insertion of the tail.

contain twenty-three species of the cat genus (*felis*); which comprehends lions, tigers, leopards, lynxes, etc. The greater number of these animals lived in the menagerie, and several bred there. The most remarkable amongst them are, the lion and panther of Africa, the tiger of India, the hunting tiger, trained by the Indians for the chace, the caracal, which is the true lynx of the ancients, the jaguar and the couguar of America, the European lynx, which is the lynx of the furriers, and the lynx of the United States of America. By the side of the lioness we see three cubs, which were born in our menagerie, and there lived until the period of dentition.

After the cats are the numerous family of the *didelphis*, or animals with a pouch; it comprehends the opossums, kanguroos, etc. There are thirty-three species of them in the Museum (1).

(1) The females of these animals present a very remarkable phenomenon; which is, that their young are born in the state of a fœtus, possessing only the rudiments of their members and exterior organs: they are then received into a pouch, which is under the belly of the mother, and formed by the skin of the abdomen folded round the mammæ. The young ones fix to the mammæ by instinct: they are preserved in that pouch from external accident, and even when able to walk, they occasionally retire into it.

In several species, when the young ones become too large to be any longer contained in the pouch, they fix on the back of the mother, twisting their tails to hers; and holding that situation even while she runs, in the manner we see here in the *didelphis murina* and *d. marsupialis*. The *opossum*, with party-coloured ears (the opossum of the Americans)

Passing to the right side of the room, we see those of the genus *didelphis,* which belong to the old world. The largest of them all are the kanguroos of New Holland. These animals, having the fore feet very short, and the tarsi extremely long, are almost always on their hind feet and leaning on their tails; and instead of walking, they jump, without the help of the fore feet. MM. Peron and Lesueur brought almost all the species possessed by the Museum. The kanguroo with red and woolly hair comes from the Blue mountains, and was brought by MM. Quoy and Gaimard, surgeons to the expedition of captain Freycinet.

Near the kanguroos we see the *dasyura,* the *peramèles,* and the *phalangers,* genera established by M. Geoffroy Saint-Hilaire. Amongst the phalangers we may observe several species from New Holland, which have the skin on each side extended from one paw to the other like flying squirrels. One of the species is scarcely as large as a mouse.

The rodentia, to the number of one hundred

is nearly the size of a cat; its young, at the time of their birth, only weigh one grain : they remain fixed to the mammæ in the pouch until they have acquired the size of a mouse. A particularity in the species of opossum is, that they have fifty teeth; a greater number than has hitherto been observed in any other quadruped. They live on trees, and hunt at night.

species, occupy the three following cases. The most worthy of attention are the *beavers*, which live in companies on the borders of rivers in Canada. With their teeth they cut down the trees, which they use in the construction of dams, to keep the water always the same height; and they build huts of two stories; the lower apartment, being under water, serves as a store-room, and the upper they inhabit during winter. The industry of these animals appears the more extraordinary, when we consider their outward form. Their number decreases, as so many are killed for the sake of their fur. The substance employed in medicine, under the name of *castoreum*, lies in two glands under the belly of this animal. After the beavers, we find the *dormouse*, remarkable for sleeping during winter; the *hamster*, so destructive to corn, which it buries in its hole, and is sometimes more than seven feet deep. The *chinchilla*, so valuable for its fur; the *alactaga*, a species of gerboa, given by M. Gamba: and the gerboa of the Cape brought by M. Delalande. Near them are twenty-three species of squirrels; amongst which we see the flying squirrel (*pteromys*), the skin of whose flanks, extending into a membrane between the fore and hind legs, enables it to remain a short time in the air, and to leap with facility from

one tree to another. We then see the *aye-aye,*
from Madagascar, so named from its cry: this
singular animal, unique in European collections,
was discovered by Sonnerat. On the lower
shelves we see porcupines, remarkable for the
long black and white spikes which cover their
bodies. We have four species of them: one from
the Brazils (*histrix prehensilis*), which has a prehensile tail, and is often found on trees.

The numerous species and varieties of hares
and rabbits occupy several shelves in the last case
but one. The order of the rodentia is terminated
by the guinea-pigs (*anœma,* F. Cuvier), of which
the *aperea* of Brazil is the original type.

The last case of this room is filled by sloths
(*bradypus*), which are at the head of the order
edentata. The unau, or two-fingered sloth (*bradypus didactylus*), and the aï, or three-fingered
sloth (*b. tridactylus*), are the two known species of this extraordinary genus, and are from
South America. Their fore-legs are much longer
than those behind; so that when walking they
are obliged to drag themselves along on their
elbows; and their *pelvis* is so large, that they
cannot bring their knees close together: their
hair is coarse and brittle. The female brings
forth but one young at a time, which she carries
on her back. These animals live on trees, and

feed on the leaves whilst one remains; and it is said, that when they pass from one tree to another, they let themselves fall to the ground, rather than be at the trouble of descending the trunk. When they sleep, they sit down, and cross the fore paws round the forehead, which hangs down on the breast; this attitude has been given to one of them, which we see on the middle shelf.

Returning to the second room, in the case to the left of the door, we see the armadillo of America (*dasypus*), covered with hard horny plates, united in the front and hind part of the body in the form of a shield. These plates are disposed in transverse zones on the middle of the back, and move one upon another, thus permitting the animal to roll itself into a ball like the hedge-hog. The three lower shelves contain the manis, originally from India, where they in some degree represent the armadillos; but their bodies are covered with imbricated scales. This animal also rolls itself into a ball, folding its tail under its belly, the skin of which is naked.

The first case, on the same side of the room, contains the ant-eaters. The very long muzzle of the ant-eater is terminated by a mouth without teeth, whence issues a filiform tongue, which it can considerably elongate, and which it introduces into the nests of termites and ants, and draws it back covered with these insects, which become fixed to it by means of the viscous matter with which it is wetted. The largest of these species, all of which are originally from America, is the *myrmecophaga jubata* (1). By the side of the ant-eaters is the *orycteropus*, vulgarly called *ground-hog*, because it lives in holes which it hollows with great facility. In the bottom of this case we see the two-horned rhinoceros of Africa, the American tapir, and another species of the same genus, sent from Malacca by MM. Diard and Duvaucel.

In the second case are two genera of quadrupeds, differing in several respects from all others: these are the *ornithorynchus*, the large flattened muzzle of which much resembles the bill of a duck ; and the *echidna*, which has a long muzzle, terminated by a small mouth like that of the anteater, and the body covered with spines like those of the hedge-hog. These animals have hitherto been found only in the rivers and marshes of New Holland, near Port Jackson. (2)

(1) The female brings but one young at a birth, which she carries on her back.

(2) M. Blumenback having made known, in 1800, the first of these animals under the name of *ornithorynchus paradoxus,* and sir Everard Home having since described the echidna, M. Geoffroy Saint-Hilaire formed of these two genera a separate order, which he named *monotrema.* This professor, having re-examined them, is of opinion, that they ought to form an intermediate class between the mammalia

The four following cases contain nineteen species of the order pachydermata. The Arabian horse, the baskir horse covered with long hair, the zebra, and the quagga, are remarkable for their beautiful form or variety of colours. The different species of wild boar are placed between the legs of these larger quadrupeds, and amongst them the *Pecary* of America, which has a glandulous opening in the back, whence issues a fœtid humour.

In the last case are the *cetaceæ*, vulgarly called *blowers*. A fœtus of the whale, a porpesse, a large dolphin, and the dolphin of the Ganges, a very rare species sent by MM. Diard and Duvaucel, are the most remarkable animals of this order in the Museum.

We have been obliged to place in the middle of this room, on account of their enormous size, the male and female elephants, which lived in our menagerie; as also the one-horned rhinoceros of India, which lived at Versailles: the two-horned rhinoceros of Sumatra, and the unicorn of Java, which we owe to MM. Diard and Duvaucel; and lastly, the two-horned rhinoceros and the hippopotamus, brought from the Cape by M. Delalande.

and the birds. The monotrema in fact differ from mammalia in the want of mammæ, and by being oviparous, but they approach them in the other organic systems.

After having gone through the gallery where the birds are placed, we enter the room which contains the order *ruminantia*. We first see the animals which, like those in the first room, being too large to be placed in the cases, stand in the middle of the room: they are, the giraffe (*camelopardalis*), which lives in the deserts in the south of Africa; and is the tallest of them all, its head being 18 feet from the ground; we have had the male ever since the journey of M. Levaillant, and the female has been recently brought by M. Delalande: the buffalo (*bos bubalus*), originally from India, whence it was taken to Egypt, and thence into Greece and Italy, during the middle ages: the aurochs (*bos urus*), from the marshy forests of Lithuania and Caucasus, which has been erroneously considered as the primitive stock of our large cattle (1): the camel with two humps (*camelus bactrianus*), from the centre of Asia, and the camel with one hump (*camelus dromedarius*); two species which are completely domesticated, and are the only medium of communication between certain nations separated by deserts: and the elk (*cervus alces*) which is found in the marshy forests to the

(1) By the side of the aurochs is a cow without horns, and a bull of a race which is half wild in the plains of Camargue, in Provence. This last was given to the Museum by baron Laugier.

north of both continents, living in small herds ; its horns widen into triangular plates, and increase with age to the weight of fifty or sixty pounds.

We will now make the circuit of this room, beginning with the case to the right of the window. We there find a young camel, born in our menagerie, which lived only three days. By its side is the *vicuna,* a wild animal of Peru, whose tawny wool, of an admirable texture, is employed in making the finest cloth: it was presented by baron Larrey. Below it, is the *lama,* the only beast of burthen in Peru, at the time of the conquest ; it is now solely employed in the service of the mines. By the side of the *lama* is the musk-deer (*moschus moschiferus*), remarkable for the long canine teeth, projecting from its upper jaw : it is found in Thibet, Tonquin, and other countries of Asia. The perfume so well known under the name of musk, is furnished by this animal, and secreted under the belly of the male. After the musk, we see the *moschus prgmæus,* the smallest and most elegant of all ruminating animals.

The second case contains the common deer, and a species one third larger (*cervus canadensis*), from North America. Before them is the *muntjac,* of different ages, from Java and Sumatra, which were sent by MM. Diard and Duvaucel. These travellers sent us, from the same countries, the *hippelaphos,* which was only known from the description by Aristotle. It is in the third case, with the axis or deer of the Ganges, whose skin is beautifully spotted.

In the fourth case we see the Louisianan or Virginian deer, of which some are red and others brown; which shews the difference of colours in the same species according to the season of the year in which they were killed. In the bottom of the same case is the white deer of Cayenne, presented by M. Poiteau. The roebuck and its black and white varieties are placed in the fifth case. By their side we see the male and female rein-deer, given to the Museum by the marshal duke of Treviso, who procured it alive from Stockholm (1). Before the rein-deer is the common roe (*cervus capreolus*), whose flesh is much esteemed.

On the upper shelf of the sixth case are the American deers: the first is the *gouazoubira,* from Buenos Ayres, which was given to the

(1) The rein-deer inhabits the frozen regions of the two continents. It is well known for speed, and for feeding by simply browzing on the lickens concealed under the snow; but it is still more celebrated for its utility to the polar countries. The Laplanders have large herds, which they lead up the mountains in summer, and bring down to the plains in winter. They live upon the milk and flesh of these animals, and make their clothes of the skins. They also use them to draw sledges and carry burdens.

Museum by M. Baillon. The second is the red-deer of Cayenne, of which we have specimens of the two races, differing in size. The lower part of the case is occupied by the *bubalus,* or Barbary cow, and the caama of the Cape; and with them begins the numerous genus of antelopes, of which we have twenty-two species.

In the seventh case is the Barbary antelope (*antilope dorcas*), a species celebrated for the elegance of its form and the sweetness of its countenance. They live in innumerable herds in the north of Africa, and are the usual food of lions and panthers. In the eighth case, which is on the other side of the door, are the steenbock, the duiker or plunging goat of the Cape, so called from its habit of darting head downwards into the covers where it lives in small herds; the stone-leaper, the gries-bock in its different ages, and the woolly antelope of M. Cuvier. All these were procured by M. Delalande.

The pasan of Buffon (*antelope oryx*) is in the ninth case. Its size, colour, straight annulated horns, and the contrary direction of the hair on its back and neck, exactly agree with the description given us by the ancients of the unicorn (1). By the side of the pasan we see the

(1) M. Cuvier is of opinion, that the description, which the ancients have given us of the unicorn, is that of a pasan, which had lost one of its

algazel of Buffon, from Senegal; the blue antelope of the Cape; and the guevei, pygmy antelope (*antilope pygmæa*), a beautiful little animal only nine inches high, and wonderfully alert. It inhabits the warmest regions of Africa.

In the tenth case are the two largest species of antelope: their size almost equals that of the horse : they are the *antilope equina* of India, and the striped antelope (*antilope strepsiceros*) of the Cape. By the side of them is the gnu (*antilope gnu*), of a very singular form, apparently borrowed from other animals: it has the body, rump, and tail of a small horse, and an upright mane ; its horns, drawn close together, resemble those of the buffalo of Caffraria; projecting bristles surround its flat muzzle, a second black mane descending under the neck and dewlap, and its feet and legs are as slender as those of the deer. It inhabits the mountains northward of the Cape, where it is somewhat rare.

The Nilghau (1) of India (*antilope picta*), and the European chamois (*antilope rupicapra*), are in the eleventh case, with several varieties of the goat, which also fill the twelfth case. Amongst these varieties we find the one that

horns, or rather of a drawing, which represented the animal in profile, shewing then but one of its horns.

(1) This is a compound of two Persian words : *nil*, which means *blue*, and *ghau*, which signifies a *bull*.

furnishes the wool of which the Cashmere shawls are made. In the twelfth case is the Caucasan ibex (*capra ægagrus*), which lives in herds on the mountains of Persia, where it is known by the name of *paseng*; it appears to be the parent of all our varieties of domestic goats (1). After it, comes the *ibex*, an inhabitant of the highest mountains, and remarkable for the uncommon size of its horns.

The thirteenth case contains the various races of sheep of which the muffoli of Corsica and Sardinia might be considered the parent. Beneath it is the African muffoli (*ovis tragelaphus*), to which a long mane hanging under the neck, and another which forms ruffles round each ancle, give a very singular appearance. This species inhabits the rocky countries of Barbary and Upper Egypt.

On the higher shelf of the fourteenth and last case, we see a race of sheep originally from Persia and Tartary. The tail of this race enlarges from the insertion, and gradually transforms itself into a double lobe of fat, weighing from fifteen to twenty pounds (2).

Here terminates the collection of mammalia. The numerous family of the ruminantia, which

(1) In the intestines of the ægagrus is found the concretion called egagropyla, or oriental bezoar.

(2) The Astracan sheep, and that from Upper Egypt, with a very short tail, belong to this race. This last was given to the Museum by H. R. H. the duke of Orleans.

we have just seen, is the most useful to mankind. Several of them were reduced to a domestic state at a period anterior to history, and have produced varieties, the primitive type of which we can with no small difficulty recognize.

§ V. COLLECTION OF BIRDS.

ON leaving the gallery of ruminating animals,
we re-enter that of the birds. The species
nearest to each other from their natural affinities,
form the genera and sub-genera of which M. Cuvier has traced the characters in the first volume
of his *Regne Animal.* To avoid the loss of space,
which the different sizes of birds of analogous genera would have caused in the cases, had they been
placed together, we have frequently been obliged
to lay aside the order adopted by M. Cuvier for the
distribution of his genera ; but they can easily be
referred to, by means of tickets on black pedestals, placed before each group : those on the red
pedestals indicate the subdivisions of the genera.
To the support of each bird is attached a label,
the first line of which is the French name with a
reference to a figure. The name given by Buffon
to the species has been prefered, for all those
which he described, as well as the coloured
plate in which it is represented by him. The

beautiful works of MM. Levaillant and Vieillot, and the rich collection of coloured plates of MM. Laugier and Temminck have served for the species unknown to Buffon. The latin name, written in the second line, is that given by Gmelin in his edition of the *Systema Naturæ*, or of the author who has since described it. The third line gives the name of the country whence the individual comes, with that of the traveller who brought it, or of the person who presented it to the Museum.

The collection comprehends upwards of six thousand individuals belonging to more than two thousand three hundred different species. Almost all are in a perfect state of preservation; and such means have been found of preparing them, that they never change. There is not so numerous a collection existing any where else, and nevertheless it has been formed in a few years. At the death of Buffon it consisted of only eight hundred species : but it has since been successively enriched by the purchase of M. Levaillant's collection; by the addition of that of the Stadtholder, the possession of which was confirmed, when we gave in exchange our duplicates in every branch of natural history ; and lastly, it has been every year increased by those sent from different countries by travellers.

This collection is of infinite use and advantage, from its being so well adapted for study, by the methodical distribution of the genera and species: the males and females and the varieties being placed close to each other.

A great number of birds, especially those remarkable for the beauty of their colours, have a totally different plumage according to their age, and even sometimes according to the season. Thus the same bird has often been described and drawn several times under different names. It is only after many researches that the different varieties and the passage from one to the other can be determined. We frequently see ten or twelve individuals of one species, presenting the same essential characters, but whose colours are totally different. Besides the diversity of the males and females, the same bird is quite different at one, two, or three years of age ; as also, if it has been killed in summer or winter. All this may be observed in the collection, which for the future will fix the type for the species. We will now resume the succinct description of what is most worthy of remark in this collection.

The gallery, which contains it, is divided into fifty-seven cases with shelves, on which the birds are arranged in the manner best adapted to their display. Care has been taken to avoid the waste of room, and at the same time the confusion that might arise from that economy of space.

We will begin with the case to the left, on entering the gallery from that of the ruminating animals, and we will make the circuit, proceeding from left to right.

Ten species of the genus *vulture* occupy the two first cases. On the top shelf of the first are the different ages of the king of vultures (1) (*vultur papa*). This bird inhabits South America; it has a beautiful plumage; the naked parts of its head and neck are covered with red and yellow, which fade after death. On the second shelf of the second case is the *percnopterus* of Egypt, vulgarly called *Pharaoh's chicken*, a bird most common over the old continent. Large flights of them follow the caravans, devour what dies, and purify the country of the dead bodies which otherwise would infect it. The ancient Egyptians held this bird in veneration, and even now some devout mussulmen leave legacies for the maintenance of a certain number. Above the *percnoptere* is the *vultur fulvus*, whose organ of smell is so acute, that it can distinguish

(1) The reason of its receiving this title is worth recording. When the male and female, which keep always together, fall upon prey, already attacked by a large party of the *aura* (a species which we see on the second shelf of the second case), these latter fly away, leaving the king and his female to finish quietly their repast.

at several leagues the prey best suited to its voracity.

At the bottom of the case is the *lœmmer-geyer,* or *gypaëtos* of the Alps (*vultur barbarus*), the largest bird of prey on our continent ; it measures ten feet between the tip of each wing. It lives solitary on the steep rocks of the Swiss mountains, carries away sheep, goats and chamois, and it is said sometimes to have attacked children. The cases from the third to the tenth contain the numerous species of diurnal birds of prey, which Linnæus united under the generic name of *falco.*

Six species of eagles commence the series. The royal eagle, the largest and most courageous is the first. It hunts in the mountains for goats, roes, and other quadrupeds of that size, and feeds on dead animals only when pressed by hunger. After this are the ospreys or fishing eagles, which keep on the borders of the sea, or near the great lakes, and the bald-buzzard, which lays waste the fish-ponds.

In the fifth case we see the great American harpy of a size larger than the common eagle ; and it is considered as having the claws and beak stronger than any other bird. It generally feeds upon the sloth, can carry away a fawn ; but the rapidity of its flight being greatly diminished by

the shortness of its wings renders it less destructive. By the side of this is the *falco ecaudatus*,
from Africa, which has a shorter tail than any
other of this genus. Lastly, the secretary of the
Cape (*falco serpentarius*), one of the most remarkable birds of prey for the length and strength
of its legs (1). It inhabits the burning and sandy
deserts of Africa, where it feeds on serpents and
other venomous reptiles (2).

In the sixth case are the male and female
astur, the sparrow-hawk, which was formerly
trained for the chace, and the male of which,
from being one third less than the female, was
named the *tiercelet*. We must remark the *falco
musicus*, the only bird of prey which sings agreeably. The buzzards, the kites, the pernis, and the
pygargi, occupy the seventh and eighth cases.
These birds pursue insects or reptiles. The *py-
gargus* deserves a peculiar attention, as the Egyptians worshipped it and embalmed it after death.
We may see the feathers of one perfectly preserved, that were taken from a mummy brought
from Egypt by M. Geoffroy Saint-Hilaire. The

(1) When very young it drags itself along on its belly. It is exhibited
in the various attitudes it assumes according to its age.

(2) This species, if we naturalized it to the climate of the French
equatorial settlements, would be of great service to the inhabitants,
by destroying the dangerous reptiles with which they abound.

honey-buzzard was sent us from Java by M. Leschenault.

In the ninth case are the common falcon and the jerfalcon (*hiero falco*), celebrated for their docility, and the rapidity of their flight; the first of these has given its name to a peculiar art, that of training these birds to pounce upon the game, either in the air or on the ground, and bringing it to their master. This sport was very much in use during the middle ages. Both these birds inhabit northern countries, and build their nests upon rocks.

In the same case we find the smallest of the birds of prey, the *falco cærulescens* from Sumatra.

The tenth case, which projects and forms a separation in the gallery, contains the hobby-falcon and the *falco tinnunculus* in its different ages. And here terminates the diurnal birds of prey, of which there are one hundred and twenty species in the Museum.

The eleventh and twelfth cases contain thirty-four species of the nocturnal birds of prey : the grand duke, the lesser duke, the *ulula,* the common owl, the little duke or *scops;* all of them inhabiting Europe. Amongst the foreign species are the Cape owl, the great American owl, and the owl with naked feet (*strix leschenaultii,* Tem.).

This last was discovered and sent us from Pondicherry, by M. Leschenault.

The thirteenth and fourteenth cases contain the beautiful and numerous family of parrots, which is divided into cockatoos, lorys, aras, parrots, and parrakeets. The cockatoos have a tuft on the head which they raise and lower at will; the greater number of them have a white plumage, that of the lorys is red. The aras are sought after on account of the brilliant and various colours which adorn them (1). Green is the prevailing colour of the parrakeets and parrots, properly so called. There are however some exceptions; the common grey or African parrot, called *jaco* (*psittacus erithacus*), which can articulate with so much facility, is of an ash grey all over the body and wings, and the tail is of a fine red. That of the Moluccas, which Buffon calls parrakeet, with the spotted blue face, is of varied colours (2). All climb the trees with the assistance of their beak (3). The species most

(1) One of them is entirely black, the *ara à trompe* of Levaillant. M. Geoffroy Saint-Hilaire calls it the *microglossa*, on account of the smallness of its tongue; the functions of which are supplied by the larynx, which it projects beyond its beak.

(2) Those colours are said to be accidental, and produced by an operation called *tapirer*; some feathers are taken away, and the naked skin of the bird is rubbed with the blood of a frog called *rana tinctoria*; the new feathers which shoot out after this operation change their colour.

(3) The *perruche ingambe* of Levaillant, which we received from New

anciently known in Europe is the Alexandrine par-rakeet (*psittacus Alexandri*), so named because it was brought from India by that conqueror.

On the first shelves of the nineteenth case are the different species of *toucan*. The enormous bills of these birds would weigh more than their body, if they were not of a cellular and light substance. They belong to the equatorial regions of America, and live upon insects and fruits. The structure of their beak prevents them from chewing their food; they throw it in the air and catch it as it falls. The brilliant feathers, which cover their breast, were formerly employed for a peculiar sort of embroidery.

On the third shelf are the wrynecks (*yunx*), small birds which owe their name to the habit of turning their neck in different ways. The woodpeckers (*picus*) are placed on the lower shelves. To them the name of climbers is best adapted, as they climb in all directions on the bark of trees, striking it with their long and flattened bills, and taking from underneath the larvæ of insects, which they seize with their tongue, armed with curved spines, and susceptible of a consi-derable elongation.

Holland, is remarkable for the length of its legs. It is the only one which runs on the ground, and seeks its food in herbage; wherefore Illiger made a separate genus of it.

The different species of cuckoos occupy the upper shelves of the sixteenth case. The European cuckoo, so named from its cry, is celebrated for its singular habit of laying its eggs in the nests of other insectivorous birds. These bring up the young cuckoo with as much care as if it was their own, even when its introduction into the nest has been preceded by the destruction of their own eggs. Amongst the foreign species we will notice the blue cuckoo of Madagascar, sent by baron Milius; the copper coloured cuckoo of the Cape; the golden and the klaas cuckoos, remarkable for the beauty of their plumage. On the sixth shelf are birds, which have been set apart to make a separate genus. Sparmann, who observed them at the Cape, gave them the name of *indicators;* because, feeding on honey, they fly at a great distance in quest of wild bee hives, and utter a loud cry when they have found them; thus serving as guides to the inhabitants, and saving them a tedious search. We have four species of this genus: the two first were brought by Levaillant, and the others by M. Delalande. The lower shelves of this case are covered by the barbets (*bucco,* Lin.), so named from the bundle of stiff bristles on each side of their beak; and with the couroucous, solitary birds, which fly only at twilight. Several species of this genus attract attention by the beautiful colours of their plumage. The three most brilliant amongst them are the blue-faced barbet (*bucco cyanops*, Cuvier). The red-bellied couroucou of Sumatra, and the couroucou narina from the south of Africa. The two first were brought by M. Duvaucel, and the other by M. Delalande.

In the seventeenth case, is the numerous genus of shrikes (*lanius*). These birds live in families; their attachment to their young is such that, although not larger than a common blackbird, the female will fight with the crow in defence of her covey, and is frequently victorious. The grey shrike (*lanius excubitor*), of the size of a thrush, remains in France all the year long. The *l. rufus,* and the red-backed shrike (*l. collurio*), quit us in winter. The latter, which is the smallest of all, pursues insects, and fixes them in the bushes, to find them when it needs food. It imitates the voice of other birds.

Amongst the foreign species, those most remarkable for their colour are the *bacbakiri* (*turdus zeylonicus*) from the Cape, the blue shrike of Madagascar, and the shrike with a red throat (*lanius gutturalis*, Daud., Ann. du Mus.), from the coast of Angola. The *vanga* is a species of shrike with a compressed beak. The most curious are the *blanchot* of Senegal, the tufted vanga of Java and Sumatra, and the striped vanga of the Brazils. The *baritæ*, which come naturally after the shrikes, are shrieking birds of New Holland and New Guinea: one of them, the *chalybean*, has such brilliant colours, that it was formerly ranked amongst the birds of paradise (*paradisea viridis*); another (the *coracias streptera*) has so strong a voice, that it has been named the wakener; this is from New Holland, as well as the musician (*coracias tibicen*), whose voice is so agreeable. Below the shrikes are placed the *breves,* from India, adorned with the most beautiful colours. Buffon knew but two species of them; there are now six in the Museum. Two of the most beautiful, the one with a red belly, and the other with a black head, were presented by M. Dussumier, who brought them from the Philippine isles.

After the *breves* come the ant-thrushes (*myothera*); they live on the enormous ant-hills in the forests and deserts of America; their plumage is brown and their voice very sonorous. We possess twenty-seven species of them; the largest is the king of the ant-thrushes, of the size of the blackbird. It is solitary, and lives in the forests of Cayenne.

The eighteenth case contains the merlins (*tur-*

dus) (1), of which we have one hundred and sixty species. By the side of the common blackbird (*turdus merula*), which is the first, we may observe a white variety; then the rose-coloured thrush of the south of France, which is so useful in destroying grasshoppers; next comes the mockbird (*turdus polyglottus*), famous for the astonishing facility with which it imitates the chirping of other birds, and all the voices it hears. Below are the singing thrushes; the largest is the *turdus viscivorus*, which feeds on the misletoe berries, and propagates this parasitic plant by sowing the seeds of it on the branches of trees (2). The smallest is the mavis, which arrives in large flights about the time the grapes ripen; and which is when fat very delicate eating (3).

(1) The merlins and thrushes are of the same genus. We give the name of *merlins,* to those species whose colour is uniform; and that of *thrush,* to those whose plumage is marked with small black or brown spots.

(2) The *turdus pilaris* differs but little from the *turdus viscivorus;* but the white variety is very rare; that in the Museum, was given to Buffon by Lewis XVI, who had killed it when hunting.

(3) The Romans thought much of this species, which they called *turdus,* and is still called *tourdre* in the South of France. Horace, speaking of the presents to be made to one whose property it was desirable to inherit, says:

> . . . . *turdus*
> *Sive aliud privum dabitur tibi; devolet illuc*
> *Res ubi magna nitet domino sene.*

Of all the birds of this numerous genus, the azure thrush of Java attracts most attention: its breast is of a velvet black, and its back of an ultramarine blue. This beautiful bird was sent us by MM. Diard and Duvaucel. We have also the white-breasted thrush from Senegal (*turdus leucogaster*), the back of which is of the most vivid crimson; and the New Guinea thrush, commonly known under the name of the magpie of paradise, beautiful on account of the magnificence of its plumage; its tail is three times longer than the body, its head has a double tuft, and the colours of its throat and neck shine with metallic lustre.

The lyra (*menura magnifica*), which inhabits the rocky parts of New Holland, is placed in this case. The tail of this singular bird is composed of three sorts of feathers: twelve of these, which are very long with slender and scanty beards, form the principal part; the two in the middle longer than the others, are stiff, and only bearded on one side; and the two outer are curved like the frame of a lyre. The female does not present the same characters. Both are of the size of a pheasant.

On the two last shelves in this case are the grakle genus (*gracula*). The most common species (*paradisea tristis*, Gm.) is famous for the service it performed in the Isle of France, in destroying the grasshoppers. Near the grakles are the orioles, which have a beautiful yellow plumage, and the different species of which are distinguished by the variety of tints on a small portion of the body. The French oriole constructs its nest very skilfully, suspending it from the extremities of the largest branches of trees.

The first shelf of the nineteenth case contains a genus established by M. Cuvier, under the name of *philedon*. The species, each of which possesses some remarkable singularity, are united by one common character, that of having the tongue terminated with a brush of hairs. The carunculated philedon (*corvus paradoxus*, Lath.) has two fleshy caruncles, which hang under its throat; the cravat philedon (*merops Novæ Hollandiæ*, Brown) has two small bunches of curled feathers, the whiteness of which forms a strong contrast to the green of the body. The monk philedon (*merops monachus*, Lath.), and the philedon corbicalao, which, as well as the two preceding, are natives of New Holland, have a tubercle on the beak, and during life the naked parts of the head and neck are of a fine blue.

Below the philedons we find the *motacillæ*, a very numerous family, characterized by a straight slender beak, and comprehending the

stonefinches, the warblers, the bullfinches, the
wrens, the wagtails, etc. There are a hundred
and seventy-two species in the Museum. The
most celebrated, not for its plumage, but for its
singing, is the nightingale. Amongst the foreign
species, the *motacilla superba, m. cyanea,* and the
malachura, all three from New Holland, ought
to be particularly noticed, the two first, for the
beauty of their colours, the third, for the delicacy and slenderness of the feathers of its tail.

Amongst the indigenous species, we will only
mention the most interesting, such as the wheat-
ear (*motacilla ænanthe*), which follows the la-
bourers in the fields, to feed on the worms
turned up by the plough. The robin-redbreast,
which seeks shelter in our dwellings during
winter: in some provinces they assemble in
such numerous flights, that the sky seems covered by them. The reed-warbler (*motacilla
salicaria*), which fastens its nest to three reed-
stalks, so that it rises and falls with the surface of
the water, upon which it reposes. The *motacilla modularis,* the only species which remains
with us during winter, and enlivens this season
with its agreeable notes: it builds its nest twice
in the year, and feeds on corn, when insects are
not to be had. The golden crowned wren (*motacilla regulus*) is the smallest of all European

birds; it weighs sixty-seven grains, and its heart, no larger than a pea, weighs from four to five. This pretty little bird makes its nest in the shape of a ball in the firs; the entry is on one side, and contains from eight to ten eggs of the size of a pea. On the last shelf but one are the wagtails, so called from the continual movement of their tail; some of them are called water-wagtails, because they live on the banks of ponds and rivers. Near them we see the *budytes*, a sort of wagtails which follow the sheep, perch on their back, and search for insects in their wool. Below the case are the meadow larks (*anthus*), known in the southern provinces of France by the name of *bec-figues*. One of the most remarkable exotic species is the sentinel lark, which lives at the Cape among the flocks.

The drongos (*edolii*) are placed in the twentieth case. We have eight species, some of which come from Africa, others from the countries on the borders of the Indian ocean; some sing as sweetly as the nightingale. The most remarkable is the racketted drongo, which has the two outer feathers of the tail three times longer than the others, and destitute of beards, except towards the end, where they form a little palette.

The cotingas, or chatterers (*ampelis*), are placed below the drongos; they inhabit the swamps of South America. At the pairing season, the plumage of the males is coloured with crimson and azure, and is generally grey during the rest of the year. Some species with a strong pointed beak feed on insects, and those with a weak depressed one feed on berries. Amongst seventeen species in the Museum, the most beautiful are the *ampelis carnifex, pompadora*, and *cotinga ;* the purple cotinga (*coracias militaris,* Sh.), the white cotinga (*ampelis caruncu- lata*), which has a caruncle upon its head. The *ampelis variegata,* whose plumage is green during the first year, and ash-grey when the bird is full grown, is besides remarkable for the bundle of fleshy caruncles which hang under its throat.

The numerous family of the flycatchers, divided into several genera, and of which the Museum possesses one hundred and fifty species, occupies the lower shelf of this case. The species with a wide depressed beak belong to the genus gnatsnapper (*muscipeta,* Cuv.); those with a narrower beak, to the *muscicapa,* Cuv. The flycatcher of Lorraine (*muscicapa atricapilla*), which nidificates in the trunk of trees, presents the same phenomenon as the cotingas. During the winter the male is of an uniform grey. but towards the pairing season a part of its plumage becomes of a beautiful black, and the rest of a pure white. Amongst the exotic species, the prettiest are the *muscicapa oranor* from Java, the red-breasted flycatcher and the *azurou*, both from Timor. Some of the gnatsnappers of Madagascar, the Isle of France, the Cape of Good Hope, and the East Indies, have very long tails, on account of which they have been named the paradise gnatsnappers. A species from Cayenne, which has been erroneously referred to the todies (*todus platyrhincos*), is remarkable for its beak being widened in the form of a spoon. On the last shelf of this case are several birds well worthy of attention from their rarity and beauty: 1st. A species near the cotingas (*coracias scutata*, Lath.), which M. d'Azzara has described, under the name of the pie with a bloody throat. 2d. The *cephalopterus ornatus*, Geoff., the base of whose beak is furnished with upright plumes, which form a large bunch over the head. 3d. The gymnocephalus (*corvus calvus*), which the negroes of the French settlements call *oiseau mon père;* its head is covered with feathers when young, but becomes altogether bald when of an advanced age: young individuals of this species are very rare in collections. 4th. Two species of a genus lately described by Dr. Horsefield, under the name of *eureylamus,* sent from Java by MM. Diard and Duvaucel. 5th. The *rupicola* of Cayenne, the male of which, when adult, is of a beautiful orange colour, whilst the young males are as brown as the females. 6th. A new species of this same genus, to which M. Cuvier gave the name of *rupicola smaragdina,* on account of its beautiful green colour: this species was sent us from Java by MM. Diard and Duvaucel, and is the more interesting, as all the others are natives of America.

At the top of the twenty-first case are many species of the genus *tyrannus.* These American birds have the same habits with the shrikes, and their courage is still greater; the females defending their young even against the eagles, and finding the means of driving away all birds of prey from their nests. The spoon-billed tyrant (*lanius pitangua*) of the Brazils, the yellow tyrant from Cayenne (*lanius sulfuraceus*), and many other species have their plumage of a sulphur yellow, and a red tuft on the head. Below the tyrants are the *euphones,* from the warmer countries of America. One of the most common species in the West Indies, is called the musician (*pipra musica*), because it articulates the seven notes of the gamut. After the euphones come the *tanagers,* American birds, very agreeably varied in colours: the prettiest species are the *tanagra septicolor* of Cayenne, the tricolor of the Brazils,

the *t. mexicana,* the *t. punctata,* the *t. archiepis-
copus,* the *t. episcopus,* the red tanager from the
Mississipi, the scarlet one from Brazil, and the
silver beaked tanager from Cayenne; they live in
the woods like some of our sparrows, and feed
equally on seeds, berries and insects. Below the
tanagers are the manakins (*pipra*), small birds
which live on insects in the forests of equinoctial
America. They are all adorned with brilliant colours. That with a long tail (*pipra caudata,* Sh.),
makes a noise like the barking of a moderate sized
dog. The titmice are placed after the manakins.
These birds of a very lively nature, are incessantly
suspended to the branches of trees, busied in split-
ting the bark to find the larvæ of insects, or in
breaking hard seeds, on which they feed. They
line their nest with down, and lay from sixteen
to eighteen eggs: the great titmouse (*parus major*), the marsh titmouse (*p. palustris*), the blue
titmouse (*p. cœruleus*), the long tailed titmouse
(*p. caudatus*), are natives of France, the rest are
foreign; by the side of the titmice is the *remiz*
(*parus pendulinus*): this little bird from the
south of Europe, builds its nest in the form of a
purse, with the down of the catkins of the willow or poplar, and suspends it to the flexible
branches of aquatic trees. Another species from
the Cape of Good Hope, builds its nest with

cotton, and having given it the form of a bottle, adds a small cupola outside for the male to rest in. In the bottom of this case are the goat-suckers, nineteen in number; they have the light soft plumage of the nocturnal birds, and their mouth is so wide that they can swallow the largest insects. They only fly about in the evening. The female lays its eggs on the ground, or on a stone, and only sits on them a very short time. An American species (*caprimulgus grandis*), is the size of an owl. One from Africa (*caprimulgus longipennis*), is remarkable from a feather twice as long as its body, which springs from the carpus of each wing, and is only feathered at the extremity.

The twenty-second case contains, first, the numerous genus of the swallows, of which the Museum has twenty-seven species. The first is the *hirundo apus,* of all birds best formed for flight; its feet are so short, and its wings so long, that when it is on the ground it cannot rise again; it therefore passes the greater part of its life in the air, and when it has rested for a short while on a wall or on the trees, it falls to recommence its flight. The swallows which immediately follow it, have longer feet, which enable them to take their flight from the ground, and rest on it longer. The *hirundo urbica,* or window swallow, is universally known. It quits us in the autumn, and returns with the spring to the nest it had formerly occupied. There is a white variety in this case; near it is the *h. riparia,* which builds its nest in the banks by the water side: it does not quit us in the winter, but plunges deep into the mud, where it remains torpid until the return of warm weather. Amongst the exotic swallows, we should observe the *hirundo esculenta,* which inhabits the Indian archipelago; it builds its nest, which has been placed by its side, on the highest shores, with the spawn of fish and other gelatinous substances, picked up from the sea. These nests are an object of commerce: considerable number of them are sent to China and Japan, where they are considered a very agreeable and nourishing food. Below the swallows are the larks; beginning with the field lark, of which we have a white variety. Immediately after come the starlings; they assemble in large flights, and are very useful to cattle, by destroying the insects which torment them. The five lower shelves are filled with the genus *cassicus,* of which we have thirty-four species; some of the size of a crow, others of that of a thrush. They are American birds, live in numbers like our starlings; most of them are brilliantly tinted with yellow, red and black. Their nest (of which we can see several in the two

frames upon the cornice) are worthy of attention; they are made with tufts of grass, of an
oval form, and from six to twelve placed one
after the other, all united by a single tube with
only one opening, like so many rooms in a passage. The tube, which is commonly from four
to six feet long, is suspended by its upper end to
the branch of a tree, and its opening is at the
lower extremity; through it each couple ascends
to that part of the nest which it has adopted.
These nests being suspended, and continually
blown about by the wind, protect the cassiques
from their constant enemies, the serpents (1).

The twenty-third case contains the numerous
family of buntings (*emberiza*) and sparrows,
which has been subdivided into several genera,
and of which we have seven hundred individuals
belonging to one hundred and fifty species. The
emberiza are on the first shelf; to these belong
the *ortolan*, which is reckoned so great a dainty.
The sparrows, properly so called, (*pyrgita*, Cuv.),
occupy the three following shelves; one of the
prettiest is the painted emberiza (*emberiza ciris*,
Lin.) of New Orleans. Four individuals shew
the different colours assumed by this bird at different ages. The linnets are on the fifth, sixth,

(1) These birds have been named *republicans*, from their habit of
living many together in the same nest with only one communication.

seventh and eighth shelves. To this group belong the goldfinch, the linnet of the vines (*fringilla cannabina*), the *fringilla linaria*, and the
Canary bird (*f. canarina*) (1). The widow birds
follow the goldfinches; the long feathers of
their tail give them a peculiar appearance; they
are all from Africa. On the six following shelves
are the numerous species of grosbeaks (*cocco-
thraustes*, Cuv.). On the fifteenth is the bullfinch
and its congeners. All these birds feed upon
seeds, and are much sought for the sweetness of
their voice, the beauty of their colours, and the
facility with which they are tamed. The cross-
bills (*loxia curvirostra,* Lin.), which inhabit the
pine forests of the northern parts of the two continents, are at the bottom of this case. The European crossbill is very familiar; it takes its food
in the claws, and carries it to the mouth as the
parrots do: it builds its nest, and hatches in January. The greater bullfinch of the north (*loxia
enucleator*), and the *colies* of the Cape, are on
the same shelf; the latter live in numbers; they
sleep suspended to the branches of trees, the head
downwards, close to each other. The last bird
of this shelf is the beef-eater (*bufaga africana*),
so called because it takes out and feeds upon

(1) This species pairs with its congeners, and its produce is endowed
with the same fecundity.

the larvæ of insects which lodge in the skin of oxen.

In the twenty-fourth case are the rollers, which resemble the jays in form, but whose colours are more brilliant. We have seven species of them. The *gracula religiosa,* or Indian grakle of Java, is on the second shelf; it is said to imitate the human voice better than any other bird. Next to it are the jays. The third and fourth shelves are occupied by a magnificent series of birds of paradise (*paradisea*), of which there are nine species. These birds live in New Guinea, and in the neighbouring islands. As their feathers are employed in making plumes, aigrettes, and different other ornaments of dress, the natives sell them very dear, and it is even difficult to procure perfect individuals. Thus it was for a long time believed, that they had no feet nor wings, and always lived in the air, supported by the very long feathers which are placed under their wings. The velvet black, the emerald green, the sapphire blue, and the most vivid red, all adorn the plumage of these birds (1). The lower shelves of this case are furnished with the different species of pies and crows. We shall only mention

(1) M. Regnault de la Susse, who has seen these birds alive at the governor of the Philippine islands, tells us they have no voice, are destitute of intelligence, and feed on berries.

the sky-blue pie of Paraguay, and the pie from the Brazils, whose colours are beautiful, and agreeably distributed. This species was sent to the Museum by M. Auguste de Saint-Hilaire.

The twenty-fifth case contains the birds which bear the name of *tenuirostris,* from their slender beaks, which are very long, and more or less arched. This family has been divided into three great genera. The hoopoes (*upupa*), the creepers (*certhia*), and the humming birds (*trochilus*). We have eight species of the first genus, sixty-four of the second, and fifty-three of the third. Two genera which belong to another division, occupy the lower part of the case. The humming birds are placed in the front and middle part of the case, in order to be better seen.

On the first shelf are, 1st. The hoopoes (*fregilus,* Cuv.), from the Alps and Pyrennees, which build in the clefts of the highest rocks; their red beak and feet form a strong contrast to their black plumage. 2d. The hoopoes, properly so called, and so named from the double row of feathers with which their heads are adorned, and which they can raise at their will. They live on insects, and lay their eggs in the holes of trees and walls. 3d. The promerops of the Cape, and the *epimachus* of New Guinea. The epimachus promefil, whose breast is like the

most beautiful burnished steel, and which has
the feathers of its flanks elongated like those of
the bird of paradise, is one of the rarest and most
beautiful birds of the collection. On the second
shelf are the species of American creepers. These
little birds resemble the pies in their habits, and
in the feathers of their tails, which are stiff and
worn at the end. The humming birds fill up
the third, fourth, and fifth shelves. They excite
attention by their diminutive size, the beauty of
their colours, and the elegance of their forms;
several of them are not an inch long; their nests
are placed by their side. The topaz, the garnet,
the red topaz, the ruby, the sapphire, and the
emerald humming birds, have been so named
from the colours of these precious stones which
they display on their breasts and necks. They
are all natives of America, and fly rapidly round
the flowers, the nectar of which they suck by in-
troducing their tongue into the corolla, which
is bifid like that of the pies and capable of the
same elongation; they also feed on small insects.
The crested-neck of Cayenne and the white
crested-neck of Brazil are not so lively, but their
size is scarcely larger than that of a hornet. The
sugar-eaters (*cinnyris*, Cuv.), which are the humming birds of the ancient continent, fill the three
following shelves. They are equally brilliant,

but less remarkable for the smallness of their size. The malachite and shining sugar-eaters present the most brilliant colours. The red sugar-eater, or heorotaire of Levaillant (*certhia vestiaria*), from the Sandwich islands, furnishes those scarlet feathers with which the islanders make mantles in high esteem among them.

Amongst the sugar-eaters on the ninth and tenth shelves, we should remark that of the West Indies, which lives in sugar plantations, climbs up the stems of the canes to feed on the insects which it finds in the axillæ of the leaves; the *certhia cœrulea* and the *c. cyanea*, blue creepers of the most beautiful ultramarine blue: the *merops rufus*, commonly called the baker, because its nest which it builds on bushes has the shape of an oven; one of these nests is at the bottom of the case. The wall-creepers (*certhia muraria*) are on the eleventh shelf; they inhabit the south of France, but are sometimes seen further north. The specimen with extended wings was killed in the Jardin du Roi. The bee-eaters are placed on the twelfth shelf; they are all adorned with the most beautiful colours. We will only remark that with a lilac head, recently sent us from Sumatra by M. Duvaucel, and that of Europe (*merops apiaster*), which also inhabits Africa in all its extent from Egypt to the Cape of

Good Hope : it flies rapidly in pursuit of bees, wasps, and other insects, and builds its nest in holes on the banks of rivers. By the side of the bee-eaters are the momots, which may be well considered as the bee-eaters of America; they have the same habits with the merops.

We will now pass on to the twenty-sixth case. The five upper shelves contain thirty-four species of king-fishers (*alcedo*), which belong to both continents; their plumage is in every species shaded with green and blue; some of them live on the borders of rivers and lakes, and feed on fish which they catch by diving; others resort to the forests and marshes, where they pursue and feed upon insects. The bottom of this case is filled with horn-bills (*buceros*), large birds from Africa and the East Indies, remarkable for the size and form of their beak. In some species it is surmounted by a straight or arched projection, which changes its form and increases with age, becoming even as large as the beak itself.

On the first shelf of the twenty-seventh case are the touracos (*corythaix*, Ill.), and the musophaga or plantain-eater, African birds which appear to form a passage from the climbers to the gallinaceæ. We have four species of them ; that which has been longest known is the touraco of Buffon (*cuculus persa*, Lin.), which lives in the neighbourhood of the Cape; the three others have only been discovered lately. The most remarkable is the *corythaix paulina*, Cuv., of Sierra Leone, which is valued by the negroes, who tame it for the sake of its voice. The rest of the case is filled by the numerous varieties of the domestic pigeon and the cognate species. We will only call the attention to the *columba muscadivora*, from the country of the Papous; that with a white cap, from the West Indies; the bronze winged pigeon (*columba lumachella*), whose changing colours are extremely brilliant; the kurukuru from New Holland, and the jamboos from Sumatra.

The remainder of this genus entirely fills the twenty-eighth case. The species with long graduated tails are on the upper shelves; the *columba phasianella* feeds on all-spice. On the fourth shelf are the green pigeons with strong beaks, which Levaillant names colombars (*vinago*, Cuv.). At the bottom are the columbigallinæ of the same ornithologist. Amongst the most remarkable of this genus are, 1st. The talpacoti from America (*columba passerina*), scarcely the size of a sparrow. 2d. The goura, or crowned pigeon of the Moluccas (*columba coronata*), the size of a cock. 3d. The bleeding dove, which has a red spot on the white plumage of its breast.

It is caught in the Philippine islands in the same manner as partridges are caught in France. 4th. The *columbi-gallina* of the Moluccas, whose green plumage has a metallic lustre, and which is ornamented with a long ruffle round the neck. The genus *columba* forms the transition from the passeres to the gallinaceæ. There are eighty-four species in the Museum.

Although the peacocks are generally known, a few moments' attention should be given to the twenty-eighth case, which contains varieties prepared so as to display the magnificence of their plumage. This superb bird, which is now domesticated, is originally from India, and the individual to the left was killed in a wild state in the mountains of Bengal. To the right is another species from Java, which differs from the preceding in the feathers of its crest, and its neck is green, spotted with blue ; it was sent to the Museum by M. Diard.

The thirtieth case, which projects and forms an angle, is the last on that side of the room, and contains the turkeys. The common turkey has been spread over Europe since the discovery of America. By comparing the individuals in our farms with those killed wild in the forests of Virginia, and which were sent us by M. Milbert, we find that domestication has deprived these birds of the metallic lustre of their plumage. At the bottom of the case is a new species, described by M. Cuvier, in the sixth volume of the Memoirs of the Museum, under the name of *meleagris ocellata;* the eyes of its iridescent plumage offer a diversity of colours of a metallic lustre, which change according to their position. It is one of the most beautiful birds known, comes from the bay of Honduras, in the gulph of Mexico, and is the only spcimen in Europe.

The thirty-first case, corresponding to that we have just seen, is the first in our return along the other side of the room, and is filled with hoccos, which come from the warm countries of America, and are analogous to the turkeys. The *pauxi,* called the stone-bird, on account of a very hard tubercle at the base of its beak, is placed on the last shelf (1).

On the first shelf of the thirty-second case are the quans or jacoos (*penelope*), American birds with a dark plumage, a tuft on their head and a part of the throat naked. On the second is the napaul or horned pheasant from Bengal, a very rare bird; the male of which has two fleshy horns behind the eyes, its plumage is crimson spotted with white. On the third and fourth shelves are the different races of domestic fowls,

(1) It has the longest trachea of all known birds.

and near them several wild species from India and the Moluccas. It cannot yet be decided from which of them our common yard fowls have sprung. The genus pheasant commences at the bottom of this case, and presents ten species. Among these we should notice the golden pheasant from China; its gold crest, its plumage, enamelled with the most brilliant colours, and its elegant shape make it esteemed as the most beautiful of the gallinaceæ: to this bird we may refer Pliny's description of the phœnix.

In the following case is a superb bird from Sumatra; its size is nearly that of a common cock, but its wings are extremely large; it is called the argus pheasant from the number of eyes on its wings and tail. It does not excite admiration from the brilliancy of its colours, but from the regular distribution of the circles on each plume and the delicate gradation of the tints. The Museum had only three feathers of this bird at the time of Buffon; it now possesses six individuals sent from Sumatra by MM. Diard and Duvaucel, four males and two females; the plumage of the latter is in nowise remarkable. Below the argus is the impeyan pheasant (*phasianus impeyanus*), which has a very elegant crest, and whose plumage presents the brilliant colours of gold, malachite or lapislazuli, according to the reflexion of the light. The crested pheasant (*p. ignitus*), from the Sonda islands, is next to it, and is not less remarkable for the singular form of its crest, than for the colour of its plumage. After it comes the rouloul from Malacca, a very rare species discovered by Sonnerat; the male is black and the female green. The lower part of the case is filled with the pintadoes (*numida*, Lin.), commonly called guinea-fowls. The most common species is gregarious in the marshes of Africa. As its flesh is very good eating, it would be advantageously multiplied in our farm-yards, were its cry less disagreeable.

The numerous family of the grouse, of which we have fifty-nine species, entirely fills the thirty-fourth case. To it belongs the heath-cock, the largest of the gallinaceæ; the water-hen; the lagopus or ptarmigan, which is of a tawny colour in summer and white in winter. This bird lives on high mountains, and passes the winter under the snow; the several varieties of partridges; and lastly, the quail, the white variety of which was sent to Buffon by Louis XV, who killed it when sporting. A great number of foreign species with varied plumage are grouped around the European, which serve as types to the different genera of this family, so celebrated for furnishing excellent game.

We now come to the *grallæ,* or waders, so called from the length of their legs; they form the fifth order in the general system of classification. Those of the two first genera, which occupy the thirty-fifth and thirty-sixth cases, differ from all others in being deprived of the power of flight. The first is the ostrich (*struthio camelus*); this bird, celebrated in the remotest ages, is sometimes eight feet high; it lives in herds in the sandy deserts of Africa, and feeds on grains and herbage; no animal can run so fast. The female lays her eggs, which weigh three or four pounds, on the sand. She abandons them to be hatched by the heat of the sun in the tropical regions, but sits upon them in the colder climates; when the young are hatched she holds them between her legs. There are some tame ostriches in Senegal, which are ridden like horses, but they are not to be trained or guided at will. Their feathers form a considerable article of commerce; they are light and waving, because their plumes are not hooked to each other as in almost all other birds. Above the female ostrich, which is accompanied by its eggs and young of different ages, is the *nandou,* or American ostrich, about half the size of that of the ancient continent. Its feathers, which are made into brooms, are sold for vulture feathers. Several females are said to lay their eggs in the same nest, and leave them to be hatched by the male. Above the male ostrich are the two species of cassowary. That of Asia has a prominence on the head, and the naked skin of its neck is tinted during life with red and blue; it feeds on fruit and eggs, but not on grains. The second species comes from New Holland, and is as rapid in its course as the fleetest greyhound; its feathers are used for ornamental purposes, and its flesh is good food.

The bustards (*otis*, Lin.) fill the thirty-seventh case. We have nine species; three of which have not yet been described. That of Europe lives in large plains, and builds its nests in the corn; it flies but little, and generally uses its wings only to accelerate its course. The male, which is double the size of the female, is very rare, and is the largest of European birds. The specimen in the Museum was presented by Viscount de Riocour. Among the foreign species we may cite the houbara, brought from Barbary by M. Desfontaines, remarkable for the cloak of long feathers which adorns its neck. Next comes the cariama of Brazil, described by M. Geoffroy Saint-Hilaire in the Annals of the Museum; its flesh being esteemed, it has been domesticated.

The *grallæ*, in the following cases, have been called shore-birds, on account of their habits. The plovers (*charadrius*, Linn.), to the number of thirty species, are arranged on the three first shelves of the thirty-eighth case. These birds live in numbers on humid soils, and strike the ground with their feet, to make the worms, on which they feed, come out; their flesh is generally esteemed. The golden plover is common in our climates. The lapwings are on the fourth shelf; they very nearly approach the plovers; that of Europe is pretty, its head is adorned with a loose elegant crest; its eggs are considered a great delicacy. Several species of lapwings and plovers have their naked face furnished with long caruncles; others have a very long and pointed spur to the joint of their wings, with which they defend themselves against the birds of prey. Below the lapwings are the oyster-catchers (*hæmatopus*), so called because they open the shells of oysters with their bill, which is strong and square at the end; they are also called sea-pies, because their plumage is varied black and white like that of the magpie.

The bottom of the case is occupied by the ibis. The most celebrated species is that worshipped by the Egyptians, and which M. Cuvier has named the sacred ibis, after having compared those which live in Africa with the mummies brought from Egypt by M. Geoffroy Saint-Hilaire. Two

of these are placed here ; the one has still its envelopes on, and they have been removed from
the other to exhibit the feathers, which are well
preserved as to their form and colour. Some
species nearly allied to the sacred ibis are found
in America ; the most remarkable on account of
its scarlet colour inhabits the borders of the sea
at Cayenne and Surinam.

The thirty-ninth case contains fifty species of
the genera analogous to the woodcocks (*scolopax*). These birds live all in the same way on
the swamps, thrusting their long bills into the
mud to seek for worms. The extremity of their
bill is soft, which renders their sense of feeling
very delicate. They moult twice a year, when
their plumage changes from a dull red which
it assumes in summer, to a deep grey during
the winter season. It is this complete change
that has induced several naturalists to multiply
the species. Care has been taken to bring here
together those specimens which differ most in
their plumage. On the first shelf are the god-
wits (*limosæ*, Bechst.) one species of which
meets in innumerable flights on the plains of the
Low Countries in the summer. On the second
are the woodcocks proper. The common wood-
cock, which lives on the mountains in summer,
and descends in the plains during the autumn.

is too well known to be particularly noticed. By the side of them are the snipes, which live in marshes, and rise to a great height, uttering a cry much like the bleat of a goat. On the last shelves are the ruffs and reeves (*machetes*, Cuv.), of which the Museum has twenty-three varieties. These birds are famous in the north of Europe for their combats to get possession of the females, which are much less numerous than the males. During the spring the head of the male is covered with red caruncles, and the neck is ornamented with a large ruffle which varies in form and colour. The turn-stones (*tringa interpres*, Lath.) fill the bottom of the case; they live on the sea-shore, where they turn up the stones with their short conical bills, and eat the worms which they find underneath.

The sand-pipers occupy the top of the fortieth case. The white-tail (*tringa ochropus*), which is very common on the borders of streams and rivers, although it lives solitary, belongs to this genus. This series is terminated by the avocets (*recurvirostra*), which are distinguished from all other birds by the strong upward curve of their bill. They run upon the mud, which they furrow with their bills in search of insects; in other respects they have the habits of the snipes. At the bottom of this case is the boat-bill (*can-*

croma cochlearia), which lives in the warm parts of South America; it perches upon trees on the borders of rivers, whence it darts upon the fish. It is remarkable for its large bill, which is like the bowls of two spoons with their concave sides adjoining, and for the long black feathers which hang from the head of the male.

The herons (*ardea*) to the number of thirty-nine species fill the forty-first case. These birds, which are of a melancholy disposition, feed on fish, and pass their lives on the borders of rivers. They advance into the water up to the body, and remain for hours with the neck drawn in between the shoulders in the most complete inactivity; if a fish, such as they like, passes within their reach, they extend their neck and dart their bill down with such rapidity, that they never fail to catch it. The common heron is of an iron-grey with a black crest, out of which three feathers, longer than the rest and very flexible, hang behind the head. They are much esteemed as ornaments, and are sold at a considerable price. There is another species equally esteemed for the pretty feathers, from which it is named the egret; the bird itself is entirely white; in the pairing season those feathers with slender stems and loose plumes are very long on its back. The bittern (*ardea stellaris*) keeps constantly amongst the reeds, whence its strong voice is heard, similar to that of a bull.

The crane and its congeners fill the three shelves of the forty-second case. On the first is the *sun bird*, also called the small rose peacock. This beautiful bird, which is nearly the size of a partridge, lives on the banks of rivers in Guyana. The different colours of its plumage are rather dark, but shaded with such delicacy that they can be compared to those of the most beautiful moths. By its side is the agami of South America (*psophia crepitans*), also called the trumpeter, from the hollow sounds it utters, which seem to come from its abdomen; its plumage is blackish with metallic shades of blue and violet; it is so easily tamed, that it conducts the farm-yard fowls as dogs do the flocks. Below it, is the royal or crowned crane (*ardea pavonina*); its voice is like the sound of the trumpet; its shape is slender, its cheeks coloured with the brightest rose and white, and the bunch of light feathers which crowns its head expands or keeps close at pleasure; it inhabits the western regions of Africa, as well as the *ardea virgo*, which is brought up as an ornament for parks, on account of its elegant forms and singular movements. The common European crane and that of southern Africa, are at the bottom of the case.

The first has been long celebrated for its regular migrations in large and regular flights, from the south to the north in the autumn, and in spring from north to south.

The storks (*ciconia*, Cuv.) are placed in the forty-third case. The first is the European, só common in Holland and the north of Europe. They migrate like the cranes, and in the spring return to the nest which they had built the preceding year on steeples and chimneys. It is a common belief with the lower class of people in some countries, and particularly in Flanders, that there is an impending calamity upon the house to which the storks do not resort. Immediately after follow seven other species, amongst which we will only cite the *ardea dubia* and *a. crumenifera;* the former lives in great numbers in the Philippine islands and Bengal, where it is called the *adjutant.* They are so useful for purging the towns from all sorts of animal remains, that no one is allowed to kill them. They have however been much sought for, since the custom has been introduced of wearing as an ornament the under feathers of their tail, which are well known under the name of maraboo plumes, as well as those of the latter species which is from Senegal.

On the upper shelf of the forty-fourth case

are two species of the open-beak genus (1).
The one (*ardea ponticeriana*), which is white,
comes from India; the other, which is black,
lives in the interior of Africa, north of the Cape.
This last is extremely remarkable for the feathers
of its breast and belly, which do not resemble
those of any other bird; their stems flattened
into thin and brilliant thongs are prolonged
much beyond the beards, and curl or twist into
a spire from their elasticity: it is a new species recently brought by M. Leschenault de la
Tour.   On the following shelves are three species of *tantalus;* one from America, one from
Ceylon, and the other from Senegal.   Previous
to the researches of MM. Cuvier and Savigny, the
Senegal species, or the *tantalus ibis,* was looked
upon as the true ibis of the Egyptians; it is not
even found in Egypt.   Below the tantali are the
jabirus (*mycteria,* Linn.).   These birds have the
same habits as the storks, from which they differ
chiefly in being of a larger size: they are found in
America and Africa living by the side of ponds,
and feeding on reptiles and fish.

In the forty-fifth case we see, first, the spoon-
bills (*platalea,* Linn.), which are so named from

(1) The open-beak (*hians*) has been thus named, because its two
mandibles, which form a crescent, touch each other only at the base
and point of the beak

the form of their bill : they live in the same
way as the storks, but their large and flattened
bill, being weaker than that of the stork, seems
rather calculated to seize small fishes, or to seek
insects and worms in the mud. The European
spoon-bill is white, that of America of a beautiful rose colour, which becomes brighter with
age. The lower shelves of this case, and all
those of the following, are occupied by the numerous family *macrodactyla,* which comprehends the rails, the jacanas, the screamers or
kamichis, the water fowls, sea partridges, and
flamingos. Some have very long and slender
claws, which enable them to walk on marshy
ground, and support themselves on the grass;
others have their feet furnished with membranes, often dentated, which facilitate their
swimming. They form the passage from the
grallæ to the *palmipedes.* Thirty species of rails
furnish the lower shelves of the forty-fifth case·
That which is called *rallus crex* lives and nidificates in the plains (1); the others remain in
marshes, run rapidly on the grass and swim.
Such are the water rails which feed on shrimps;
the spotted rail (*rallus porzana*), which builds

(1) This bird has been also called the king of quails, because it
arrives and sets off with them, which led to the belief that it served
them as a guide.

its nest of rushes in the form of a small boat, and fastens it to some aquatic plant, so that it rises and falls with the stem of the plant according to the level of the water. The jacanas (*parra,* Lin.) have longer nails than any other bird; that of the great toe is considerably longer and sharper than the rest: they inhabit the warmer regions of America and India. Their wing at the first joint is armed with a spur or lancet more or less pointed, from which they have been vulgarly called surgeons. This characteristic is still more evident in the kamichi (*palamedea cornuta*). This bird has a horny stem at the base of its bill, which adheres only to the skin; it lives in the swamps of South America. The admirable pages written by Buffon on this bird, have given it a great celebrity. Next to it has been placed the Chaïa (*parra chavaria*), a very fine species which had not yet been seen in collections, and was lately brought from Paraguay by M. Auguste Saint-Hilaire.

The coots (*fulica,* Briss.) and the sultans (*porphyrio,* Briss.), which we see at the top of the forty-sixth case, are remarkable for the beauty of their plumage, shaded with violet, blue, and aquamarine. They stand on one leg and bring their food to the bill as parrots do. One species only is found in the south of Europe, and more particularly in Sicily, others in southern Africa, the East Indies, New Holland, and South America. Next to the porphyrions are the coots, whose feet are furnished with dentated membranes; their flesh is esteemed. By the side of the coots is a very rare bird which forms a genus by itself under the name of the scabbard-beak (*vaginalis*, Lath.), on account of the singular form of its beak, the base of which is encircled by a horny substance. We are perfectly ignorant of the habits of this bird, which is found in the Malouin islands, whence it was brought by the naturalists attached to M. Freycinet's expedition. The lower part of this case is occupied by the flamingos (*phœnicopterus*), remarkable for the excessive length of their legs and neck, and the curious shape of their beak; they live in troops on marshes and sea-shores; they sometimes undertake long voyages; they feed on shell-fish, insects and fish-spawn; they build a pyramidal nest elevated above the water, and place themselves astride upon it to hatch their eggs. The European species, when adult, is white with rose-coloured wings; when young it is entirely grey. That of America has a scarlet plumage, but the young ones are grey, spotted with black.

We here terminate the *grallæ*, and proceed to

the *palmipedes,* so called from the membranes which unite their toes. The legs in all of them are placed very far back, which fits them the better for swimming. This order comprehends four families, the *brachyptera* or divers, the *longipennes,* the *totipalmes,* and the *lamellirostres.* Twenty-seven species of the first family fill up the forty-seventh, forty-eighth, and forty-ninth cases. On the first shelves of the forty-seventh are the *colymbi;* although their wings are very short, they go very far inland from one pond to another; their close, smooth, and silvery plumage is employed as a fur. Some species carry their young under their wings when swimming. The divers in the lower part of the case breed in the north, but come to the coasts of France in the winter. The guillemots (*uria,* Lath.), on the second shelf of the forty-eighth case, are stupid birds which live on fish and crabs, and build their nest in the clefts of steep rocks. The *colymbus grylle* changes its colour more decidedly than has been yet observed in any other bird, being white in winter, and quite black in summer. We have here an individual killed during the interme- diate season, whose plumage is white, speckled with black. By its side is the small *colymbus,* known to travellers by the name of the Green- land pigeon, which lives in the north, and builds

its nest under ground. Immediately after comes
the *alca cristatella,* a very rare bird, found in
the Aleutian islands by M. Choris, who presented
it to the Museum. After these are the auks
(*alca,* Lath., *fratercula,* Briss.), remarkable for
their large flat beak, and the pinguins, whose
wings are so small that they cannot support
themselves in the air for an instant; they always
remain in the waters, and are excellent divers;
they build on steep rocks, which they climb,
using their feet and wings equally.

The *aptenodytes,* Forst., fill the greater part of
the forty-ninth case, they are from the Antarctic
seas; the shortness of the feathers on their
wings is such that they may be taken for scales.
They seldom get on shore except when they
breed, and to reach the spot they drag themselves
along with difficulty. The largest, the *Patago-
nian pinguin,* which is as large as a goose, lives
in great numbers in the straits of Magellan; the
plumage of that part of its skin which lines the
belly is silvery, and much sought for by furriers.
The *aptenodytes chrysocoma* inhabits the Malouin
islands and New Holland; it bounds along the
surface of the water.

The fiftieth and fifty-first cases contain the *longipennes.* They live in the open seas, are found
in all countries; some are met with six hundred

leagues from land, and they rest upon the waves. The first genus is that of the *procellaria,* Lin., which are also called tempest birds, because they are seldom seen, except on the approach of a storm. They build in the holes of rocks. The largest of them (*p. gigantea*) has been called the bone-breaker, from the strength of its beak. That most spoken of by sailors is the *procellaria capensis,* whose back is spotted with black and white; it is seen in great numbers in the environs of the Cape. The smaller is more particularly called the stormy petrel (*p. pelagica*); some have been found carried by the wind more than forty leagues inland. Below these are the *stercorarii,* which eat the dung of sea-mews and gulls. They are northern birds which visit our coasts in winter. At the bottom of the case is the genus albatross (*diomedea*), which inhabits the Austral ocean ; the largest has been called the *Cape sheep,* on account of its size, its colour, and gregarious habits ; its voice is said to resemble that of an ass.

The gulls and sea-mews, of which we have twenty-two species in the fifty-first case, are found in all latitudes ; they fly with rapidity near the coasts, never stretching so far out as the petrels; they live on fish; their plumage varies according to the season, but is generally of an ashy blue on the back and white on the belly, in the adults; when young they are of a greyish colour. Twenty-three species of sea-swallows (*sterna*) fill the lower shelves of this case; they owe their name to the extraordinary length of their wings and tails, and to the rapidity of their flight. They skim the surface of the water to catch molluscas and small fish. Some species ascend large rivers, others keep on the lakes. The cut-waters (*rhyncops*), which are at the bottom of the case, are distinguished from all other birds by the extraordinary form of their beaks; the upper mandible of which is like the blade of a knife, and much shorter than the inferior. This singular form renders it impossible for the bird to pierce any thing with its beak, but he uses it when flying over the surface of the waves, to seize the molluscas which float on them. One of these species, the *rhyncops nigra*, is commonly found in the seas of the West Indies; the other, but recently known, inhabits the Austral ocean.

The fifty-second and fifty-third cases contain the *totipalmes*, so named because their great toe is united to the others by a membrane; notwithstanding this formation they perch upon trees. The largest of them is the pelican, which is remarkable for the length of its beak, whose lower mandible supports a naked dilatable membrane, which forms a bag capable of containing fish and water. The pelican inhabits the sea-shores and lakes, it feeds on living fish. At the time of the pairing season, the hooked extremity of its beak becomes of the liveliest red. When it feeds its young, it ejects the fish kept in reserve in the above mentioned bag, and cuts them to pieces; the blood that then drops on its breast has given rise to the belief that it tore its breast to nourish its young. Below the pelicans are the cormorants, which destroy a great quantity of fish; they build on trees or in the hollows of rocks.

The frigate birds (*pelecanus*, Lath.) are in the fifty-third case. Their wings, which measure from 10 to 12 feet, are so powerful that they fly to an immense distance from land, especially between the tropics; they dart upon flying-fish, and strike the boobies (*morus*, Vieill.) to make them quit their prey. These boobies, which we see on the lower shelves, have been so called from the stupid manner in which they suffer themselves to be attacked by sea birds and by men. They sometimes rest on the rigging of vessels, and are then easily taken with the hand. The tropic birds (*phaeton*) occupy the bottom of the case; they are called straw-tails on account of the two long beardless feathers in their tail;

they keep constantly in the tropical latitudes, the approach of which they announce to sailors.

The four cases which terminate the gallery are filled with the *lamellirostres*; to this family belong the swans, geese, and the numerous family of ducks and the *mergus*, Lath. The swans are well known for the beauty and elegance of their form, and for their down which is so useful. The beak of the wild swan is yellow at the base, and black at the extremity, it is a distinct species from the domestic swan, which has a red beak. The black swan of New Holland, and that with a black neck sent from the Brazils by M. Saint-Hilaire, are remarkable species. Above the swan is the bernacle goose (*anas erythropus*, Lath.), celebrated for having long been thought to spring from trees like fruit; it passes the summer in the north, and visits our climates in winter. Next to this is the Egyptian goose, which is very common in Africa. We see it often represented on ancient Egyptian monuments; it was worshipped for its attachment to its young, and the Egyptians called it *chenalopex* (fox-goose).

Among the ducks (of which we have seventy-eight species) we will only cite: 1st. The eyder (*anas mollissima*), a bird common in the north of Europe, and which only visits the coasts of France in the severest winters; it furnishes a very precious down.   2d. The musk-duck (*anas moschata*), native of America, and which is now common in our farm-yards, where it is known under the name of Barbary duck.   3d. The Carolina duck (*anas sponsa*), much esteemed for its elegance and the colour of its plumage. 4th. The fan-water-fowl from China (*anas galericulata*), the male of which has some of its wing feathers standing up over the back, and widened in the form of a fan.

The mergansers (*mergus*), which we see at the bottom of the case, live on the lakes of northern countries, and winter on the coasts of France, etc.; they vary much in their colour, and have a crest on the head.

Here terminates the collection of birds, which for elegance, richness, variety of forms and brilliancy of colours, equals all that the imagination can conceive as beautiful.

The centre of the gallery is occupied by a set of cases in which the animals without vertebræ are arranged.   We will examine them after we have seen the collections of reptiles and fishes which are on the first floor. On the walls of the staircase, which leads to the two rooms below, are expanded the skins of large serpents of the *boa* genus, the colours and scales of which are very well preserved.

§ VI. COLLECTION OF REPTILES.

REPTILES do not arrest our attention in an equal
degree with birds, either by their elegance of
form or variety of colours; most of these animals are of an unpleasant or repulsive shape;
and the brilliant speckles, which embellished
many of them whilst they were living, have
completely faded since their death. But the singularity and variety of their forms, and their
different properties, some fatal to life, and others
capable of being rendered subservient to the
wants of man, give to the animals comprised in
this collection, at least an equal degree of interest. Who has not heard of the sea tortoise,
of the crocodiles of the Nile, of the pithons of
India, of the boas of America? And who would
disdain the examination of these animals, many
of whose species are celebrated, either for their
peculiar habits, for the phenomena which they
exhibit, for the terror which they inspire, or for
the fables of which they have been the subject?
That the study of reptiles has become almost

a passion amongst naturalists, is a fact attested by the great perfection to which this branch of natural history is brought at the present day; notwithstanding that the habits of these animals, and their abode in damp and unhealthy places, seem alike calculated to make them elude the research of man (1).

The collection of reptiles in the cabinet of natural history, is unquestionably the richest in the world; it consists of eighteen hundred individuals, belonging to more than five hundred species. But what renders it of incalculable value to the student is, that it contains almost all the individuals from which the beautiful plates of Seba were copied, and that it was from them that Linnæus composed his descriptions; here also are to be found the originals which served for the work of M. de la Cépède.

Reptiles are divided into four orders, namely: *chelonians,* or tortoises; *saurians,* which comprehend the crocodiles, lizards, etc.; *ophidians,* or serpents; and *batracians,* to which the toads, the frogs, the salamanders, etc. are referred.

---

(1) Of the species known, scarcely a fifteenth part are dangerous to man, either on account of their size and voracity, as the crocodile; or on account of their venom, as the viper. A great number are serviceable to him, either from their flesh and their eggs which afford an excellent food, or from their furnishing materials which are employed in the arts.

We shall follow this division ; but the place which the specimens occupy in the cabinet, renders it necessary to examine them twice (1).

Amongst the reptiles of the three first orders, as amongst the quadrupeds, there are some specimens too large to be placed in the cases; these have been suspended from the ceiling or attached to the wall; and as they first attract our attention, we shall describe them before those which are in the cases.

The species of the genus tortoise, suspended from the ceiling, are : the *leather tortoise,* or the lute of the Mediterranean seas. This species is the largest of all; and the one we here see is seven feet long; its weight is often more than twelve hundred pounds. It has no apparent breastplate; its upper shell or carapace (2) has five projecting bones or ridges, and is covered

(1) The want of space not having permitted the arrangement of all the reptiles to be made according to the degree of affinity of the genera, we shall not, in pointing out the objects which are more worthy of notice in this collection, designate them by the cases in which they now are contained, as we have done for the birds; because as the fishes are about to be removed into the room occupied by the library, the reptiles will then be distributed more at large and in their natural order, when the description we give will be found consistent with the new arrangement.

(2) The two bones which envelope the body of the tortoise are called the *bucklers;* the one on the back is named *the upper shell,* or *carapace;* and the under one, *the breastplate.* Both are in most tortoises covered with scales, fitted together like inlaid work. The characters fo-

with a brown skin which resembles leather; its flesh is good food.

The true tortoise, or green tortoise, which is nearly as large as the preceding, and which in general weighs eight hundred pounds, feeds on sea-weeds, and inhabits in great numbers all parts of the ocean near the equator. Its eggs, which it deposits in the sand, and its flesh are considered a wholesome food, and are an important resource for the navigators in the torrid zones. This species is caught with nets, and dragged under water to the shore; when brought to land it is fastened to clumps, and four men are often requisite to lift one of them.

The caretta (*testudo imbricata*), of nearly the same size as the former, furnishes the tortoise shell employed in the arts; it is found in the seas which border on warm countries; but the greater number are obtained from Ascension island, where, after traversing the ocean for two or three hundred leagues, they resort in great numbers to lay their eggs. When they come on shore to deposit them in the sand, the people engaged in catching them, turn them on their backs, and as the tortoise in that position is ut-

---

ascertaining the species are taken from the number, the form, and the colour of these scales. The carapace of the tortoise is very strong, and can support a prodigious weight without breaking.

terly unable to move, they are sure to be found in the same place; and it is not until a considerable number of them are caught in the like manner, that they are collected and carried to the ship.

The great land tortoise of India: the feet of this species are rounded off like a stump, instead of being long and flat like those of the sea tortoise; it is the largest of all the land tortoises, its carapace being sometimes three feet long. Of the three species which are here, the two largest were given by M. de l'Etang, and the smallest which weighed one hundred pounds was kept sometime alive in our menagerie.

Amongst the tortoises which are fastened to the wall or placed on the cornice, we shall only remark the following: the *testudo radiata* from New Holland; the great emyd with a large back, from Cayenne, (*emys expansa*); an enormous carapace of the *tyrse* or soft tortoise of the Nile (*trionyx ægyptiacus*), which renders the most essential services to the Egyptians, by devouring the young crocodiles immediately after they are hatched; it was brought from Egypt by M. Geoffroy Saint-Hilaire; the *testudo fimbria*, the carapace of which is covered with pyramidal points, it inhabits the fresh water ponds and rivers near Cayenne, where it conceals itself under the leaves of aquatic plants, exposing only the extremity of its nose, which from its length resembles a proboscis; in this position it awaits the approach of young birds and small aquatic animals, which it seizes as they pass near it. The *testudo fimbria* differs from all other species in this respect, that its mouth, which is very wide, instead of being terminated by a horny substance, is formed with fleshy lips as in *batracians*.

Let us now examine the reptiles belonging to the order of the *saurians*, which are attached to the ceiling or the wall. The largest is the crocodile of the Nile, which it appears, inhabits also the rivers of Africa, and even those of Madagascar. The length of the specimen we see here is 13 feet, but they are sometimes found 25 feet long. Both on account of its strength and its voracity the crocodile is the most dangerous of all the animals of this order; its enormous mouth, extremely wide and extending beyond the ears, is armed with conical teeth; it is covered all over with an impenetrable coat of mail, and it is impossible to wound it, except in the belly, or in the intervals between the scales of its cuirass. They are seen in great numbers on the banks of rivers, sometimes they lie on the shore, and sometimes keep concealed under the water with their nose out, and ready to dart with rapidity upon the animals which pass near them. The females lay their eggs on the sand, and cover them with leaves ; they are hatched by the heat of the sun ; the young ones betake themselves to the water as soon as they quit the shell. One method employed for catching the crocodile is, to dig across its path a deep ditch, and to cover it with branches and leaves; it can also be caught by means of a bait, under which is concealed a strong hook which sticks into its palate ; the bait is attached to a long rope, with which the animal, when weakened by the loss of blood, is drawn out of the water. It is said that the negroes eat the flesh of the crocodile; but it smells so strongly of musk, that it is quite revolting to the stomach of an European. The largest of the crocodiles, after that of the Nile (1), is the one

(1) The crocodile can be tamed when abundantly supplied with food. Worship was offered to this reptile in three of the principal cities of ancient Egypt, Memphis, Thebes, and Arsinoë ; which last was surnamed Crocodilopolis. They used to bring up one of these animals in a lake; it was fed by the priests, and called *souchi,* as the sacred ox was called *apis.* Precious stones were fastened to its ears, and after death it was embalmed, and placed in the tombs of the kings. Throughout the rest of Egypt it was held in detestation, and the ichneumon (*viverra ichneumon*) was worshipped on account of the war it made on the crocodile by destroying its eggs. It is difficult to explain the cause of this absurd superstition, which continued as late as the third century of the christian æra. But what is worthy of remark is, that a similar superstition is practised in Java. The inhabitants of that country go out to meet the crocodile, offer him presents, and

with a slender muzzle (*crocodilus acutus*). This species is peculiar to the Antilles and South America. M. de Humboldt saw a prodigious number of them in the river Orinoco, and he measured one which proved to be 23 feet long. It is a singular fact, that in extreme heat they become torpid like our lizards in a cold winter. M. Descourtils, who observed this species in St. Domingo, says, that the female deposits its eggs in the sand, comes in quest of them when they are ready to be hatched, and defends its young with great courage.

At the side of the crocodiles we observe the gavial (*crocodilus gangeticus*), which is remarkable on account of the extreme length of its muzzle; it inhabits the Ganges, lives on fish, and is perfectly inoffensive to man.

The *saurians* attached to the wall are: the bicarinated crocodile from India: the pike muzzled caïman, which inhabits the waters of the Missouri and Mississipi: the caïman with bony eyelids, from Cayenne: the ouaran of the Nile (*lacerta nilotica*), which was brought from Egypt by M. Geoffroy Saint-Hilaire; this great lizard feeds on the eggs of the crocodile, and we

crown him with owers. M. Leschenault witnessed this singular ceremony. The species which is the object of it is the bicarinated crocodile.

see it often represented on Egyptian monuments:
the elegant tupinambis of Java and the neighbouring islands, brought by M. Leschenault; this
traveller states, that these animals assemble in
considerable number on the banks of rivers,
where awaiting the approach of quadrupeds that
come to quench their thirst, they make a joint
attack upon them, drown and devour them:
the *dragonne*, sent from Cayenne by M. Martin;
it is an excellent swimmer, and retires into burrows by the waterside: the safeguard of America, a beautiful lizard spotted with blue on a
black ground, with rings of the same colour on
its tail; the flesh of this and of the former is
much esteemed: lastly, the iguana of South America, so remarkable on account of the crest which
it has on its back, and for its indented dewlap.
This animal is most always found on trees, where
it attracts attention by the brilliancy of its colours; it lives on insects and vegetables, is of a
gentle nature and easy to tame; but during the
pairing season the male is fierce, watches constantly over the female, and becomes furious if
any one approaches her. The iguanas are highly
esteemed, and are hunted for the sake of their
flesh and eggs, which are considered an excellent
food; one of the species we see here has on this
account been named *iguana delicatissima*. The

cases contain both the young and eggs of the species we have just mentioned; but before describing them we must examine the serpents, which like the crocodiles and tortoises, have been placed against the wall.

The largest species are: the *boa anacundo*, sent from Cayenne by M. Banon; the *amethistine python*, or the *ular-sawa* of Java, brought by M. Leschenault; and the pithon of Senegal.

It is said that these serpents grow to the length of 30 feet; the longest of those in the cabinet is only 19. They inhabit swamps and marshes, and feed upon quadrupeds, which they swallow entire, after having crushed them to death by writhing themselves round them; they smear the body all over with a glutinous slime, and then take it little by little into their mouth, which is susceptible of great dilatation, their jaw bones not being articulated, but simply connected by elastic ligaments. In this manner they swallow antelopes, goats and deers; and their skin is capable of expanding sufficiently to contain them. When the animal they have thus taken in, exceeds their own size, they remain in a torpid state, and are several days without eating.

Amongst the serpents of an inferior size which are stuffed, we shall mention, as worthy of remark, the *rattle snake,* or *boiquira,* sent from New York by M. Milbert; it is five feet long. This serpent is considered as more venomous than any other; it is called the rattle snake, because its tail is terminated by hollow scaly and horny processes, loosely jointed together, and which on being shaken, occasion a sound very much like that produced by dry seeds in their envelope. The yellow or spear-headed viper (*trigonocephalus*) from Martinico, a very dangerous serpent, given by M. Servile; this specimen is 8 feet 9 inches long. The *lachesis* of Cayenne (*crotalus mutus*), brought by M. Poiteau, the tail of which is terminated by a horny, very hard, and sharp point.

Let us now go round the room from left to right, in order to point out the principal objects contained in the cases. We shall follow the scientific arrangement, which is also that observed in the distribution of the genera in the cabinet. We shall state the number of species in our possession belonging to each genus, and dwell upon those only which appear to deserve peculiar attention.

The tortoises are divided into *land tortoises,* or *emyds; box* tortoises; and trionyx, or soft tortoises.

In the Museum are sixteen species of the land

tortoises. The most common is the *Grecian,* which is found in Italy, Sardinia, etc. It is of this species that broth is made for consumptive persons; its carapace is much arched, and is seldom more than one foot in length; it feeds on leaves, fruits, and snails; passes the winter under ground, pairs in the spring, and lays from four to six eggs. Amongst the exotics we may mention the *geometrical* from the Cape, and the *spotted* from South America. The carapace of the latter is speckled with a variety of colours.

The *emyds,* or fresh water tortoises, are web-footed, which enables them to swim with great facility. We have twenty species of them; amongst which the *yellow* and the *mud* tortoises are European, and found in the temperate regions as far north as Berlin. They assemble in numbers, and feed upon frogs and fish. They are kept on account of the quality of their flesh, which is excellent food. The most remarkable amongst the exotic species, for its varied and beautiful colours, is the *testudo pulchella.*

The box tortoise is so called from the peculiar formation of its breastplate; the front part of which, and sometimes also the hind part, move backwards and forwards upon the centre as if they were hinged on it; so that when the animal has drawn its head and feet under the carapace, the two moveable parts of the breastplate are brought close to it, and the tortoise is shut in as completely as if it were in a box. The Museum possesses five species of this kind of tortoise: we shall only mention two; the striated tortoise (*t. virgulata*) from South America, and the *couro*, brought from Java by M. Leschenault. We have already seen the most remarkable species of the sea tortoises, or *cheloniæ*. Some young and eggs of the *caretta mydas* and *coriacea* or leather tortoise are placed in their regular order under the *emyds*.

The last genus of this family is that of the tortoises whose carapace and breastplate are covered with a soft skin; we have five species of them; they live in fresh water and feed on aquatic animals. M. Geoffroy Saint-Hilaire has named them *trionyx*, because they have only three claws on each foot. There are several young of the soft tortoise of the Nile, called *tyrse;* a large carapace of which species is placed against the wall. The soft tortoise of America, called the ferocious, lies in ambush in order to seize on ducks and other water fowl, which it drowns, and devours under water. Here ends the first order of reptiles, *chelonians* or tortoises, of which there are a great number in the cabinet to exemplify the differences resulting from age and the varieties.

We proceed to the *saurians*. The first genus is that of the crocodile, of which we have twelve species, belonging to the same family as those we have already noticed in our inspection of the large specimens attached to the ceiling; but here they are of a smaller size and of different ages; some which were just hatched are preserved in spirits of wine. We also see on the shelves the eggs of the crocodile: these eggs are of a regular form, their shell is granulated, and their smallness, when compared to the bulk of the full grown animal, is quite surprising. In the second genus are comprised the *monitors*, or tupinambis, of which we have already spoken; the Museum possesses fifteen species of them.

After the tupinambis come the *lizards;* of these there are in the Museum fifty-three species; nearly one fourth of which belong to the climates of Europe. The greater number have been preserved in spirits. This tribe is remarkable for its swiftness and agility; they retire to the clefts of rocks and walls, or to holes under ground; they feed on flies and other insects, which they will not eat unless they are living. The least shock breaks off their tail, and when this happens a new one grows out, and soon acquires the length of the old one. The lizards are in general speckled with vivid and varied colours,

which become dull after death. The *lacerta
agilis*, found on our walls in France, is the most
common; it feeds on worms and small insects,
shaking them rapidly before it devours them.
The female lays from eight to ten eggs, which
are hatched eleven or twelve days afterwards.
But the most beautiful amongst the lizards of
Europe, is the *lacerta viridis*, or green lizard,
commonly found in Spain and the south of
France; its length is more than one foot; its
body is regularly striated, and speckled with
black on a fine glossy green ground. There
exists still a sort of confusion in the systematical
arrangement of the lizards, which are very numerous, and the species of which are so nearly
allied, that it is sometimes difficult to ascertain
them. It would be desirable that a monograph
on this family were written; for the completing
of which, the collection now in the Museum
might afford no inconsiderable help.

The fourth genus is that of the *stellio*, of which
there are thirteen species in the Museum. They
are distinguished by rings formed of erected and
sharp scales on their tail. The stellion of the
Levant (*lacerta stellio*), so common in Egypt, furnishes, according to Belon, the matter used for a
cosmetic, known under the name of *cordylea*, or
stercus lacerti. The mahometans kill this animal

wherever they find it, being displeased with the motion of its head downwards, which they consider as a gross imitation of them when they pray. The species named *cordyla* is covered with very hard scales, and inhabits the Cape; that which is called the *whiptail* of Egypt, is found in the deserts of that country, and reaches sometimes two or three feet in length.

The *agames,* the name given to the fifth genus, are found in every part of the world; there are twenty-one species of them in the collection; some are covered with small rhomboïdal scales. The varying or changeable *agame,* discovered in Egypt by M. Geoffroy Saint-Hilaire, changes its colour with still more rapidity than the chamæleon.

Next to the *agames* are placed the *basilisks*, or long-tailed lizards (1), the only species of the saurian tribe with radiated dorsal and caudal fins. We know but two species of this genus, living in fresh water ponds and rivers of the East India islands; they are remarkable for their form, but we are totally ignorant of their habits; there is some reason to believe that they feed on vegetables.

(1) The basilisk, properly so called, is in no respect the basilisk of the ancients. This was the serpent of which so many ridiculous stories have been reported.

The seventh genus is that of the *dragon* (*draco volans*), to whom a lateral extended membrane gives a very different aspect from that of other lizards. This kind of wing, of a peculiar mechanism, although it does not adhere to the legs of the animal, as in the flying mammalia, can nevertheless be extended at pleasure, and must aid considerably in the act of springing from one tree or branch to another, or when it pursues insects. The three species in the cabinet were sent from the East Indies; their eggs are quite round, and not larger than a common pea.

The eighth genus is formed of the *iguanas*. They are very large lizards, some of which measure more than six feet from the tip of the nose to the extremity of the tail, which is sometimes twice the length of the trunk. They climb trees with an amazing agility, and feed on leaves and seeds. Eight species of them have been sent from India and America.

After the iguanas come the *anolis*, of which there are in the cabinet fourteen species, all natives of America. The striated swelling they have under each toe, and which is formed by the widening of the second phalanx, enables them to walk on the most polished surface, and even to run on the ceilings of apartments. They have under the throat a sort of wen or swelling, which when inflated, especially at the pairing season, shows the most vivid colours, and is as large as a cherry in the *a. bullaris.* These lizards change their colour in the same way as the chamæleon, and when two males meet they are said to fight with great fury.

M. Cuvier has made several important divisions in the very numerous genus of the *geckos,* which comes next.   They inhabit the warm regions of both continents, and are allied to the anolis in this respect, that they have the same striated swelling under their toes, but with this remarkable difference, that it occupies their whole length, and enables them to walk with still more facility on ceilings; they differ also from the *anolis* by the flattened and elongated shape of their body, which is more like that of the lizards. They have a kind of heavy and crawling walk ; their eyes are large, and are affected by light in the same way as are the eyes of cats, and their claws are also retractile like theirs.   There are forty-six species of *geckos* in the collection.   The two most commonly found in the southern countries of Europe, and which are considered as venomous in a high degree, are the *lacerta mauritanica* and the *l. gecko.*   We are assured that the skin becomes red from a mere contact with them, and that on this account the second species is, in Egypt, vulgarly called *the father of leprosy*.

The eleventh genus in the present systematical arrangement of the saurians is composed of the chamæleons. These reptiles are celebrated from the facility with which they can change their colour. It was long believed that they could receive and transmit the colour of the object they stood nearest to; but it is now well ascertained, that they owe this faculty to the capacity of their lungs; and that according to the quantity of air they take in, their blood acquires a more or less vivid hue, by which their transparency is augmented or lessened. This phenomenon takes place by the simple act of breathing, but is more striking in the pairing season. The chamæleons have a very extraordinary appearance, if we consider their flat body, angular and sharp on the back; the pyramidal helmet on their head; their long prehensile tail; their fingers, five in number, but separated, three on one side, and two on the other; and the mobility of their eyes, which they can move in all directions, and keep open or shut the one independent of the other. Their tongue, which they can stretch and move with vivacity, has a glutinous covering, by means of which they catch insects. These animals keep constantly on the trees. Fourteen species of them are in the collection, some from the south of Europe and from Africa, others from the East Indies. The European species is nearly a foot long, whilst that of the Cape is very small, and the Molucca species has two long prominences on the muzzle.

The *scinks*, which are the last in the order of the saurians, have very short legs, their body shaped as a spindle, is covered with fish-like scales, and has no contraction at the head and tail. The family of the scinks, of which there are forty-five species in the collection, is very numerous, and inhabits the warm countries of both continents. It has been distributed according to the number and situation of their feet into five genera : the scinks proper, the seps, the two-footed scinks, the *chalcides*, and the scinks with two hands.

The Nubian and Abyssinian, or common offici-nal scink, is celebrated for the quickness with which it burrows in the sand to conceal itself from its pursuers, and also for its medicinal powers. It was long employed as a specific remedy for restoring the enfeebled vital functions, and enters into the compound medicine known by the name of *theriaca*. The species of scink found in the West India islands is more than a foot long, and is as thick as the arm ; the planters name it the *land-pike*. Out of the four species in the collection, one is African, and has five fingers to each foot; the second, which was brought from New Holland by Peron, has only four; the third species found in Italy is viviparous, and has but three; and the fourth, which is from the Cape, has only one finger to each foot. The species belonging to the genus *chalcides* present the same irregularity.

The two-handed scinks, of which we know but one single species from Mexico, want the hind feet, whilst that with two feet wants the fore ones. The only species we know of this last genus are: the *lepidopode*, brought from the Cape by Peron, with two scales at each foot: a species discovered in the Brazils by M. Auguste Saint-Hilaire; and the *sheltopusik of Pallas*, from the banks of the Volga, the feet of which are reduced to a very short scaly and scarce visible appendage. This specimen was given by M. d'Urville; it is the largest of the three species, and is said to be sometimes six feet long.

The mere inspection of these animals will readily satisfy us, as to their forming the passage from the saurians or lizards to the ophidians or serpents, which come immediately after in the general classification adopted by M. Cuvier.

The two first genera of this order, those that

are more nearly allied to the saurians, are the slow-worms (*ophisaurus* and *orvet*, Daud.). The glass slow-worm (*anguis ventralis*, Sh.) inhabits the southern states of North America: it is remarkable for the deep furrow which separates its dorsal and ventral scales; its tail is much longer than the abdomen, and it breaks with such facility, that it has been named on this account the *glass snake;* and the same remark is made of the common slow-worm (*anguis fragilis,* Sh.). These animals feed on worms, insects, and young snails. There are in the collection several foreign species of this family; that which comes next is composed of the ophidians or serpents. The *amphisbœnæ,* which constitute the first order amongst them, are much allied to the *chalcides* by the verticillated arrangement of their scales. The Museum possesses several varieties of amphisbæne; all are inhabitants of the warmer regions of America, where they are found concealed in the nests of ants.

We shall now describe the ophidians, the collection of which amounts to five hundred and fifty specimens belonging to upwards of two hundred species. The greater number have been preserved in spirits of wine, and are contained in long cylindrical glasses, through which they can be perfectly well seen. This mode of preservation is undoubtedly preferable to the stuffing of the skins, as it maintains the natural form of the animal, whilst it injures its colour but little.

The *ophidians* are divided into two tribes; those that are not poisonous and those which are; the upper jaw of which last is furnished with fangs implanted into a gland which contains the venom: these fangs being hollow admit the liquid, and carry it into the wound they make. We shall only cite the species which are most deserving our notice. The largest amongst them are the *boas* and the *pithons.* We possess fourteen species of the first, and three only of the last. Young individuals of the species we have seen on the wall are placed in the cases.

The genus *coluber* is the most numerous of all. We shall mention only five species amongst those that inhabit our climate. 1st. The ringed snake (*coluber natrix*), lives in the meadows, and feeds upon frogs; it is kept in Sardinia for the purpose of destroying the mice; women and children often play with it, and in some parts of Italy it is dressed for eating under the name of the *bush-eel.* 2d. The French snake (*c. atrovirens*), which acquires sometimes four feet in length, and is easily tamed. 3d. The smooth snake (*c. austriacus*). 4th. The dun snake (*c. elaphis*), which is from five to six feet long. 5th. The æsculapian snake, which in all probability is the same with the serpent of Epidaurus of the ancients. This species is still more easily tamed than the preceding. Near it has been placed an old skin of one of them, to shew the manner in which they cast it, rolling it outward from the head to the tail.

Amongst the foreign species we shall cite the *ibiboca* of India; the coach-whip (*c. flagellum*), whose length is three feet, and its diameter only a few lines; the iridescent snake (*c. ahœtulla*). No reptiles have such vivid colours as these two, their scales show the varied hues of gold and precious stones; the iridescent is of a surprising agility; children in India play with it, and put it round their neck. And lastly, we shall cite the long-snouted snakes (*c. mycterizans* and *c. nasutus*), very remarkable for their very sharp and long snout.

We shall now pass on to the poisonous serpents, whose bite is so dreadful and mortal. Some have a flattened tail, and inhabit the Indian seas, where they prove most fatal to fishermen, when they happen to take them in their nets. They are named *hydrus*, or water snake. There are two species in the collection; one of which, the black-backed hydrus (*anguis platura*, Lin.), although it is extremely poisonous, is eaten by the natives of Otaheite. The most renowned amongst the land serpents is the rattle snake (*crotalus*), of which the Museum possesses four species. The banded rattle snake (*c. horridus*), common in the United States, and the striped rattle snake (*c. durissus*) of Guyana, are the largest, and measure sometimes six feet; their bite proves mortal in a few minutes. The vipers and analogous genera, such as the *trigonocephales,* twelve species of which are in the cabinet, are classed immediately after the crotalus. The specimen of the *megæra,* which we see here, was caught in the act of swallowing a very large frog which it had not yet entirely taken in, it shows at once the length of its fangs, and the disproportionate size of the prey it can eat. Next to these is the *platurus,* much resembling the *hydrus* from its flattened tail and poisonous fangs. The only species we know inhabits the Indian seas. The *naja* is a very remarkable snake, from the thickness of its neck and the straightness of its cervical ribs. We have two species of these serpents: they can draw their head in and out of their neck, and exhibit the most fantastical attitudes. The *coluber naja,* or spectacle snake, is so named from the large spot on its neck much resembling the figure of a pair of spectacles. The Portuguese call it *cobra de capello.* The Indian jugglers, after they have deprived it of its fangs, play with it, and manage it so as to make it assume a sort of dancing motion. The other species is the *coluber haje* from Egypt; it was procured by M. Geoffroy Saint-Hilaire. It is said that by simply pressing on its head, this snake can be deprived of all motion. This species is thought to be the asp of the ancients.

The common viper (*c. berus*) is found at Fontainebleau, near Paris. The individuals of this species vary so much, as to the arrangement of the black and white spots and stripes on their skins, that they have been taken for as many species, whilst they constitute but one and the same. The *cerastes,* or horned vipers, are distinguished by a pair of curved processes, situated above the eyes; they are frequently seen represented on Egyptian monuments. This snake has a great resemblance in form and colour to the *erix* of the Turks; but this being deprived of horns and venom, the Egyptian jugglers engraft the spurs of birds on its head, and exhibit it afterwards for a *cerastes,* to show how they can prevent the effect of its poison on themselves. A specimen arranged in this way is seen in the collection. The last genus amongst the ophidians is the genus *cæcilia;* thus named from the smallness of their eyes. The two species in the cabinet inhabit Guyana, and are found in the nests of ants. The

reptiles comprised in the third and last division
are the batracians (1). Their body is destitute
of scales: their form and habits are for the first
age, those of the fish, and their feet appear but at
a second period. They are divided into two
families, and each of these into four genera. The
batracians belonging to the first family have no
tail; they are the frogs (*rana*), the tree frogs
(*hylæ*), the toads (*bufo*), and the *pipas*. The salamanders, the tritons, the sirens, and the proteus,
compose the second family. All the species belonging to the first are oviparous; the young or
tadpoles have no feet, live in the water, and
respire by means of *branchiæ* (2); their head is
very large. As they grow in size the *branchiæ*
become obliterated, and drop off; the feet shoot
out, and the tail shortens and disappears totally.
We may observe here tadpoles of almost every
species, and in every degree of their developement. Their size varies according to the state
of metamorphosis, and according as the time at
which that takes place approaches the period of
their birth. The vital strength with which these
animals are endowed is very great, and has fur-

(1) From the Greek *batrachos*, which signifies a frog.

(2) *Branchiæ* is the name given to the lamellated organ, situated on
each side of the head of fishes, and partly exterior, by means of which
they separate and respire the air contained in the water.

nished the most interesting experiments on the phenomena of the animal economy. It was by dissecting frogs that galvanism was discovered. The Museum possesses more than twenty-five species of frogs. The two most common in France are the green and brown frogs: the thighs of the first are esteemed good food. The American species, called the bull-frog, is four times larger than the green, and feeds on water-fowl; which it seizes by the feet, and drags under water. A specimen of this voracious frog is here seen, preserved in spirits of wine, into which it was thrown when taken, at the time it was endeavouring to swallow a duck, one half of which had not yet been gorged. The Cayenne frog, which is called *jackie*, presents the singularity of being smaller when at the complete state of frog, than when at that of tadpole, owing to the transformation; for when it takes place, the *branchiæ*, skin and tail, fall off at once, and consequently the animal is reduced to a smaller volume. To this difference of size, existing between the perfect animal and its tadpole, may be attributed the errors some people have fallen into, in taking the tadpole for the full grown frog, and believing this last was changed into a fish. The tree frogs (*hylæ*), which come next, differ from the *rana*; they have the last phalanx of the fingers widened into a disk, by means of which they can walk and fix on smooth surfaces with their body downwards. More than thirty species of this genus are in the collection. That which is more commonly found in France is of a green colour, with a yellow and black stripe on each side; it climbs on trees and fixes on the leaves, and it does not pair before four years old. The colours in the foreign species are generally more varied; the most celebrated amongst them is the frog, the blood of which is employed for colouring the feathers of birds, and which we mentioned in speaking of the parrot of the Moluccas.

The toads (*bufo*) are more squabby than the frogs; their body is covered with a sort of fœtid humour; they are the most hideous and disgusting of the reptiles, but are not venomous. Of the thirty species in the collection, we shall only mention the four following amongst the European. The common toad (*b. vulgaris*), which is found in dark recesses, and pairs in the water: its female is of a surprising fecundity; the spawn is enveloped in two gelatinous strings from 20 to 30 feet long, which the male draws off with its legs. The *bufo calamita*, or rush toad, which has a strong smell of gunpowder; lives in the clefts of walls, and takes to the water only in the pairing time. The alliaceous toad (*rana alliacea*), which smells of garlick; its tadpoles are very large, and are eaten as fish in some countries. The *bufo obstetricans*, common in the rocky soil near Paris, helps the female to exclude its ova: collects them round its legs, after having covered them with gluten, and carries them about until the time of their birth draws near, when it seeks for a pond or ditch of stagnant water; the young then come forth and swim instantaneously. Amongst the foreign species are: the *b. agua*, its body is from 8 to 10 inches long, and covered with warts as large as beans; the horned frog (*b. cornutus*), which has a kind of callous process, resembling a horn, on each eyelid; and the crested toad (*b. margaritifera*). We know but one species of the genus *pipa*, or toad of Surinam; it inhabits the dark and humid parts of houses; it has a flattened body, and its way of reproduction is so different from that of the whole tribe of frogs, that it ought to be mentioned here. When the spawn is milted, the male places it on the back of the female, which goes instantly to the water; there its skin swells, and a great number of cellular tubercles are formed in it; the young are hatched and undergo their metamorphosis in these cells, in which they remain. When their legs are well formed.

the female leaves the water. We have several specimens of this toad preserved in spirits of wine.

The first genus of the second family is that of the salamander. This animal, when in its perfect state, has a cylindric tail; when young, the branchiæ are loosely floating on the neck, and the fore feet are formed before there is any trace of the hind ones. The land salamander is viviparous; it lives in holes, and deposits its young in some neighbouring pond. The salamander is furnished with pores and small foraminæ through which exudes a fœtid liquor, considered as venomous. It has long been believed that it could resist the action of fire; but it is now ascertained, that it can be consumed in the flames as well as any other animal.

The tritons belong to the second genus; their tail is compressed, and they inhabit the water. If a limb, or the portion of a limb, is severed from a triton, it is replaced by a new one. There is in the collection a specimen, which lived four months after its head had been cut off: it was kept in a bottle filled with water, which was renewed every day, and the wound was completely healed. When a triton happens to be enclosed in the ice, it remains in it as long as the winter lasts, and swims off the instant the ice is melted. There are more than twelve species of this genus in the cabinet; the largest of which is from 15 to 18 inches in length, of a deep blue colour, and inhabits America. Amongst the tadpoles of the tritons we shall observe that of the crested triton; the branchiæ of which remained on the animal, although its metamorphosis had been in other respects completed.

We see close to the salamanders and tritons the axolotl of Mexico, presented by M. de Humbolt. M. Cuvier, who published a description of this animal, after having dissected every part of it, could not decide whether it was a perfect animal, or a large species of salamander. Next to the axolotl, is the anguine siren (*proteus anguinus*); it retains during life the external branchiæ, which the salamanders preserve only when they are young. This singular animal inhabits the subterraneous waters of the duchy of Carniola in Austria. Dr. Schreibers, director of the imperial museum of Vienna, made our Museum a present of a good representation in wax of the proteus anguinus, several specimens of which are preserved in spirits of wine. The last genus of the siren, is the eel-shaped siren; its branchiæ are exactly like those of the proteus; but it has no hind legs; it inhabits the rivers of Carolina, and feeds on insects. The specimens we see here were sent by M. L'Herminier.

§ VII. COLLECTION OF FISH.

COLLECTIONS of natural history are generally designed to unite the most interesting and the most curious amongst the productions of nature; to show in each family the objects best calculated to give an idea of its characteristic forms, and those which indicate the link between one family and another; to establish the principles of classification; and also to show the natural state and the origin of the different substances circulated by means of commerce. To this end they must be well classed, but it is not necessary that they should be very numerous. A garden containing fifteen hundred well chosen plants is sufficient to give an idea of the vegetable kingdom; and all that is essential to be examined, in order to have a notion of zoology, may be contained in one room.

But the collections of the Museum are destined to a much more extended and important purpose: namely, the progress of science. The object for which they were formed, that towards which they ought to tend, is to assemble the greatest possible number of species, taken at different ages and in different places, in order to determine the essential characters which, being common to several species, unite them into genera; to afford, in short, the means of considering beings under every relation they bear to each other and to man; and to enable us to write their history.

Considered in this point of view, the collection under our inspection is not only the most considerable that ever existed in this class of animals, but also the most complete of zoology in general. It comprehends about five thousand specimens, belonging to more than two thousand two hundred species; that is to say, a number nearly double that of the species distinctly described and figured by naturalists. It offers the elements of the classification which M. Cuvier has established in his *Règne Animal;* the type of the ichthyological memoirs which he has inserted in our Annals; the far greater part of the fishes which M. de Lacépède has described or figured in his great work, and almost all the known genera. Of each species, it possesses generally one preserved in spirits of wine, which affords the facility of examining its interior organization in case of necessity. The greater

number of those which are dried have been covered with a varnish which has revived the
colours, and they appear almost as brilliant as
they were some hours after being taken out of
the water.

This collection has been newly arranged according to the method of M. Cuvier, and all the
species have been ticketed with the greatest
exactness. The oldest specimens are those found
by Commerson, at Madagascar, the island of
Bourbon, and the isle of France. When, after
the death of that traveller, they arrived at the
King's Garden, there was no place in which to
exhibit them, and they remained shut up in the
boxes in which they had been sent, and were
in a manner forgotten : fortunately Commerson
had made drawings of them, and it was from
those drawings, and the notes which accompanied them, that M. de Lacépède described them
in his history of fishes. We feel the advantage
of possessing the originals in studying the work
of the historian. The collection was afterwards
enriched by MM. Peron and Lesueur, and by the
other naturalists who accompanied captain Baudin to the Pacific ocean. More recently it has
been considerably augmented by collections sent
from New York by M. Milbert, from South Carolina and from Guadalupe by M. L'Herminier,

from Martinique by general Donzelot and M. Plée,
from Brazil and the Cape by M. Delalande, from
Coromandel and from Bengal by M. Leschenault,
and from different places, particularly the Ma-
rian islands, the Sandwich islands and New
Guinea, by the naturalists embarked with captain
Freycinet. Through all these channels a great
number of new species have been obtained.
Whilst other travellers were sending us the
rarest productions of the seas, lakes and rivers
they had visited in the most distant countries,
we have not neglected on our part to procure
all the species of France, Germany and Italy, from
the Northern ocean and from the Mediterranean; and although the fish of these seas had
long been studied, they have afforded us many
species which naturalists had confounded, because they had not been able to collect in order
to compare them. This collection has only been
formed a few years, and we have no doubt but
it will soon be augmented. What we have said
is sufficient to show its importance; but it is impossible on a superficial view to appreciate its
value; for its greatest utility is not that which
attracts the eye, it can only be felt by the naturalist. We shall confine ourselves to pointing
out the order in which it is disposed, showing
the principal differences which exist between

the forms of different families, which forms are more varied and more singular than in the reptiles; and noticing the rarest species, and those which are most deserving of attention either on account of their singularity or their utility.

Although this collection is extremely numerous, it is not convenient for those who are beginning the study of ichthyology. The objects cannot be all placed at the proper distance from the eye; and of one genus, which sometimes comprehends a hundred species and two or three hundred individuals, we cannot at first sight distinguish those which present the most essential characters. The greater number of the specimens are preserved in spirits of wine, in which they may be seen very well, but cannot be examined without holding in the hand the bottle which contains them. As the Museum ought to facilitate the study of the elements as well as that of the details, a selection of one individual of each genus has been made from the general collection; this, as it is arranged and named after the method published by M. Cuvier, presents the genera and sub-genera which have the greatest affinity with each other. It occupies twelve large frames, placed in the passage to the great room. All the specimens are dried with the greatest care, and so as to facilitate the examination of them.

On the floor of the great room is placed the basking shark (*squalus maximus*); several other specimens of which of considerable dimensions are attached to the ceiling; we shall point them out when we have occasion to speak of them in their regular order. In this as in the preceding rooms we shall follow the cases from left to right.

The fishes have been divided into two great series and into eight orders by M. Cuvier. We shall not touch upon this classification, of which full information may be obtained by examining the collection destined for study. We must follow the natural families, calling the attention to those genera and species which seem to us to merit it the most.

The first family is composed of two genera: the lamprey (*petromyzon*, Lin.,) and gastrobranchus (*myxine*, Lin.), united under the denomination of suckers, because they attach themselves closely to different bodies, fixing upon them their round and fleshy mouth, and their tongue which acts as a piston. We have eight species of lamprey. The only one which is sought after is the great lamprey; it is caught at the mouth of rivers, which it ascends in the spring: the other species live in fresh water. The fishes of this genus fix themselves upon others in the manner we have described, and penetrate into their bodies, which they devour by degrees with the sharp and pointed teeth placed at the bottom of their mouth. There are but two species of *myxine* known : the one, the blind gastrobranchus (*myxine glutinosa,* Lin.,) inhabits the North seas; the other, which M. de Lacépède has named dombeyan, because Dombey discovered it, is very rare. M. Delalande however has brought us several specimens from the Cape of Good Hope.

The second family, that of the selacians, is composed of the sharks, rays, and fishes allied to them. It comprehends a great number of species which deserve our attention, either from their gigantic size and their voracity, or from the use which is made of them in the arts. The first genus is that of the sharks commonly known by the name of dog-fish; their rough skin is employed in polishing different surfaces, such as wood, ivory, etc. M. Cuvier has divided them into twelve sub-genera. We have forty-one species of them; the largest is the basking shark (*squalus maximus*), from the North seas, which is sometimes upwards of 3o feet in length. The specimen in the middle of the room was cast upon our shores by a north west storm. The species which is most celebrated for its voracity is the white shark (*squalus carcarias*); its mouth is armed with a great number of teeth placed in five or six rows; these teeth, which are extremely hard and white, are in a triangular form terminating in a sharp point, having the sides cut like a saw. After these follows the sub-genus of the hammer headed sharks (*s. zygaena,* Cuv.), very remarkable for the form of their head, of which there is not another example in the animal kingdom; it is flattened, and as it were truncated, with the two sides extended like the head of a hammer, and the eyes at the extremities. We have five species of the *zygaena;* that which is most common in our seas is sometimes 12 feet long.

The genus most allied to the sharks is that of the saw-fish, whose distinguishing character consists in a very long depressed snout in the form of a beak, furnished at each side with strong prickles fixed in the manner of teeth. This beak is a powerful weapon, with which the saw-fish attacks the largest fish without fear. We have five species of them; on the ceiling is a large specimen of the most common species (*pristis antiquorum,* Lath.). The saw-fish connects the sharks and the rays, which M. Cuvier has divided into seven sub-genera, and of which we possess fifty-seven species. The first sub-genus is that of the lengthened shape ray (*raia rhinobatos*), whose body resembles that of the saw-fish, and of which a species is found in the Mediterranean; the others inhabit the seas of South America, the Cape of Good Hope, and Coromandel. The torpedoes or cramp rays, which form the second sub-genus, have the head flattened in the form of a quoit, the mouth large, short and fleshy. These forms ally them on one side to the rhinobatos, and on the other to the rays properly so called.

The torpedo rays are celebrated for the faculty which they possess of giving violent shocks to those who touch them. The organ which performs the part of an electrifying machine is an apparatus of tubes divided by diaphragms into little hexagonal cells situated near the head, before the pectoral fins.

The torpedoes give shocks to the animals which disturb them; and it appears that in this manner they benumb those which they intend to make their prey; they have no other weapons. Torpedoes are to be found in almost all the seas. The largest, and that which gives the most violent shocks, lives upon the coast of the Cape of Good Hope. The torpedo with five spots, the marbled torpedo, and the torpedo of Galvani inhabit our coasts, but it is only lately that they have been distinguished from their congeners.

The other species of the genus ray have the body extremely depressed, and very wide, with a long filiform tail. Their teeth, which are of different forms, serve to distinguish the sub-genera; in certain species the tail is smooth, in others it is armed with long prickles, furnished on each side with very deep notches. These weapons are not venomous, but they inflict wounds which it is very difficult to cure.

The thorn-back (*raja clavata*), and the rough ray (*r. rubus*), which are common in our markets, are the species most esteemed as food.

Of those which inhabit our seas, the skate (*raja batis*, Lin.,) arrives at the greatest size; it has been known to weigh more than two hundred pounds. Amongst the foreign species, one of the most remarkable is the pearled ray (*r. syrhen*), which belongs to the sub-genus of the *pastinacæ;* its body is covered with close set and hard tubercles; after they have been worn on a millstone, to polish them and render the skin smooth, it is used for covering various kinds of cases, and called shagreen. This species of ray is found on the coast of Coromandel, from whence it was sent by M. Leschenault.

The Mediterranean furnishes also a gigantic

species (*r. cephaloptera*), which from the singular form of its fins has established the sub-genus cephaloptera ; its head is truncated, and the pectoral fins, instead of uniting with it, are extended and give the animal the appearance of having two horns; it is the giorna ray of Risso, *Ichtyol. of Nice*. On the ceiling is another species from Brazil.

The last species of the selacians is that of the chimæras, of which we have two species: the one from our seas (*c. borealis*), which is called the king of the herrings, because it is found at the head of those wandering fish, is remarkable for having its tail continued into a long and slender filament, and for the short upright process with a fringed tip on its head ; the other from the Antarctic seas (*c .australis*), much resembling the former, but having its upper lip extended into a lengthened cartilaginous appendage bent downwards.

After the chimæras comes the family of the sturgeons (*accipenser*), which is divided into two genera. We have four species of sturgeon: two large specimens of the most common species are attached to the ceiling; of the swimming bladder of this fish isinglass is fabricated, and *caviar*, a dish much sought for amongst the inhabitants of the north, is made of its spawn.

Near the sturgeons is a very rare fish which inhabits the fresh water in North America, whence it was sent to us by M. Lesueur, the foliated polyodon, so called because the edges of its snout are furnished with a small membrane reticulated by a great number of vessels which resemble the fibres of leaves.

The fishes which we have hitherto seen are cartilaginous; those which follow belong to another series, the osseous. The first family is that of the *gymnodontes*, which consists of three genera: *diodon, tetrodon*, and *mole*. The first commonly called spiny orbs, or sea porcupines, have the body covered with long quills; they have the power of inflating themselves into the form of a ball, and float upon the water; the second have their body covered with shorter prickles. We have fifty-four species of them. One of the most remarkable is the *fahaca* of the Arabs (*tetrodon physsa*, Geoff. St.-Hilaire): during the inundations of the Nile a great number are cast upon its banks. The third genus comprehends the species vulgarly called moon-fish: their body is compressed and without prickles, and their tail is very short. Amongst the four species in the Museum, that of our seas, which is of a silvery colour, weighs sometimes three hundred pounds; there are two specimens of it attached to the ceiling.

The next family, that of the *sclerodermata*, comprehends two genera : the file-fish and trunk-fish. We have sixty-six species of the first genus and eighteen of the second : they all inhabit the equatorial seas, except the Mediterranean file-fish, vulgarly called *pesce balestra*. The colours of the file-fishes are varied and brilliant during life, but disappear after death. The trunk-fishes, instead of scales have a long crust, whose hexagonal compartments are soldered into a cuirasse, covering their head and body; their form is like that of a triangular or square box; they are all, except one, armed with prickles differently situated according to the species.

The order of the *lophobranchia* is very remarkable on account of its gills, which, instead of being fashioned like the teeth of a comb, are in the form of a little tuft. This order comprehends three genera : that of the pipe-fish (*syngnathus*), of which we have ten species; that of the sea-horse (*hippocampus*), of which we have four species; and that of the *pegasus,* of which we have only two. Almost all the fishes of this order have the body from one end to the other covered with laminæ or escutcheons which give it an angular form. The swimming pegasus, or the pegasus of the Indian seas, derives its name from its singular form, occasioned by the size of its pectoral fins.

From the *lophobranchia* we pass to the fifth order, which comprehends the greatest number of the fresh water fish, and of which the first family is that of the salmon. This very numerous genus has been divided into many sub-genera, according to the form of their teeth. We have forty-four species in the cabinet: every one knows the salmon, the trout, the smelt, the grayling salmon, etc. Amongst the foreign species we shall only notice the *piraya* (*salmo rhombeus*), which lives in the rivers of South America, and seizes with its sharp teeth the animals which bathe in the rivers.

The second family, that of the *clupeæ*, comprehends a great number of sea and river fish. The principal part of those which inhabit the sea go up the rivers to deposit their spawn. This family has been divided into eight genera; we possess forty-three species. The first genus is that of the herrings, divided into seven sub-genera by M. Cuvier. We have nineteen species of the first division, to which belong the shad, the herring, the sprat, etc. Among the six others is the anchovy, of which we have six species, differing from the herrings in the lengthened form of the .snout. The most remarkable of the *migalopes* is the king-fish of the Carribee islands (*clupea cyprinoides*, Bl.); the beautiful

specimen in the cabinet was sent from Guadaloupe by M. L'Herminier.

The second genus comprehends only two species: one from the East Indies, sent by MM. Sonnerat and Leschenault; the other from South
America, by M. Milbert.

The genus *chirocentrum* consists of only one
species, from the Indian seas, called *sabre* fish on
account of its form; it was presented by M. Leschenault.

We have four species of the fourth genus,
erythrinus, which are commonly found in the
western and eastern seas; and only one of the
genus *amia,* a small fresh water fish, brought
from Carolina by M. Bosc. The genus *vastrea*
also consists of fresh water fish; we have two
undescribed species, the first of which was
brought from Senegal by Adanson, and the other,
a very large specimen, called the gigantic vastrea
by M. Cuvier, was sent from the Brazils.

The two species in the Museum, belonging to
the genus *lepisosteum,* inhabit the lakes and rivers
of America. The first, *esox osseus,* the bony-
scaled pike, is known in North America by the
name of *caiman;* the other is found in the more
southern seas of America, and it has been described by M. de Lacépède.

The eighth and last genus of the family *clupea*

is that of the *polypteri,* a species of which was discovered in the Nile by M. Geoffroy, who named it *p. bicher (nilotic polypterus)*; it is described and figured in the Annals of the Museum, vol. I, p. 57, fig. 5.

The third family, that of the pikes (*esoces*), is divided into three genera. The *pikes proper,* the flying fishes, and the *mormyri.* These are all extremely voracious; the greater part of them live in the sea. We have twenty-five species in the cabinet, which M. Cuvier has subdivided into nine sub-genera; one only of which (the *chauliodus*) is wanting in the collection.

The European pike is universally known. Another fresh water species, from North America, sent to us by M. Lesueur, bears the name of *reticulated pike.*

The *orphies* are remarkable for their long snout and their green bones. We have four species of them; one of which, the gar pike (*esox belone*), is a native of the European seas; the three others, from the American coasts, have been described by M. Lesueur. The *hemiramphi,* Cuv., have a remarkable character; the lower jaw being considerably prolonged beyond the upper, into a long point without teeth. One of the three species we possess is from Brazil, the others are from India.

The flying fishes are to be found in every sea, but more particularly near the tropics. The size of their pectoral fins enables them to support themselves some minutes in the air; when they spring out of the water to escape the fishes which pursue them, they often become the prey of sea birds. We possess three species of them. The genus of the *mormyri* comprehends eight species, all inhabitants of the Nile, from whence they were brought by M. Geoffroy. One of them, which has a very long snout, was known and revered by the ancient Egyptians under the name of *oxyrhynchus,* and it is often seen depicted on their monuments.

The fourth family, that of the carps (*cyprini*), consists of fresh water fish, found in all latitudes, of which we have thirty-five species. To the first genus, *cyprinus,* belong the barbel, the bream, the tench, and several white fish which have served as types of various sub-genera established by M. Cuvier. The gold-fish of China, which adorns our ponds, is a species of this genus, the domesticated varieties of which are very numerous. Next to the carps is the *gonorhynchus,* of which only one very rare species is known, which was brought from the Cape by M. Delalande. The loches come next; we have four species of them: the three first inhabit our rivers and ponds, and the last, which has not been described, was sent us from North America by M. Milbert.

The genus *anableps* is very distinct from the loches, to which it has been united: the eyes are very projecting; the cornea and iris are divided into two parts by transverse bands, thus forming two pupils, although there is but one crystalline humour and one retina; this conformation is the only example amongst vertebrated animals. These species inhabit the rivers of the warmer regions of America, and are said to be viviparous.

The genera *pœcilia*, *cyprinodon*, and *lebia* complete the family of the *cyprini*. We have six species of the first, small fish which are found in the rivers of North and South America; only one of the second, and two of the third, lately described by M. Valenciennes.

Next to the *cyprini* are the *siluroïdeæ*: their skin, destitute of scales, is bare or covered with large bony plates: they almost all inhabit the rivers of hot climates. We have fifty-seven species of them. The *siluri*, which are the first genus of this family, have the first ray of the dorsal and pectoral fins large, strong, and serrated; they erect it at pleasure, and it is a dangerous weapon: their mouth is furnished with numerous and long cirri, and their bony head is not covered by the skin. The European silurus (*s. glanis*) is the only species found in northern climates. It is common in the Danube, and is the largest of our fresh water fish. The want of spiny processes to the dorsal fin has caused the separation from this genus of the electric silurus of the Nile, under the name of *melapterurus*. M. Geoffroy brought it from Egypt; the Arabs call it *raasch*, or thunder, and like the torpedo it is capable of giving shocks. The last genera of this family are the *aspredos* and *loricaria*, whose body is covered with large and angular plates. We have eight species of them.

The first family of the fourth order is almost entirely composed of the genus *gadus*, which comprehends the cod, the whiting, the hake, etc. These fishes live in immense shoals in the European seas. A considerable number of vessels resort to the North, in search of cod-fish, which are salted and dried. This is a great article of commerce, particularly with the Dutch. We have twenty-six species of cod. The burbot (*gadus lota*) is the only species amongst them which inhabits the rivers. The *grenadiers* (*lepidoleprus*, Risso,) are nearly allied to the *gadi*. We have two species, which live in the deep waters of the Mediterranean; they were given us by M. Risso.

The second family of the same order, vulgarly called *flat-fish*, is composed of the great genus of the flounders (*pleuronectes*), of which we have fifty-nine species. To this genus belong the turbot, the dab, the sole, etc. These are the only vertebrated animals that are not symmetrical; both the eyes and both the nostrils are placed on the same side of the head, and the mouth is unequally divided. A very large *hippoglossus*, Cuv., from the North seas, is fixed to the ceiling; it was caught at St. Valery, and sent to us by M. Baillan.

The third family, that of the *discolobi*, comprehends four genera not very numerous in species. The two first (*lepadogaster*, Gouan, and *cyclopterus*, Lin.,) are most of them very small fish, which hide themselves under the stones; there are eight species in the collection. The *echeneides*, Lin., are remarkable for the lamellated and flat discus on their head, by means of which they attach themselves to different bodies, either to large fishes or to the bottom of vessels, and are thus carried with them in their course. We have four species: the two most common are the *e. remora* and the *e. naucrates*; the first from the Mediterranean, and the other from the Indian seas. It has been said of the *remora*, that it can arrest the progress of a ship. We have three species of the genus *ophicephalus*, all from the Indian seas.

The seventh order of fishes forms only one family, the eel-shaped, of which we have sixty-five species divided into several sub-genera. The body of these fishes is linear, like that of the serpents; they have no ventral fins, in consequence of which they have been named *apodes*, and the number of their other fins differ according to the genus. The eels have them all, the *ophisuri* want those of the tail, the *murænophies* are without the pectoral fins, and the *apterichti* have no fin whatever. The first genus is that of the eels: the common eel (*muræna anguilla*) is found in the rivers in every latitude; the one which is attached to the ceiling is 5 feet long. The Romans were particularly fond of a species of murænophis (*murena helena*), and kept it in reservoirs appropriated to the purpose. A specimen of this genus, 4 feet long, is attached to the ceiling. The *gymnoti* are distinguished from the eels by the want of the dorsal fins; they are fresh water fish, and inhabit the warmer regions of Africa and America. We have six species of them; one of which, the electrical gymnotus is much celebrated for its electric powers, which are so great that it can give the most violent shocks at pleasure, at a considerable distance and in any direction, thus killing the small fish and other animals on which it feeds. If horses go into a pond where there are *gymnoti,* they immediately fall down, and are unable to rise. This power is exhausted if too much exercised, but it is renewed by repose: the organ in which it exists occupies the lower part of the tail. One of these gymnoti was brought alive to our menagerie, but it did not live long enough to enable us to repeat all the experiments which M. de Humboldt had made in America.

We are now arrived at the eighth and last order of fishes, which is the most numerous of all: it has been divided into eight families.

The *tenioïdeæ* have been thus named from the resemblance of their long and flattened bodies to a ribbon. We have five species of them, amongst which are the *lophotes* of Lacépède (*giorna*), a rare and beautiful fish from the gulph of Genoa, sent us by M. Martial Duvaucel; the Cepedian gymnetrus of the Mediterranean, the colour of whose body is bright silver, and that of the fins, red: and the garter fish (*hi idojus*), which was caught at La Rochelle. A specimen of each of these is attached to the ceiling.

The family of the *gobioïdeæ* comprehends the

blennies and the gobies : we have seventy-six species of the first genus and forty-six of the second. The fish of this family are generally small, and live in shoals on rocky shores. They exist a long time out of the water. Some species of the blennies from the Indian seas, remarkable for the number and smallness of their teeth, which are moveable as the keys of a pianoforte, are distinguished from the others by the name of *salarias*.

The genus *anarrhichas* (wolf-fish) is nearly allied to the blennies; those which are named sea-wolf or sea-cat are a great resource to the Irish, who eat them dry, make shagreen of the skin, and use the liver as soap. A large specimen is suspended from the ceiling. The *periophthalmi*, of which we have five species, are akin to the gobies. Their eyes are furnished with moveable eyelids, which meet at the top of the head. These fish can live a long time out of the water, and by the help of their pectoral fins can run upon the mud. The specimen from Senegal was given to the cabinet by M. Delcambre, who took it for a lizard, and shot it. The other species are from the Moluccas. The *sillagones* belong to the same family; the *s. acuta*, Cuv., and the *s. domina* Cuv., are the best fish of the Indian seas. The *callionymi*, Lin., the last of the family of the gobies, are very pretty fish with a smooth skin. We have four species of them.

The third family, that of the *labroideæ* comprehends several genera; they have long and scaly bodies. The first genus, that of the *labrus*, is characterized by the thickness of its fleshy lips. M. Cuvier has divided it into seven subgenera. The first, that of the *labrus proper* (rainbow fish), to which belong the varying labrus (*l. turdus*), the *louche* of the Mediterranean, and the anile labrus (*l. vetula*) from the North seas, a beautiful fish striped with blue and orange. 2d. The *julis*, Cuv., which differs from the labrus in not having scales on its head. We have forty-seven species; one of which from the Mediterranean (*l. julis*), jurella labrus, is distinguished by its colour, which is a fine blue or violet on the upper parts with an undulating orange stripe on each side. 3d. The *crenilabri*, Cuv., of which we have forty-nine species, mostly from the Mediterranean, and all beautifully coloured. 4th. The *ephibuli*, Cuv.; amongst which is the insidious sparus, (*sparus insidiator*), remarkable for the great elasticity of its snout, to which it suddenly gives the form of a tube to catch the small fish within its reach. This species is rarely found in collections of natural history. 5th. The *elopes*, Lin., which have a long snout, like the *ephi-*

buli, but a very small mouth, incapable of extension. The only two species known are from the Indian seas, and were brought home by Commerson.

The second genus of this family (*novacula*, Cuv.,) are distinguished by the sharpness of their forehead, which descends almost vertically to the mouth. We have six species; one of which is from the Mediterranean, and the others from the equatorial seas. Next comes the genus *chromis*, Cuv. We have seventeen species, one of which, from the North American seas, has been named *sparus auritus*, because its skin is prolonged beyond the gills. The last genus is that of the *scarus*. We have seven species which inhabit the seas of hot climates. The singular form of their snout and the brilliancy of their colours has caused them to be called parrot-fish.

The next family is that of the *sparoïdeæ*, of which we have about three hundred and fifty species. M. Cuvier has divided them into twenty-five genera, which are characterized by the form of the jaws, teeth, and operculum. We have twenty-two species of the genus *smaris*, Cuv., to which belong the smare sparus (*sparus smaris*, Lin.) and the mendole sparus (*smæna*, Lin.), which are found in the Mediterranean. The most remarkable of the others are: 1st. The *spari*, Cuv., which have round teeth placed close together in the manner of a pavement. We have eighteen species. That which is most celebrated for the flavour of its flesh and the beauty of its colours is the gilt-headed sparus (*s. auratus*, Lin.), which is found in almost all seas. 2d. The *bodiani*, Cuv., which have a spiny operculum without notches, and the *serrani*, Cuv., whose operculum is both spiny and notched. These two genera comprehend the greater number of the sparoïdeæ. We have thirty-one species of the first and fifty-four of the second. During life they have all very brilliant colours, and their body is generally spotted or marked with transverse stripes. The *s. anthias*, Cuv., is remarkable for its beautiful red colour. 3d. The *scorpœnæ*, Lin., are divided into four sub-genera, and we have twenty-two species of them. The spines with which their head is covered, and the cirri which hang from their body give them a hideous appearance; they are vulgarly called sea-hogs. Two species of them are found in our climates, the others are exotic. The sub-genus *pteroïs*, Cuv., comprehends fishes which inhabit the fresh waters of India and the Moluccas; they are of a very singular form, but elegantly coloured; they are particularly remarkable for the extension of their pectoral fins, which has been the cause of the surnames of *volitans* and *evolans* being given to some of them; there are two species in the cabinet which have not been described.

The family of the perch (*perca*) is as numerous as the preceding: its distinguishing character is two fins on the back. There are in the collection more than two hundred species, which M. Cuvier has distributed into eighteen genera and many sub-genera. We shall only mention those which contain the most remarkable species. The first are the surmullets (*mullus*), of which we have fifteen species. The most celebrated is the red surmullet from the Mediterranean, its back is of a beautiful red, its belly silvery, and its flesh is very delicate. 2d. The perch (*perca*), the most common species of which inhabits fresh waters. M. Cuvier has divided it into six sub-genera. We will only mention the sandre perch (*perca lucio-perca*, Lin.), a very delicate fish which, from the flattened form of its snout resembles the pike. A very fine specimen, given to the Museum by the marquis de Bonnay, is on the ceiling. We have thirty-four species of the genus *perca*. 3d. The *sciænæ*, Cuv., have an obtuse and scaly snout. Of the thirty-one species in the Museum some are from the Indian seas, others from the Mediterranean. The most remarkable for its size and the excellence of its flesh is the *fegaro* of the Genoese (*sciæna aquila*, Cuv.); it is often 6 feet long, is frequently caught in the Mediterranean, and rarely found in the British Channel. We see three specimens of this fish attached to the ceiling. M. Lesueur has recently sent from North America a species of fegaro caught in the lake Erie, and to which he gave the name of *s. oscula.* 4th. The *pogonias*, Lacép., which differs from the *sciænæ* in having a great number of cirri under the throat. The only known species of this genus inhabits the North American seas, whence M. Milbert has sent us large specimens. 5th. The gurnards (*trigla*, Lin.), remarkable for their mailed and angular heads, and the loose rays placed before the pectoral fins which are extremely developed. We have eighteen species. The most common in our markets is the cuckow gurnard (*trigla cuculus*, Lin.); its colour is a bright red. The most-remarkable is the flying gurnard (*t. volitans*, Lin.); its pectoral fins are so large that it can fly for some minutes above the surface of the water; it inhabits the Ocean and the Mediterranean. 6th, The *lophii*, Lin., of which we have twelve species, divided into three sub-genera by M. Cuvier: the *lophius* proper, the *antenna-*

rius, Com., and the *malthe,* Cuv. These fishes are too remarkable to be passed over in silence. The European frog-fish (*lophius piscatorius*, Lin.), inhabits our seas; it is from 4 to 5 feet long; the cavity of the gills is very large, while the aperture is small, and it can live a long time out of the water. Its skin is without scales, and its pectoral fins are supported as though it had arms; the rays on the head are long and moveable; and they are kept in motion while it lies imbedded in the mud, and thus serve as a bait for the small fish, which, being deceived by the similitude, mistake them for worms, and become a prey to the frog-fish. We have eight species of the genus *antennarius,* described and figured by M. Cuvier in the Memoirs of the Museum. They are small fish, whose pectoral fins, supported by a long pedicle, enable them to crawl upon the mud. We have three species of the genus *malthe,* two of which are very rare; the third. vulgarly called the *sea-bat,* inhabits the Atlantic.

The next family, that of the *scomberoïdeæ,* Cuv., of which we have more than seventy species, is divided into fourteen genera. The first, which has given the name to the family, has been separated into six sub-genera. 1st. The common mackrel (*sc. scomber,* Lin.), very common in the summer on our coasts. 2d. The tunny (*s. thynnus*), the fishery of which is of great importance to the inhabitants of the Mediterranean coasts. 3d. The German mackrel (*s. germon*) found in large shoals in the bay of Biscay: the specimen in the collection was sent from La Rochelle by M. Dorbigny. 4th. The carang mackrel (*caranx*, Lacép.) remarkable for the line strongly mailed on each side of the tail. We have thirty-seven species of them; one of which, the scad mackrel (*s. trachurus*, Lin.,) inhabits the Mediterranean.

The *vomeres*, Cuv., of which there are seven species in the collection from the American and Indian seas, are distinguished by their sharp and elevated forehead, and by the compressed form of their body, which is almost as wide as it is long. M. Cuvier has divided them into four subgenera, according to the form of their fins; it is to the generally round form of these fish, and to their silvery hue, that they owe the appellation of moon-fish.

Only one species of the genus *tetragonurus*, Risso, is known under the vulgar name of *sea-crow*. M. Risso, who presented it to the Museum, has described and named it *t. Cuvieri*.

The genus stickleback (*gasterosteus*), comprehends the smallest of our fresh water fish; they are all spiny. One of the two species which inhabit our streams, the smaller stickleback, is not one inch in length. The second is somewhat larger, and the third comes from North America.

The *centronoti*, Lacép., are allied to the *gasterostei*. We have four species of them. The pilot (*g. ductor*, Lin.) belongs to this genus: it is so called because it swims before the shark, and points out its prey. We have fifteen species of the genus *zeus*, Lin., which has been divided into three sub-genera. The most common is the *z. faber*, Lin., found in the Ocean and the Mediterranean; it is of a yellow colour, with a black spot on each side: the other species belong almost all to the Indian seas. The genus *lampris*, Retz, has but two species, very rare and remarkable for the beauty of their colours. The first is the opah (*z. regius*), which has been suspended to the ceiling on account of its size; the other, as yet undescribed, was presented by count Lynch, and M. de Lacépède will soon publish it under the name of *z. argyropomus*. The sword-fish (*xyphias*), remarkable for its sword-shaped snout, which is sometimes 3 feet long. The specimen in the gallery is *xyphius gladius* from the Mediterranean; we have only the snout of the East Indian species, the broad-finned sword-fish.

The genus *coryphœna*, Lin., has been divided into four sub-genera, all in the collection, except the *pteraclis* from India. The most common is the *coryphæna hippurus*, the dolphin of the sailors, a fish from 3 to 4 feet in length, of a beautiful silvery blue, with yellow fins. It is very abundant in the seas of hot and temperate climates, and feeds mostly on the flying fish (*exocœtus*).

The principal character of the *acanthuri* is, that they are provided with one or more strong spines on each side of the tail. We have seventeen species of them. The *monoceri*, Willugh., are allied to the acanthuri, as having like them caudal spines, but they differ in the more or less prominent process which has caused them to be named sea-unicorns. We have three species from the Indian seas.

The family of the *squamipennæ* is so called because the fishes which belong to it have the greatest part of their dorsal and anal fins covered with scales. We have more than one hundred species of this family, which M. Cuvier has divided into eighteen genera. The most numerous is that of the chætodons, which have very numerous teeth, close set and setaceous; they are striped with lively colours, and are natives of the Indian and American seas. They compose eight sub-genera, of which we have forty-two species in the collection. There is only one species of the genus *osphroneme* known (*o. olfax*, Com.); its flesh is esteemed; it was brought from Java to the Isle of France, and from thence to Cayenne. This genus and that of the *trichogaster*, Schn., which is next to it, have this particular character, that one of the rays of the ventral fins is lengthened into an articulated bristle as long as the body. After these comes the *toxotes*, Cuv., which had been confounded with the labrus, but from which it materially differs. There is only one species known, which inhabits the Indian seas, viz. the *labrus jaculator* of Linnæus, so called from its habit of darting drops of water on the insects which fly near it, and on which it feeds. The *perca scandens*, Daldorf (*anthias testudineus*, Bl.), of which M. Cuvier has made the genus *anabas*, is a very extraordinary fish from the complicated apparatus of its gills as well as on account of its habits. As it can retain water for a length of time in the cavity of its gills, it is thus enabled to crawl upon the earth and to climb up the palm-trees, where it remains in the rain-water collected in the leaves. The *polynemi*, the last genus of the family of *squamipennæ*, are remarkable for setaceous processes on each side near the pectoral fins, which are longer than their bodies. This circumstance is more particularly remarked in the *p. paradisœus*, Lin.   These fish inhabit the Indian seas.   We have five species of them in the cabinet.

The last family is that of *fistularia*, so named from the cylindrical form of their snout, at the extremity of which is placed the mouth ; it comprehends two genera : *fistularia* and *centriscus ;* each of which forms two sub-genera.   The fistularia are cylindrical fish, the length of whose head is equal to one third of the body : we have two species of them from the seas of hot climates.   The *aulostoma*, Lacép., differs from the fistularia, properly so called, in having the head shorter and no bristles in the tail.   We have but one species, the Chinese fistularia (*f. chinensis*, Bl.).

The *centrisci*, known by the specific name of sea-snipes, have a tubular snout like the fistularia, but their body is compressed and much shorter. We have two species of them : the one from the Mediterranean, *c. scolopax*, is of a silvery hue ; the other which belong to the sub-genus *amphisile*, Klein, was sent from the coast of Coromandel by M. Leschenault.

§ VIII. COLLECTION OF ARTICULATED ANIMALS;

COMPREHENDING THE CRUSTACEÆ, THE ARACHNIDES, THE INSECTS, THE ANNELIDES AND THE WORMS.

THE animal kingdom naturally forms two great divisions: that of the animals with vertebræ, which have an interior skeleton, and that of the animals without vertebræ. The collections which we have hitherto seen consist of animals belonging to the first division, and those which we have yet to see are comprised in the second. In order to facilitate our examination, we shall divide them into only two sections, the articulated, and the non-articulated. As they both comprehend animals of a small size, they have been placed in cases which occupy the centre of the gallery on the second floor.

The collection of articulated animals, of which we have about twenty-five thousand species, is divided into five classes, which we shall examine successively. Its existence may be dated from the new organization of the Museum; for in the former cabinet there were only one thousand five hundred specimens, almost all belonging to the arachnides and insects; these came mostly from Reaumur's private collection, and were sent either to that celebrated entomologist or to Buffon by Cossigny, Duhamel, Poivre, Adanson, Granger, Chavalon, Commerson, Sonnerat and several other correspondents. They were exhibited with a view to give an idea of this branch of zoology, rather than as objects of curiosity. It began to present an interesting series when M. Desfontaines enriched it in the year 1796 with the insects which he had collected in Africa. It was considerably augmented from the cabinet of the Stadtholder, and was afterwards much increased by specimens belonging to all the classes which Dombey, Maugé, Peron, Lesueur, Maté, Hogard, and Olivier collected in their travels. Still more recently it has been greatly augmented by those which MM. Delalande, Leschenault, and Auguste Saint-Hilaire have found, the first at the Cape of Good Hope, the second in the Peninsula of India, and the third in the Brazils.

In consequence of the presents which have been made us, the exchanges with naturalists, and the zeal of many of our correspondents, among whom we are happy to enumerate MM. L'Herminier, Milbert, Lapillaye and Dorbigny, our collection has attained such a degree of superiority, that in respect of exotic species there are few in Europe to be compared with it.

What gives a great value to this collection, and renders it essentially classical is, that not only all the species are referred to their genera, but that they are all named with precision. M. Latreille, who has taken this task upon himself, has in the distribution of the families and genera followed the order established by M. de Lamark in his system of animals without vertebræ; and the different objects have been arranged, so as to shew their characters and to preserve their forms. by M. Dufresne, the chief of the zoological department.

We shall now examine the different parts, following the order we have indicated.

On entering the gallery of the birds by the small staircase, we must turn to the right to reach the saloon of the carnivorous quadrupeds. The crustaceous animals are placed vertically in the upper part of the cases which stand in the middle of the room. The collection is composed of about six hundred species belonging to fifty-four genera; and we may venture to assert, that there does not exist another so complete, particularly since M. Leschenault has enriched it by the specimens he has collected in India. All the species are labelled, and placed according to their affinity. Those which were too large to enter the frames are placed in twenty-seven glazed boxes on the cornices of the cases which contain the carnivorous animals.

The first boxes, those on the left of the entrance, contain a series of lobsters (*palinurus*), many of which attract attention by the variety of their colours. The most beautiful is the (*palinurus ornatus*); one specimen of which was sent to us from the Isle of France by captain Mathieu of the artillery, with several other rare and well preserved crustaceæ. In the three following boxes is the *cancer gammarus*, Lin., one of which, 3 feet long, was sent from North America by M. Milbert. Next to these is the genus *calappa*, Fabr., which cover the anterior part of their body with their fore feet, which terminate in a flat crest, and are raised perpendicularly. Lastly, the *portunus*, Fabr., which resembles our common crab, excepting that its hind feet are shaped into a sort of fin.

On the opposite cornice, near the entry into the room of the *pachydermata*, we see a very large crab which inhabits the seas of New Holland; immediately after which are the *gecarcini*, Leach. Their habits are very remarkable;

they live in burrows and in burial grounds ; but in the spawning season they go to the sea-shore, keeping always in a straight line.   In the following boxes are the *pagarus*, Latr., the largest species of the genus; the *dromia*, Latr., which having the hind feet elevated, and terminated in a hook, makes use of them to seize upon the sponges, shells, and other objects under which it takes shelter, and which it carries about with it : the *ranina*, Latr., which is said to climb even on the roofs of houses; and the *scyllarus*, Fabr., or large crawfish of the Mediterranean, the lateral antennæ of which are widened into a crest.   In the two last boxes are the *monoculi*, the cyclops of America, and the Molucca crab, which are sometimes more than 2 feet in length ; their shell is convex, rounded in front, and terminated by a long sharp tail.   The black slaves use the shell of this animal to carry water in, holding it by the tail, which answers perfectly the purpose of a handle.   One species of this genus (*l. heterodactylus*, Latr.), is a favourite dish amongst the Chinese ; and the inhabitants of Japan paint and carve it upon their monuments as a symbol of the zodiacal constellation of Cancer.

We are now going to examine the series of the crustaceæ which are in the entablement of the chest of drawers, beginning opposite to the *limuli.* As all the species are labelled, we shall mention only those which are most worthy of attention. They are, 1st., the spiny podophthalmus (*p. vigil*, Leach), whose eyes are situated at the extremity of very long moveable peduncles. 2d. The ocypode (*cancer eques*) which has the same organ surmounted by a tuft of hair, and which has been so named from the swiftness of its course. 3d. The river crab, so celebrated among the ancients. 4th. The *gelasima maracoani*, one of whose claws resembles a large pair of scissars. 5th. The masked *grapsus* of New Holland. 6th. The *cancer pelagicus*, also called *cedo-nulli*; it is distinguished by the marbled veins on its shell; and the *p. forceps*, whose claws are slender and fringed. 7th. The *parthenope*, Fabr., so extraordinary from the size of the claws, and by the inequalities and colour of the shell which give it the appearance of a stone. 8th. The hermit crab (*pagurus*, Fabr.), which enters into any vacant univalve shell and drags it along. 9th. The *phyllosoma*, which is as thin as a leaf. 10th. The *squilla*, whose claws are shaped like harpoons; and other smaller specimens of the species contained in the boxes above the cornices, many of which are eatable; such as *cancer pagurus*, large edible crab, the *c. mœnas* and the *c. puber*.

We shall now proceed to the collection of arachnides and insects. It begins at the extremity of the chest of drawers placed in the gallery of the birds. It occupies its whole length, extending to the middle of the other chest in which are the crustacea, and is composed of about fifty thousand specimens belonging to more than twenty thousand different species. As it would have required a great deal of room to exhibit so great a number of objects, and as the colours of many of the insects are liable to change when exposed to the light, this collection has been so divided as to form two. The first consists of the old collection and the duplicates of the new acquisitions, and is displayed in the vertical frames which are on the top of the drawers. The genera alone are indicated, and a number affixed to each species refers to a catalogue by M. Latreille; this is communicated to those who wish to study the nomenclature. The second collection, by far the more complete, is shut up, in order to ensure its preservation, in the drawers placed below the chests. This collection is arranged according to the most recent and approved methods; the species afford the type of those described by M. Latreille, and it is shewn to those who wish to examine the characters; but as it is not exposed to public view, we shall only oberve, that it contains almost every known genus, and a great number of very rare species.

The class of the arachnides which comprehends the genera *aranea, scorpio, acarus*, etc., is the first in order. The two first frames contain different species of scorpions, animals which in form resemble the crawfish. Their abdomen is terminated by a long jointed tail, ending in a crooked point or sting, through which the poison is evacuated. The African scorpion (n° 1.) is the largest of the tribe. The common Italian scorpion (n°' 10 and 19) will easily be recognized by the inhabitants of the southern departments of France.

The spiders, common to every part of the world, compose the next family, which being very numerous is distributed into several genera, the characters of which are taken from the disposition of the eyes. The fangs with which the animal wounds its prey, and out of which is evacuated a poisonous fluid, constitute a part of the organs of mastication. They have different habits according to the genus in which they have been classed; the skill with which the greater number of them work their delicate and regular webs, in the centre of which they place themselves; the industry with which they construct the bag (n°s 5, 85 and 87,) destined to contain their young, and their extreme care in defending and preserving it, should induce us to examine them with interest, notwithstanding the disgust which they generally excite. Some large species belonging to warm climates, are truly dangerous; such are the species (n°s 1 and 2) named after M. Leblond; the *fasciata*, Seba. (n° 7), and the bird-catcher (n° 3). This last seizes on humming birds and devours them. A very remarkable species on account of its industry is the *m. cœmentaria*, Walk. (n° 78); it digs a gallery many feet in length under ground, which it inhabits with its young; the entrance is shut by a circular door, fixed by a hinge, which falls of itself when opened. This species inhabits the neighbourhood of Montpellier; four specimens of these nests are placed by its side. The species so much celebrated in Italy by the name of *tarantula* (n° 68) belongs to the genus *lycosa*, Latr. It has been supposed that the bad effects caused by its bite were only to be dispelled by the help of music and dancing: but in fact the bite is less dangerous than has been imagined, and it can easily be cured. This species of lycosa is different from the *tarantula* of Fabricius, which constitutes another family; to this belongs the genus *phryne* of the equatorial climates. Their principal characters consist in a flattened body, the two pincers or claws furnished with numerous prickles, and the two fore feet elongated and antenniform.

The genera *solpuga* and the *chelifer* belong to the family of the arachnides. The first inhabit the warm and sandy countries of the ancient continent, and are considered as venomous; they run with astonishing rapidity, and are distinguished from the rest of their tribe by two enormous mandibles projecting beyond their large head.

We often find in our books, herbals, etc., a small insect much like a scorpion, without a tail; it is the *phalangium cancroïdes*, Lin. (n° 2), vulgarly called the book-scorpion; it feeds on the insects which destroy books.

The family of the *myriapodes* (*mitosata*, Fabr.), vulgarly called centipedes, on account of the number of their feet, equal to that of the segments of their body, which is long or depressed, are placed between the arachnides and insects properly so called. It is divided into two great genera: *julus* and *scolopendra*. The largest species of this last are exotic; their head is armed on each side with two curved fangs through which a poisonous fluid is discharged. The *sc. morsitans* (n° 2) is considered as dangerous in the West India islands: and the *sc. electrica* (n° 6) is possessed of a high degree of phosphoric splendour.

We are now come to the most numerous class of the animal kingdom, which comprises the greatest variety of forms, and is the most wonderful from the instinct peculiar to each species, and the most surprising from the metamorphosis experienced by all the animals belonging to it. Insects are equal to birds in the richness and variety of their colours; they even surpass them in some respects, particularly in that of the phosphoric light which emanates from some of them; and while they divide with them the empire of the air, they far exceed them in number, and their families are still more numerous than those of the plants.

Insects have been divided into several orders, differing from each other in their forms, in their mode of transformation and in their character. We will begin with the most extensive, the order *coleoptera*. This name has been given to those insects which like beetles, or cantharides, etc., have wing-sheaths or *elytra*.

The genus *lucanus* (stag-chafer) is at the head of this order, to which belongs the species *l. cervus* (n°' 1 and 2, stag-beetle), often seen flying in our woods on a summer's evening.

The *scarabæus, coprises*, and *geotrupes* come

immediately after the genus lucanus. The males in these three genera are often distinguished from the females by certain processes on their head and thorax. We will give as examples the *sc. hercules* (n°s 1 and 2), or hercules-beetle of South America; the *sc. actæon* (n°s 16 and 17); the *silenus* (n°s 31 and 32); and the *nasicornis* (n°s 39, 40, and 41). The males of the *sc. dichotomus* (n° 10), the *atlas* (n° 12), the *claviger* (n° 23), the *codrus* (n° 28), and the *longimanus* (n° 37), are no less remarkable. The last numbers, from 66, belong to small species, which live in excrementitious substances, and form the genus *aphodius*.

The genus *copris* is still more numerous than that of the *scarabæi;* it contains many species worthy of attention; such is the *ateuchus sacer* (n° 1), which was one of the emblems of Osyris and of the sun amongst the ancient Egyptians; it is often seen depicted in their hieroglyphics, on their monuments and sculptured stones. This insect incloses its eggs in balls of dung, and with the help of one or more of its kind, rolls these balls with its hinder feet to the hole prepared to receive them. These habits are common to the greater number of the following species, some of which are of a gigantic size; such as the *c. antenor* (n°s 35 and 36), the *bucephalus* (n°s 38 and 40), and the *lancifer* (n° 41).

The *geotrupes* and the *lethrus* make deep holes in the earth to lay their eggs in. Next to them is the genus *hexodon*, which is very rare in collections; these insects were brought by Commerson from Madagascar, although we are assured they are also found in Ceylon.

Next come the cock-chafers (*s.melolontha*, Lin.), the type of which is a common species (n° 13), very destructive to vegetables, whether in its larva or perfect state. The *s. fullo*, Lin., which inhabits the sandy and maritime shores of Europe, and which is of a brown colour spotted with white, is one of the largest species known. The males (n°ˢ 7 and 9) are remarkable for the size of the club, divided into seven leaves, which terminates their antennæ. The body of the scaly cock-chafer (n° 90) is covered with small scales of a deep and brilliant blue colour; it is very common on the ferns in the south of France. The thickness and the appendages of the hind feet of the *m. crassipes* (n° 119) give it a singular appearance. Few coleoptera equal the *cetonia* in the richness and variety of their colours. The *c. auratas* is often seen in the centre of flowers, and particularly on roses. The *c. fastuosa* (n° 22) is also found in France. The most remarkable are the Chinese cetonia (n° 41) and the *cacicus* (n° 40). This last, which is sometimes four inches long, differs from the others placed in the same frame in the form of its head; it inhabits the coasts of Guinea and Congo, and is known only in two or three Museums. This and some other species have been united by M. de Lamarck into a genus to which he has given the name of Goliath.

After the cetoniæ come the small coleoptera, which we shall notice on account of their habits. Some, as the *dermestes*, Lin., and the *anthrenus*, Geoff., are extremely noxious; as in their larva state they eat furs, and destroy collections of birds, insects, etc. Others, as the *byrrhus*, Lin., the *nitidula*, Fabr., the *silpha*, Lin., the *necrophorus*, Fabr., confine their ravages to the dead bodies of animals, and thus stop the course of the dangerous exhalations which proceed from them. Scarcely is a mole dead before the *necrophori*, which are seldom seen except on such occasions, fly around its body, insinuate themselves beneath it, dig by degrees into the earth, and bury it entirely, after having deposited their eggs in its body. The most common species with us is the *sylpha vespillo*, Lin. (n° 3). The insects of the genera *hydrophilus*, Fabr., *elophorus*, Fabr., *dytiscus*, Geoff., and the *gyrinus*, Lin., people our ponds; the last especially often attract attention, by the varied curves which they form on the surface of the water. Next to these aquatic insects comes another family of coleoptera, equally carnivorous with the last, but confining themselves to the land; it is composed of the genera *carabus*, Lin., *manticora*, Fabr., *cicindela*, Lin., *scarites*, Fabr., and *elaphrus*, Fabr. Many species of them, particularly of the first mentioned genera, which are remarkable for their size as well as their lustre and for the symmetrical elevation of their wing-sheaths, have a fœtid smell, and when caught sometimes emit an acrimonious and corrosive fluid. The *carabus auratus*, Lin. (n° 9), common in our fields, and the *c. sycophanta* (n° 24), have the wing-sheaths of a very brilliant golden green colour; the latter feeds as well as its larva upon caterpillars, and especially upon those of the oak. There are others of the same genus (n°ˢ 102 and 108) which have a very extraordinary method of attempting to escape from their enemies. They cause a caustic liquor of an ammoniacal smell, which immediately dissipates into vapour, to proceed repeatedly and with an explosion from the hinder extremity of their body; for this reason they have been named crackers, pistols, etc. The *manticora maxillosa*, Fabr., (n°ˢ 1 and 2,) comes from southern Africa; it is one of the largest insects of this family, and is provided as well as the *cicindela* with strong jaws. Amongst the species of this last genus, which are almost all spotted or striped with white on a coppery red or green ground, we will mention the *c. campestris*, Lin. (n° 21), which is equally wonderful for its habits and for the form of the larva. A perpendicular and rounded hole, excavated in a sandy soil, and exposed to the sun, forms the solitary abode of this larva. The strength of its mandibles enables it to dig, and to separate the particles of earth; an enormous head, the supeperior part of which is formed like a basket, answers the purpose of a hod, which it empties by the help of two hooked processes situated on its back, and which assist it to climb to the mouth of the hole, where it empties its load. Its habitation being thus prepared, it stands in ambuscade near the entrance, which it closes with its head. When its prey gets within reach it darts forth, seizes it with its mandibles, and precipitates it to the bottom of the cell. In the same cell, after having closed the entrance, it undergoes its last metamorphosis.

A greater length of body, very short wingsheaths, and the habit of keeping recurved the two last rings of the abdomen in the act of walking, signalize the genera *staphylinus*, Lin.

oxyporus, and *pœderus*, Fabr., which live upon decayed animal or vegetable substances. Few insects prove with time so destructive in the larva state as those which belong to the genera *ptinus*, Lin., *ptilinus*, Geoff., and *anobium*, Fabr.; some, as the *p. fur* (n° 2) and *p. scotias* (n° 3) commit their depredations on our herbals. The little holes with which the rafters of our furniture and houses are sometimes perforated, the galleries dug through books and papers, are the work of several species of anobium and ptilinus. The noise, like the ticking of a watch, which is frequently heard in wainscoated rooms, is made by the anobium. The male and female, often at a distance from each other, call and reply by knocking violently and repeatedly against the pannels.

The richest and most brilliant colours generally adorn the *coleoptera* of the genus *buprestis*, Lin. The Indians use the wing-sheaths of the *b. chrysis* (n° 1) to adorn their dresses. The *b. vittata*, Fabr. (n° 23), the *b. fastuosa* (n° 29), and many other species exhibited in the same frame are equal in beauty to the chrysis. On the wing-sheaths of the *b. fasciculata* (n° 7) are small linear bunches of yellowish hairs; the *b. gigantea* (n° 2) is remarkable for its size, being 2 inches long; and on the elytra of the *b. ocellata* (n° 24) are two large luminous spots. The next genus is that of the *elater*, Lin.;, when placed on their back these insects cannot turn themselves, on account of the shortness of their feet; they spring perpendicularly until they regain their proper position ; to do which they contract their feet close to the body, then causing the head and the abdomen to approach, they unbend themselves like a spring, striking briskly on the ground, and rebound into the air. The greatest number of the elaters are deficient in that richness of ornament which distinguishes the preceding coleoptera. Many species are nevertheless remarkable; such as the *e. flabellicornis*, Lin. (n° 1), *speciosus* (n° 7), *ocellatus* (n° 9), *ferrugineus* (n° 18), etc. This last is indigenous. The *e. noctilucus* (n° 10), called *cocuja* in South America, and the species n°ˢ 11 and 13 have luminous spots, by the light of which, it is said, that the smallest print máy be read. This luminous property which characterizes some species of the two preceding genera, the object of which appears to be the union of the two sexes, is common to all those of the genus *lampyris*, Lin., commonly called glow-worms, and by the Italians *lucciola;* but in these the phosphoric effect is produced by the last rings of the body; in some species it is peculiar to the females, which have no wings at all, or so short as to be useless for flight. The luminous part forms a whitish or yellowish spot, and the lustre which it sheds appears to vary at the pleasure of the insect. The *lampyres* are very abundant in hot countries; they fly only by night, when their swarms illuminate the sky. On traversing the Alps towards Italy, the traveller is presented with this spectacle produced by the *l. italica*, Lin. (n° 14), the males and females of which are winged. The females of our native glow-worms are without wings, and the males present scarcely any phosphoric light.

Next to the *lampyres* come the genera *malachius*, Fabr., *telephorus*, Schæff., and *lymexylon*, Fabr. The malachii, when taken in the hand, protrude on each side of the thorax, a little below the base of the elytra, irregular and red vesicles, which Geoffroy has named cockades. The telephori, when in the larva state, live gregarious under ground. The *lymexylon navale*, Oliv. (n° 1), commits great ravages in ship-building timbers.

In all the coleoptera which we have hitherto examined the extremity of the feet or tarsi have five joints; those of the following section have one joint less in the two hinder tarsi. The greatest part of these insects are peculiar to the sandy countries of the south of Europe, to Africa, and the south west part of Asia. Some of them avoid the light; such as the *opatrum*, Fabr., the *tenebrio*, Fabr., the *blaps*, the *pimelia*, the *scaurus*, and the *erodius*, Fabr., of which Linnæus formed but one genus, *tenebrio*. They generally conceal themselves in the sand or under stones, crawl slowly, and are often incapable of flight. We frequently see in our houses the *tenebrio molitor* (n° 27), the larva of which is known by the name of meal-worm. The *t. mortisagus* (n° 5) is found in the damp and obscure recesses of houses; it is of a coal-black colour, and when crushed or even handled diffuses a highly unpleasant smell. Some other coleoptera belonging to this section have wings, and varied colours; they are most of them herbivorous. Many, such as the *crodalon*, Fabr., *diaperis*, Geoff., *cistela*, Fabr., *mordella*, Lin., have some affinity with the preceding; the others differ from them in having the head large, and in the shape of a heart, in the softness of their abdomen, and wing-shells, and in their blistering properties: such is the cantharis (*meloë vesicatorius*, Lin.), n° 2 *bis* and following. In Italy and China the physicians use for the same purpose the species of the genus *mylabris*, amongst which we shall notice the *meloë cichorii*, Lin.

(n° 15). The insects belonging to the genus meloë of Linnæus were formerly considered as a specific against madness; M. Latreille is of opinion that they were the buprestis of the ancients. Those of the genus *pyrochroa*, Geoff., live in the forests, and are distinguished by their fine scarlet colour. The *chrysomela boleti*, Lin. (n° 1), is to be sought for in the fungi of trees, particularly in that called the *boletus*.

We come to a new division of the coleoptera, those which have four joints to all their tarsi : they feed upon vegetables during all the periods of their lives. They are divided into several families, the principal of which are, first, those with long antennæ : the larvæ of this family live in the inside of trees, which they hollow in such a manner as to cause them to perish. The genera *priamus*, Geoff., *cerambyx*, Lin., *callidium*, Fabr., *necydalis*, Lin., *saperda*, Fabr., *stenocorus*, Geoff., *leptura*, Lin., and *spondylis*, Fabr., belong to this family. The *cerambyx gigas* (n° 1) is one of the largest exotic insects known. The males of some species, as the *cervicornis* (n° 2) and *maxillaris* (n° 7 and 12), are distinguished from their females by the strength of their mandibles. There are but three species in France : the *scabricornis* (n° 20), the *faber* (n° 26), and the *coriarius* (n° 27). During the day they keep under the bark or at the foot of trees, and fly only by night. Amongst the exotic species some, such as the *speciosus* (n° 23), are ornamented with very brilliant metallic colours. The genus *cerambyx*, to which that of the *lamia*, Fabr., is here united, comprehends a great number of species which have their thorax tuberculated, and are in general very large insects. The *c. heros*, Fabr. (n° 2), is common in oak forests; its larva is probably that known by the Romans under the name of *cossus*, and which they esteemed a delicacy. The *c. moschatus*, Lin. (n° 8), lives on willows, etc. This species, which is of a bronze or bluish green colour, diffuses a strong odour of roses and musk which can be smelt at a considerable distance, and even after the insect is dead. Another, which has a little tuft of hair at each joint of its antennæ, the *c. rosalia* (n° 7), has the same properties, and is found on the Alps and Pyrennees. The same remark has been made of several exotic species. That which is named after M. Desfontaines (n° 26) was in the collection that professor brought over from Barbary. The *c. longimanus* (n° 45), or long-limbed cerambyx is easily recognized by its marbled wing-sheaths, the extreme length of its anterior feet, and its extraordinary size. The larva of *lamia gigas* (n° 78) lives on the baobab tree of

Senegal. Other genera of the same family present various combinations of form and colour. The *trogosita caraboïdes*, Oliv. (n° 5), or *tenebrio mauritanicus*, Lin., is very noxious; its larva, which is called *cadelle* in Provence, feeds upon corn, and the insect in its perfect state is often found in flour, bread, etc.

. The second division, that of the *cycloïdeæ*, so named from the rounded form of their body, comprehends the genera *cassida*, *hispa*, and *chrysomela*, Lin. This last has been divided to form the genera *galeruca* and *crioceris*. They are all small round insects, which feed only on the tender parts of plants; they are not the less distructive on this account, as they are equally voracious in their larva state, and assemble in great numbers on the same plant. It is thus that the *galeruca* of the elm (*chrysomela calmariensis*, Lin.,) sometimes strips the tree almost entirely of its leaves. The insects of the genus *altica*, Geoff., which devour the plants in our kitchen gardens, and which are called garden fleas, are here united to the *galeruca*, but they are placed at the end of the genus. Several coleoptera belong to this division, having longer bodies, and their wing-shells frequently spotted, compose the genus *crioceris*. Such of their larvæ as are known to us conceal themselves under a covering of their excrements. The larvæ of the *cassidæ* have similar habits; their bodies are spiny, and a posterior appendage supports the substances with which they cover themselves. This particular may easily be proved by observing the *c. viridis,* which is very common on the leaves of artichokes, and which differs but little from the *c. equestris* (n° 39). The equatorial regions of America abound most in species of this genus. Their body is almost round in the form of a buckler, elevated in the middle, and bordered all round by the edge of the thorax and wing-shells, and as they are flat beneath, they fix themselves closely to the plants on which they feed. The exterior angle of the basis of their wing-shells is sometimes lengthened into a point, which in the *c. spinifex* (n° 16) is perforated.

The *rhincophori,* so called on account of the length of their snout, are the first family of the third division of the coleoptera. The principal genus is that of the weevils (*curculio*), which has been divided into several sub-genera. Few insects are more destructive in their larva state than these. The *bruchus* eats the seeds of several leguminous plants, those of coffee, and the kernels of many stone-fruits. Two species, the *curculio granarius* (n° 15) and *c. oryzæ* (n° 14), when they get into our granaries, are amongst the greatest scourges with which we are afflicted. Others, such as the *c. palmarum* (n°s 3 and 4) live in the palm-trees, and its larva is much esteemed in the West India islands. Its nest, woven of the fibres of the tree, is placed here along with the larvæ. That of the attelabus (*rhynchites bacchus*, Herbot, n° 3), lives in the curled leaves of the vine, and if the weather is favourable to its increase, whole vineyards are in a short time deprived of their verdure through the ravages of these insects. Other larvæ of the same genus destroy the buds of flowers. The females lay their eggs in the buds while they are yet tender, and they always appear at the same time as the vegetables upon which their posterity are to be supported. They live also upon plants when in their perfect state, and we frequently see the leaves perforated by their proboscis. Many weevils remain constantly on the ground: these, such as the *brachyceri*, are almost always rough and of an ashy colour; others, like the *brenti*, have the body, and particularly the head and the thorax elongated. In some it is the snout only which arrives at these extraordinary dimensions, as in the hazel-nut-weevil (*rhynchænus*, Fabr.), n°s 191 and 192.

Some of the *rhyncophori* placed here amongst the weevils, and which form the division D, more particularly attack the composite and umbelliferous plants. These are called *lixæ* by Olivier. Many of the short snouted weevils are ornamented with the most brilliant colours, as their specific names denote: such as the *c. imperialis*, or diamond beetle (n° 93), the splendid (n° 94), the *chrysis* (n° 96), the *regalis* (n°ˢ 101 and 102), etc. This last, with gold coloured patches on a bluish green ground, is so brilliant that some persons use it for ornament instead of precious stones. The male of the curculio (n° 16) has the snout covered with down. The larvæ of the *bostrichi*, Geoff., feed on the woody parts of trees. One of these, the *destructor* (n° 11), particularly attacks the elm, and forms diverging furrows under the bark; the *curculio* (n° 160) and several analogous species do much injury to the resinous trees.

The third and last division amongst the coleoptera is composed of insects having three joints to their tarsi. They are almost all small spotted hemispherical insects, known by the vulgar name of *lady-birds*, and which naturalists have named *coccinella*. The most common is the seven-spotted lady-bird (*c. septempunctata*), n° 10. The collection of coleoptera which we have examined is contained in twenty-five frames; the three last of which exhibit their several metamorphoses.

The next order, that of the *orthoptera*, is composed of the genera earwig (*forficula*), cockroach (*blatta*), camel-cricket (*mantis*), locust (*gryllus*), and other insects which have soft wing-sheaths, and the wings folded longitudinally like a fan. Many of these orthoptera, such as the locust and camel-cricket, feed upon other insects, but the greatest number live on vegetables. Being provided with very large wings which they can keep unfolded for a great length of time, they are able to transport themselves to great distances, and they emigrate like birds of passage and with the same view. They unite in such numbers that they darken the skies. The camel-crickets, and especially the *acridium lineola* (n° 24) and *a. migratorium* (n° 28), often bring destruction upon the most fertile countries; they devour every thing, and their dead bodies sometimes even infect the air. The inhabitants of some parts of Africa and the Levant pick them up, take off their wings and feet, pickle them and sell them for food. The species bearing the first numbers are remarkable for the form of their thorax. The American cock-roaches are extremely voracious and have a fœtid smell. There are some species equally troublesome in Europe; such as the *b. orientalis* (n°s 16 and 17), the *b. lapponica* (n° 26).

which has spread itself even into the smoky cabins of the Laplanders. The *gryllus monstruosus* (n° 1) is the most extraordinary species; its wings are rolled in a spiral form, and its tarsi are much widened. The male of the *g. umbraculatus* (n° 11), a species peculiar to Spain and Barbary, has a sort of veil on its head. It would be useless to speak here of the species of our own country (n°⁵ 8, 9 and 12). To this genus is united the *gryllus gryllotalpa*, or mole-cricket, which burrows under ground in the manner of a mole. These insects are spread in every part of the world: they, together with the genera *gryllus, acridium,* and *truxalis,* form a division remarkable for two important characters, the faculty of hopping, and that of producing a monotonous sound by means of friction, which is commonly called chirping. Among the grasshoppers the males alone enjoy this property, and nature has converted a part of their wing-shells into a musical instrument; the hinder legs of the *acridium* act the part of the bow of a violin, which the animal passes rapidly over the exterior surface of its wing-sheaths. The abdomen of the female locust terminates in a long sword-shaped process, with which she pierces the ground in order to deposit her eggs. The prevailing shades among the grasshoppers and the camel-crickets (*mantis*) are green or ash-coloured. The insects of the genus *phasma* are almost all exotics; they are in general of a very singular shape, and of very great dimensions. The *mantis siccifolia* (n°ˢ 1 to 3) resembles a parcel of dried leaves; others look like the small branch of a tree, whose twigs are represented by the feet. We see here seventeen species of *mantis* or *phasma* sent from the East Indies or from South America. One of the largest species of *phasma* (the *serripes*), brought from New Holland by Peron and Lesueur, is contained in a separate frame. The camel-crickets are not less curious than the *phasmata*; one of them (*m. oratoria*, n° 27) well known in the South of France, has the fore legs much longer than the others, serrated, and armed with a claw, with which it seizes the small insects on which it feeds. The proportions and forms of the different parts of the body are very various in this genus. The female lays a great number of eggs symmetrically arranged in the manner of a honey-comb; an idea of which may be formed from the specimens here exhibited.

The leading characters of the insects which compose the two orders we have been considering are, the existence of jaws and of wing-sheaths, which are hard or horny in the coleoptera, and soft in the orthoptera. Those of the following orders are also provided with jaws, but their wings are four in number, of a membranaceous texture, and unprotected by a sheath. These wings are nearly of equal length, coriaceous and reticulated in the neuroptera; softer, and the two lower ones shorter, in the hymenoptera.

The family of the dragon flies (*libellulæ*) is very numerous; species of them are to be found in every country; they are remarkable for their elegant light form, for the variety of their colours, and above all for the rapidity of their flight. They seize in the air upon their prey, which consists of flies and other insects. Those which constitute the genus *agrion* have a filiform abdomen; we shall notice among them the varieties of the *l. virgo* (n°s 91 to 98), and those of the *l. puella* (n°s 100 to 106, 117 to 119). The *termes* is frequently mentioned by travellers under the denomination of white ant. The population of our European ant-hills is nothing in comparison with the dwellings of the exotic white ants. The habitations of several species are raised above the surface of the ground in the form of a pyramid or small tower, and placed near one another like houses in a village. They make the more havock from their working together in communities, and always under cover. They form subterraneous roads or galleries in order to reach the trunks of trees, the interior of which they devour. Nothing but metal or stone can arrest their progress. N° 2 is a white ant in its larva state, when it is designated by the name of labourer.

The *myrmeleons* are well known from the singular habits of the larva, described in so interesting a manner by Reaumur. Persons who have not observed these insects may form some idea of their history upon inspecting the frame in which they are placed; it contains the larvæ, the sandy funnel which serves them as a trap for other insects and particularly for ants, the cocoon in which they pass to the state of nymph, and the relicks of their transformation.

The *phryganeæ* are distinguished by the folding of their lower wings; they proceed from aquatic larvæ, which reside in tubular cases formed by agglutinating any materials that come in their way. Several specimens are seen n°ˢ 17 to 22.

The order of the *neuroptera* is terminated by the *ephemeræ*, which are extremely short-lived insects. Their existence, considered from the time their wings are developed, being of only a few hours' duration, has been the cause of their being thus called. The males and females die a few hours after they have perpetuated their race. But before they arrive at this state they live two or three years in the water, and when in their last stage myriads of them quit the liquid element, and fixing upon different bodies, divest themselves of their newly acquired pellicle and wings, and display new ones. They may then be seen fluttering in the air and over the surface of the water, pairing, laying their eggs, and falling to the ground, sometimes in such quantities as to cover it an inch thick. When the white may-flies (*eph. albipennis*) are very abundant, they present the appearance of snow on falling to the ground. The *e. longicauda*, Oliv. (n° 1), is the largest species of the genus, and on this Swammerdam made his very interesting remarks. It is frequently found at the entrance of the rivers of Holland and Germany.

The first genus of the hymenoptera is the *tenthredo*, so named from the saw-edged piercer with which the females are provided, and with which they make incisions in different parts of plants in order to deposit their eggs. The division *cimbex* of this genus comprehend the largest species of the family. We shall cite as examples the *c. femorata* (n° 1), *c. lutea* (n° 3), and *c. vitellina* (n° 6). Their larvæ when disturbed emit a greenish fluid, which they dart to the distance of a foot. The hymenoptera which follow have, for the most part, a piercer composed of three filaments, of which the two lateral ones serve as a covering to the other, which is the oviduct properly so called. The *ichneumons* deposit their eggs in the body of other insects, and particularly in that of the caterpillars and their chrysalis. In their larva state they do not injure the vital parts of the animal, and kill it only when they approach the time of their transformation. The *cynipes* make incisions in certain parts of plants, and cause those tubercles known by the name of galls, which we see on rose-trees, oaks, etc.; and which become a habitation and food for their posterity. The *chrysides*, whose brilliancy of colour is so remarkable, are associated with those hymenoptera which destroy the larvæ or nymphs of insects.

There are other hymenoptera which form a second section remarkable for their instinctive faculties, and which instead of a piercer have a sting. To this section belong several well known insects, in whose history we must necessarily feel interested, such as the ant, the wasp, the sphex, the bee, etc. We shall fix our attention however less upon the insects than upon their productions, which are here exhibited in three frames. In the first we see the honey-comb of a bee of Bengal (n° 1), the nest of the *vespa crabro* of Cayenne (n°s 2 and 3), and several other nests belonging to European species (n°s 4, 5, 6, 7). The second frame is devoted to our domestic bee. In the upper part of the third case are several nests of earth, bored with linear holes, which are the work either of the *sphex spirifex* or of the *apis lapidaria*, Lin. Below these are the galls and cocoons of different species of ichneumon and cimbex, and also a soft substance or down which is gathered by the *formica fungosa* of Cayenne. The *sphex compressa* (n° 51), known in the isle of France by the name of blue-fly, wages a continual war with the cockroaches.

The insects which we have hitherto noticed are all furnished with jaws; in those which follow the mouth is simply a tube of varied structure, calculated to suck or pump liquid substances. We shall begin with those which have been most favoured by nature, and which are decked with her richest ornaments, the butterflies, sphinxes and moths, collectively called *lepidoptera* by naturalists. Those which fly by day, and whose antennæ increase towards the point, are called butterflies; those which fly by night, and whose antennæ diminish gradually from the

base to the tip, are called moths.   The antennæ
of the sphinxes are in the form of a spindle:
M. Latreille has named them *crepuscularia*.   The
number of species which compose this order is
more than eight thousand.   The three genera of
Linnæus have been considered as so many families, which have been subdivided into genera.
The first cases contain the species of the genus
sphinx of Linnæus, from which the *sesia* and
zygæna have been detached.   The *sesiæ* have
glazed wings, and their body is terminated by a
brush of hair or scales; their caterpillars inhabit
the interior of plants.   That of the *s. apiformis*
(n° 1) feeds on the substance of the bark of the
poplar-tree.   We frequently see the *s. stellata-*
rum (n° 1), particularly in the autumn, hovering
around flowers, agitating its wings, passing with
rapidity from one to another, and dipping its
long trunk in their corolla.   The *s. tiliæ* (n° 17
and 18) is more commonly found on the trunk
of the elm; on the poplar we find the species
(n° 19) which takes its name from that tree.

A very beautiful caterpillar, that of the *s. eu-*
phorbiæ, Lin. (n° 25 and 26), lives on the spurge.
N° 54 is that of the sphinx, found upon the common privet (*s. ligustri*); the death-head sphinx
(*s. atropos*, Lin.,) owes its name to the disposition of the spots on its thorax.   Among the

European species we shall notice the *s. celerio* (n° 33), *s. elpenor*, Lin. (n° 23), *s. porcellus* (n° 41), *s. convolvuli* (n° 50), and that of the oleander (*s. nerii*, n° 60); the sphinx of the vine (n° 46) and the *s. labruscæ* (n° 59) were brought from the West Indies by M. Maugé. Most of the species which we have named are accompanied by the remains of their caterpillar and chrysalis.

The diurnal *lepidoptera*, or the genus butterfly, (*papilio*, Lin.) fill thirty cases. Almost all the names of the gods and demi-gods, as well as of the heroes and celebrated men of antiquity, have been adopted by Linnæus and Fabricius to designate the species. They have been divided into groupes which have names derived from the same sources; such as the *equites*, divided into Greeks and Trojans; the *nymphales, danaï, heliconii*, and *satyri*. Some smaller species bordering on the sphinxes compose the *plebeii*, which are divided into *urbicolæ* and *rurales*. The largest and most brilliant species come from the equatorial climates, especially from the Brazils and the Moluccas. The collection is too numerous to allow us to stop at every object worthy of attention; we shall therefore confine ourselves to the indication of a few species of the different tribes.

Among the *nymphales* may be noticed the *me-*

nelaus (n° 1), the *achilles* (n° 4), *laertes* (n° 8), *hecuba* (n° 14), and *idomeneus* (n° 21).   The species from n°ˢ 477 to 556 have been collectively designated by the denominations of pearled and chequered: the *heliconii* are distinguished by their their graceful forms.   In the beautiful division of the *equites,* or butterflies properly so called, we shall point out the *priamus* (n° 1), of which M. Godart discovered the female (n° 3), before considered as a separate species under the name of *panthous;* the *remus* (n° 5), the *helena* (n° 9), the *gambrisius* (n° 21), the *æneas* (n° 54), the *hector* (n° 64), the *ulysses* (n° 93), the *diomedes* (n° 95), which last, the same naturalist discovered to be the female of the preceding one ; the *paris* (n° 98), and lastly, two French species, the *machaon* (n° 122) and the *podalirius* (n° 120).   Another frame is filled with butterflies analogous in the form of their wings to those of the last division, and constituting the genus *urania;* among these, the *ripheus* from Madagascar (n° 154) is the most remarkable and rare.   The *leilus* (n°ˢ 152 and 153), common to Brazil and Guyana, and the *orontes* of India (n° 149) are also very beautiful.   Amongst the *danai candidi,* Lin., we shall notice two aurora coloured butterflies, the *cardamines* of our country (n°ˢ 49 and 52), and the *eupheno* (n°ˢ 57 and 58) ; the males alone are of this colour; they appear on the approach of spring, as does the *p. rhamni* (n°s 155 and 157). We may also notice four French lepidoptera, nearly allied to the *danai :* the two first, *apollo* (n°s 17 and 18) and *semi-apollo* (n°s 19 and 20), belonging to the genus *parnassius*, Latr., inhabit the Alps, and are much sought after by amateurs; the two others, *rumina* (n° 75) and *hypsipile* (n°s 76 and 77), of the genus *thais*, Fabr., are speckled with black and red on a yellow ground, and are found in the southern departments of France.

Next are the lepidoptera which, according to Linnæus, compose the section of the *plebeii*, subdivided into *urbicolœ* and *rurales*. The first, which are also named *argus*, fill two frames. We shall notice as examples those little butterflies which are common in meadows and lucerne fields (*corydon*, n° 131), and which, at least in one of the sexes, have the upper side of the wings of a beautiful blue, and the under side of an ashy colour with spots like eyes. The urbicolæ, keep their inferior wings nearly horizontal when at rest; they are here united with the *castniæ* under the generic name of *hesperiæ*, given at first by Fabricius to all the plebeii.

The third section, that of the moths of Linnæus, or nocturnal lepidoptera, is so numerous

that modern naturalists have been obliged to sub
divide it into twenty-five or thirty genera; the
principal of which are the *bombyx*, *noctua*, *pha-
læna*, *hepialus*, *pyralis*, *tinea*, and *pterophorus*.
The characters of these divisions have been derived from their simple or pectinated antennæ,
from the snout, the disposition of the wings, the
form of the caterpillars and the number of their
feet. These insects have not in general the brilliant and varied colours which strike us in the
diurnal lepidoptera, but their history is particularly interesting on account of their industry and
habits in the caterpillar state, as well as of the
injury which we experience from some, and of
the riches which we derive from others. Some
species are remarkable also, when in their perfect state, for their size, which almost equals
that of our bat. Most of the caterpillars of this
large family have within them two vessels filled
with a substance of which they spin their silk;
these terminate at the lower lip, whence proceed the delicate threads that compose the cocoon, in which they remain during the chrysalis
state.

The collection of nocturnal lepidoptera occupies eighteen frames, many of which are calculated to afford information concerning the different metamorphoses and the labours of these

insects. They contain the envelope of the caterpillar carefully preserved, or its representation
in wax, the chrysalis, the nymph drawn out of
its envelope, and lastly, the cocoons in which
the caterpillars enclose themselves, and the silky
nests which they prepare for the residence of
their community. In one of the frames is a nest
formed of a multitude of cocoons symmetrically
placed near each other like the cells of a honey-
comb; it is the work of a species, whose silk
is used in Madagascar.

The ten first frames enclose the genus *bombyx* (1), to which belong the *bombyx mori*, or
silk-worm moth (n° 123). Amongst the numerous species of this genus we shall notice the
phalœna pavonia major, Lin. (n°ˢ 30 and 31), the
largest of our indigenous lepidopterous insects,
exhibited here in all its states, together with
its work; the *p. atlas* (n°ˢ 1 and 13), a native of
China, still larger than the preceding; several
species known by the name of *luna* (n° 4), *cecropia* (n° 21), *promethei* (n° 23), and *tau* (n°ˢ 37 to

(1) The silk-worm lives in a wild state in several provinces of China;
its eggs were transported from that country to Constantinople by some
Greek missionaries under the reign of Justinian; from thence they
were conveyed to Italy, and the silk-worm was first known in France
in the middle of the fifteenth century; but it was not cultivated
till the time of Henry IV, when Sully caused plantations of mulberry-
trees to be made for its support.

40), the last only of which is indigenous; the *b. pythio-campa* (n°˙ 78 to 81), the *b. quercifolia* (n° 98); and other species (n°˙ 52 to 71) whose colours are disposed in stripes or chequers, and of which a section has been made under the name of scaly (1).

The moths (*phalenæ*), properly so called, have been collectively designated by the name *geometræ*, in consequence of the manner in which their caterpillars crawl. A great number of species have angular or denticulated wings; such as the *sambucaria* (n° 100), *alniaria* (n° 103), *syringaria* (n° 111).

Amongst the *noctuæ*, the *agrippina* (n° 1) is the largest species known; it measures more than ten inches between the tips of the extended wings. We may also notice the *odora* (n° 2), *bubo* (n° 3) and *crepuscularis* (n° 4). Several large species have the under wings red or blue with black stripes; as they are very pretty, they have received endearing names, such as *nupta* (n° 28), *electa* (n° 30), *sponsa* (n° 31). In these three species the ground colour of the wings is red, but

(1) The caterpillars of the *b. processionea*, Lin., are very remarkable; they live gregarious under a tent. When they wish to remove from one tree to another they proceed in a regular triangle; one of them keeping at the head of the column, two coming next, then three, until the column has attained a certain size, the whole band following their leader in all his windings on the way.

it is blue in that of the ash-tree (n°ˢ 25 and 26).
The *pyralis fagana* (n° 4) is one of the prettiest
species of this genus: the caterpillar of the *po-
mona* (n° 11) feeds on the seeds of apples and
pears: the cocoons of several species of pyralis
are in the form of a boat. The caterpillar of the
cossus ligniperda (n° 6) and that of the horse-
chesnut (*æsculi*, n° 7) live in the interior of
different trees and often destroy them. The *hepialus humuli* (n° 8) attacks the roots of the hop-
plant.

The name of *tineæ* is given to those lepidoptera which destroy fur and woollen cloths. The
industry of the caterpillars is very singular; of
the remains of the hairs they feed upon, they
make a tubular case to reside in, and lengthen or
enlarge it by the addition of new materials, as
they grow. The insects which are here united to
the tineæ or cloth-moths, and which are distinguished by their white wings spotted with black,
form at present the genus *yponomeuta*, Latr.;
their caterpillars, which are very small, live gregarious under a silken tent, and they often entirely strip of their leaves the trees on which
they establish themselves. N° 2 is the *tinea evo-
nymella*, Fabr. The last frame of the lepidoptera contains those species whose wings, from
their division and their down, resemble the plu-

mage of birds; they constitute the genus *ptero-phorus*, Latr.

The following order, that of the *hemiptera*, is composed of insects which resemble the coleoptera and the orthoptera in their wings, but which have a snout or sort of jointed beak. To this order belong the lantern fly (*fulgora*), the *cicada*, the bug (*cimex*), the plant louse (*aphis*), and other analogous insects. In the lantern flies the extremity of the head is more or less extended. The most remarkable in this respect is the *fulg. lanternaria*, or Peruvian lantern fly (n° 1), which is much sought after on account of its form and colour. We are assured that its snout is phosphoric, which has given rise to its name of lantern bearer. The *f. diadema* (n° 2) deserves also to be noticed.

There are a great number of cicadæ, all inhabiting hot countries, and differing much from the grasshoppers, which are called cicadæ in the north of France. The inhabitants of our southern departments are often tormented by the stridulous noise made by the males of some species, such as the *hæmatodes* (n° 12), and the *c. orni* (n° 25). This noise varies in different species. The description of the organ from which it proceeds and of the oviduct of the female is found in Reaumur's Memoirs. The Greeks eat the nymphs, which they called *tettigometræ,* and even the perfect insect.   The *c. orni* (n° 25) makes an incision in the flowering ash, whence exudes the honied and purgative juice named *manna.*   The genus *cicadella* is allied to the preceding.

The bugs are divided into two great families, the terrestrial or *geocorisæ,* and the aquatic or *hydrocorisæ.*   They are generally carnivorous and of an highly unpleasant smell.   The *reduvii,* Fabr., belonging to the family of the geocorisæ, have a sharp snout and a formidable sting.   The larva of the *cimex personatus,* Lin. (n° 5), inhabits our houses; it resembles a spider covered with dust, and feeds upon the common domestic bug.

To the family of the *hydrocorisæ* belong the genera *hydrometra, nepa* or water scorpion, *notonecta, naucoris,* and *corixa.*   The *nepa maxima* (n° 1) is a very large insect, from South America.

Those hemiptera in which the body of the females increases very considerably towards the season for laying, when they acquire the form of a gall nut, are called gall-insects.   To this family belongs the cochineal insect (*coccus cacti*), which yields so beautiful a colour, a native of South America, imported from Mexico with the cactus opuntia.   The order of the hemiptera is terminated by the genus *aphis,* or plant louse, which lives upon the leaves and tender shoots of plants and trees.

The next order, that of the *diptera,* comprehends the *musca,* the *culex,* the *tipula,* the *tabanus,* the *œstrus,* and a great number of other genera, not remarkable either in form or colour, but whose history is nevertheless worthy of attention. The gnats or musquitoes and the tabani are the scourges of men and domestic animals in hot countries; it is with difficulty that we preserve our meat from the flies which lay their eggs in it; the gadflies lay theirs on several parts of cattle; the *œstrus bovinus* (n° 5) attacks them, and draws blood by its bite; the *œ. maroccanus* (n° 7) torments the dromedary. The *bombylii,* like several other lepidoptera, hover around flowering plants, making a humming noise. The *stratiomides* are distinguished by their spiny scutellum; their larvæ, which are aquatic, as well as those of several *syrphi,* respire through the posterior extremity of their body, which has the form of a tube, and is shortened or lengthened by the insect at will. We cannot enter into more ample details, nor particularize a greater number of species; we shall content ourselves therefore with observing, that to the *tipulæ* belong those numerous swarms of flies which we often see hovering in the air on a summer's evening. The class of insects is terminated by the genus *pulex* (flea), which contains very few species, and of which M. de Lamarck has formed the order aptera.

The collection which we have just gone through occupies two hundred and four frames, 16 inches wide and 14 high, fifty-four of which are filled with crustacea, twelve with arachnides, and one hundred and thirty-eight with insects, that is to say, twenty-six of coleoptera, fourteen of orthoptera, six of neuroptera, ten of hymenoptera, seventy-four of lepidoptera, five of hemiptera and three of diptera.

The *annelides* were formerly classed among the worms, which they resemble at first sight. They differ however in their organization and in the colour of their blood; M. Cuvier made a distinct class of them, under the name of red-blooded worms, in a memoir read to the Institute in 1802. M. de Lamarck has since named them *annelides,* and his nomenclature is generally adopted.

M. Cuvier's observations upon the animals of this class led him to form a separate collection of them for the Museum, which has rapidly increased in consequence of the presents made by MM. Savigny and Dorbigny. M. Savigny has presented all the species which he collected in the Red Sea, and the series of the *lumbrici*, or common worms, upon which he has written an excellent treatise; and lastly, the collection lately brought by him from the Italian seas. M. Dorbigny has sent us a great number of species from La Rochelle.

M. de Lamarck divides the *annelides* into three orders : the first, that of the *apoda*, comprehends the family of the *hirudineæ*, to which belongs the leech, and that of the *echiureæ*. The second order, or that of the *antennatæ*, comprehends four families, the *aphroditæ*, the *nereideæ*, the *euniceæ* and the *amphinomi;* among them we may remark the *aphroditæ*, whose bodies are furnished on each side with numerous hairs with metallic reflections displaying all the colours of the rainbow. The animals of this order live in the sea; their body is not protected by a tube, and on this account they can be preserved only in spirits. They have been placed in the lower part of the cases which contain the crustaceæ, by the side of the worms; a still more numerous series of them is to be seen in the cabinet of comparative anatomy. The animals of the third order, the *sedentaria*, have their body covered with a tube which is generally calcareous. These tubes are placed next to the shells in the cases opposite to the parrots. This order is also divided into four families; that of the *dorsales* comprehends two genera, the *arenicola* and the *siliquaria*. The arenicola lives in the sand and makes no tube. There is but one species known, which has been named *piscatoria*, because fishermen use it as a bait. It is found in abundance on our shores. The siliquariæ form a calcareous tube; we have seven species of them.

The family of the *maldaneæ* comprehends the genera *dentalium* and *clymene*; we have eleven species of the first and none of the second. The family *amphitrite* consists of four genera, the *pectinaria* and the *sabellaria*, which form their tubes of grains of sand stuck together by a glutinous substance; the *terebella*, whose tube is composed of small shells, and the *amphitrite* with a membranous tube; we have two species of this genus. The last family, that of the *serpuleæ*, comprehends, 1st, the genus *spirorbis*, extremely small animals which attach their tube to shells and other marine substances: we have five species of them. 2d. The *serpula*, of which there are nineteen species in the cabinet, and among them several very rare ones, brought from the South seas by MM. Peron and Lesueur. 3d. The *vermilia*, of which we have seven species: one of them, *v. rostrata*, enveloped in a madrepore, and remarkable for the spine which projects from the exterior edge of its tube, comes from the seas of New Holland. 4th. The *galeolaria*, from the same seas. 5th. The *magilus*, of which one of the two species in the cabinet, the *m. antiquus*, from the isle of France, forms tubes several feet in length with a spiral termination.

The collection of intestinal worms was formed at Vienna by M. Bremser, who has made a particular study of this class of animals, and sent to the administration of the Museum in exchange for other objects of natural history. It consists of six hundred and six species of worms, taken from the interior of two hundred and fifty-nine species of vertebrated animals belonging to the four classes. Each bottle is numbered in reference to a catalogue, in which is found the name of the worm and that of the animal from which it was taken.

These worms are the more interesting, as each species belongs to a particular animal or to a particular part of the body, and produces a different malady according to its nature.

We observe among this numerous series, the ascarides from the human body (n° 4); those of the horse (n° 173); the tænia (n° 9); the hydatid from the sheep (n° 131); and the echinococcus (n° 12), found in the human liver, but happily very rare.

The manner in which these worms are intro-duced and propagated in the brain, liver, etc., is a problem that has long attracted the attention of naturalists, but which is not yet resolved. The examination and comparison of many species may perhaps throw some light upon the subject.

§ XI. COLLECTION OF INARTICULATED INVERTEBRATED ANIMALS:

COMPREHENDING THE SHELLS, THE ECHINI, AND THE POLYPI.

THE lovers of natural history have at all times taken pleasure in forming collections of shells, on account of the elegance of their forms and the beauty of their colours ; but these collections were objects only of luxury and of curiosity. As none but the most brilliant and rare shells were sought after, as their characters were even altered in order to augment their beauty by polishing them, and as the animals to which they belonged were not thought of, they were useless in zoology. But this is no longer the case; the researches of M. de Lamarck on conchology have proved, that the characters of a shell indicate those of the animal to which it belonged, as the genus of a quadruped is indicated by its teeth.

The distinction between terrestrial, river and sea-shells, and the comparison of those belonging to living subjects with those in a fossil state in different strata of the earth, has also led philosophers to decide upon the origin of different formations. In consequence of the numerous researches and the classification of M. de Lamarck, a knowledge of conchology has become one of the principal bases of geology.

The first shells in the cabinet were brought by Tournefort from the Levant, and presented by him to Louis XV. When Buffon had the superintendence of the Garden, he obtained permission to have them deposited there. Adanson presented those which he had collected at Senegal, and those which came from the cabinet of Reaumur were added.

Since the organization of the Museum, the acquisition of the cabinet of the Stadtholder, the shells brought by the naturalists of captain Baudin's expedition, those found by Richard at Cayenne, those collected by Olivier in the Levant and in Persia, those which have been recently sent from North America by MM. Milbert and Lesueur, and lastly, those which we have received from different correspondents, have made the collection very considerable; and the comparison of the fossil species with the analogous living ones, renders it valuable for study.

Those animals whose body is covered with a shell belong to the class of mollusca; but though

several animals of this class have only an interior
shell, and others have none at all, the affinity of
their organization does not admit of their being
separated.

The animals belonging to certain shells are
preserved in spirits of wine; those of a large
size, as also the naked mollusca, are placed at
the bottom of the third division of the chest of
drawers, immediately after the collection of
insects.

In going through the collection we shall follow the order adopted by M. de Lamarck;
deviating only by proceeding from the most
complicated animals to those which are more
simple.

The mollusca naturally form two divisions.
The first comprehends the univalves, or those
whose shell is formed of only one piece, and of
some analogous genera without shells. The second comprehends the bivalves, or those whose
shells consist of two pieces. There are aquatic
and terrestrial species belonging to the first division, but all those of the second are aquatic.

The univalve mollusca are divided into five
orders; the *heteropoda*, the *cephalopoda*, the *trachelipoda*, the *gasteropoda*, and the *pteropoda*.

The order heteropoda consists of the single
genus *carinaria*. The most celebrated species

for its delicacy and rarity is the *argonauta vi-treus,* Lin., found in the South seas. There are but three specimens of it known in Europe; we owe that in our possession to captain Huon, who commanded the expedition in search of La Peyrouse, after the death of Entrecasteaux. Near it, are the shell and the animal of a small and very rare species preserved in spirits of wine, which was discovered in the Mediterranean.

The order of the cephalopoda is divided into three sections; the *sepiaria,* the *monothalamiæ,* and the *polythalamiæ.* The *sepiæ* (cuttle-fish) have no shell, but under their mantle is a calcareous plate, which is horny in the *loligo.* Some large specimens of both these genera are placed at the bottom of the case, with the other mollusca preserved in spirits of wine. Close to their liver is a bag filled with a black or brown substance, which they make use of to darken the water around them. It is from this substance that the genuine Indian ink and the colour called sepia are made.

There is but one genus of the *monothalamia,* the *argonauta* (paper nautilus), of which we have seven species. The shell which is very thin, of an elegant form, and of a pure white, has no interior partitions; one from the Indian ocean (*a. tuberculosa*) is much sought after, under the name of tuberculated argonaut. These animals sail on the surface of the sea by means of membranous arms, which they extend in the manner of sails.

The nautilus, the ammonites, and the belemnites belong to the order polythalamia. We possess both species of nautilus. The umbilicated nautilus is very rare; the *n. pompilius* has been sawn to shew its internal compartments. The ammonites are found only in a fossil state; a prodigious quantity of them of different sizes, from one line to six feet in diameter, are contained in chalk and the first strata of secondary formation; we have more than twenty species. A fragment of one has preserved its delicate and pearly surface. The belemnites may be considered as nautili elongated into cones; they also are known only in a fossil state, as are likewise the other genera, whose shells are for the most part extremely small.

The order of the trachelipoda is very numerous: it comprehends the animals which crawl by means of a foot fixed to the neck. We shall notice the thirteen families of which it is composed.

The 1st is that of the *involutes*, to which belong the genera *conus, oliva, cypræa, terebellum,* and *ovula.* These shells are smooth, of an agreeable form and of varied colours.   We have one hundred and thirty-two species of cone, several of which bear a high price, such as the *c. cedonulli*, the *c. omaïcus*, the *c. obesus*, the *c. araneosus*, and a new species from the Chinese seas, presented to the cabinet by M. Dussumier. We have seventy-eight species of olive and seventy-four of cypræa; their colours are still more varied than those of the cones. The largests pecies is the *c. cervus;* the rarest is the *cypræa aurora* from New Zealand. A small species, found on the coast of Guinea, is used as money by the inhabitants of that country. There are five fossil species of cypræa. The genus *terebellum* contains only three species, two of which are fossil; that from the Indian ocean is very remarkable. We have ten species of ovula; the *o. birostris* and the *o. volva* are very rare: the pink variety of this last comes from the Indian seas, and was given us by M. Dussumier.

2d. The family of the *columellaria* comprehends five genera, all very numerous. They are generally small shells, the volutes alone are of large dimensions; at the bottom of the case there are several more than a foot in length, such as the *v. neptuni*, the *v. cymbium*. We have ninety-eight specimens of them belonging to forty-three species. One of the most expensive is the *v. vexillum*. We shall also point out the *v. diadema, javanica, junonia*, and *magnifica* of New Holland. There are several species marked with small spots on transversal and parallel lines, which have been named music shells. We have eighty-five species of the genus *mitra*, ten of which are found in a fossil state in France and Italy. The *mitra papalis* is remarkable for its size and its red spots. The *m. cardinalis* and the *m. sanguinolenta* are very rare species.

3d. The family of the *purpurifera* is divided into eleven genera, which comprehend two hundred species. Those of which the animals have been dissected have a vesicle filled with a colouring liquid. The species which has served for a type to the genus *purpura*, the *p. patula*, is found in the Mediterranean. It was from this animal that the purple of the ancients was made; but since the introduction of cochineal it is no longer used. The rarest of these shells is the concho lepas, which was brought from Peru by Dombey; that which is most commonly found on our coasts is the *buccinum undatum*.

4th. The family of the *alata* contains four genera which are remarkable for the extension of the opening of the shell. In the strombi this edge is entire, it is cut in the pterocera, and extended into a canal which is often as long as the shell itself in the rostellaria. The last genus, that of the struthiolaria, of which we have but two species, is very rare.

5th. The family of the *canalifera* is composed of three hundred and fourteen species, divided into eight genera, the most numerous of which are the cerithium and the murex. We have one hundred and twenty-three species of cerithium, sixty-four of which are fossil, and fifty species of murex. The *murex cervicornis* is the rarest of this last genus. The *m. cornutus* is known by the name of Hercules's club. The shells of the genus murex, which are carved and cut like leaves, are called endives. The rarest of this division is the *murex radix*. The calcareous stone, out of which the greater part of the houses in Paris are built, has been named *calcaire à cerites* because it contains a prodigious quantity of the *cerithium coronatum*. The *cer. ebeninum* is a very rare shell from the seas of New Zealand. The *fusi* are remarkable for their pyramidal form, and the length of their tube. Of the sixty-four species in the collection, thirty are fossil.

6th. The family of the *turbinacea* comprehends eight genera, the greatest part of which are worthy of attention either on account of their rarity or the beauty of their shells. We have twenty-eight species of turritella, eight of which are fossil : these last are the more interesting to geologists, as they partly characterize the jura limestone. The genus phasianella consists of fifteen species; one of which, the *bulimoïdes*, was so scarce forty years ago, that three thousand francs were given for a single specimen. The naturalists who accompanied captain Baudin's expedition, brought so many from Mary's Isle, that it is now to be procured at a very moderate price. Of the thirty-eight species of turbo which we have in the cabinet, we shall only point out the *t. marmoratus*, *t. coronatus*, and the *t. sarmaticus*. Next to this genus are ninety-four species of monodonta, several of which, from New Holland, are ornamented with the most lively colours. Next come ninety-two species of trochus, the rarest of which are the *t. solaris*, the *t. indicus*, the *t. imperialis*, and the *t. granatum*. The last genus of this family is the solarium; we have fifteen species, seven of which are fossil. The most singular is that known by the name of *perspectivum*.

7th. The family of the *scalaria* is composed of the genera *delphinula*, *scalaria*, and *vermetus*. We have eleven species of the delphinula, three of which are fossil. The *scalaria pretiosa* is one of the rarest and most esteemed shells; we have two specimens of it three inches long. The only known species of the genus vermetus was brought from Senegal by Adanson.

8th. The four genera which compose the family *macrostoma* are not very numerous; that of the haliotideæ consists of seventeen species, all very beautiful, amongst which the *h. iris* is undoubtedly the most brilliant of shells.

9th. The genera *natica, nerita, neritina, navicella,* and *janthina,* belong to the family of the *neritacea*: each of the three first comprehends more than thirty species. The neritinæ and the navicellæ are found in rivers; the last are from isles of France and Bourbon. We have three species of janthina of a lilac colour. The animals belonging to this genus have a cellular bladder, which they draw into their shell or extend beyond it, as they wish to remain under the water or to float upon its surface. We have a specimen preserved in spirits.

10th, 11th, and 12th. The families of the *peristomatæ,* the *melaniæ,* and the *lymnææ,* comprehend nine genera which inhabit fresh waters. The rarest species is the *cerithium fluviatile,* Lin., from Madagascar. We have twenty-seven species of the planorbis and twenty-five of the lymnææ. These shells are interesting, because when met with in the fossil state, they prove the strata to beof fresh water formation.

13th. The animals composing the numerous family *helicoidea* are very noxious on cultivated ground. The species most esteemed by conchologists belong to the genera *bulimus, achatina,* and *auricula.* We have one hundred and ninety-two species of helices; these are terrestrial shells from all countries, the most curious of which is the *h. vesicalis.*

Our knowledge of the next order, *gasteropoda,* is due to the anatomical labours of M. Cuvier, and from his observations M. de Lamarck has di-divided it into families. The greater number of the animals which compose it are destitute of a shell; we shall notice only those which are provided with one. The limax and other genera which have no shell are preserved in spirits of wine among the naked mollusca.

The family of the *bullœæ* is very numerous, but owing to the lightness and fragility of their shell, they are rare in collections; on this account that of the Museum which has twenty-four species is valuable to the student.

The family *calyptreæ* is composed of eight genera: the *crepidulæ* to the number of fourteen species, and the *calyptrææ* to that of twenty, are distinguished by the interior partition of their shell. In the *fissurellæ* there exists no partition, but the apex is pierced.

To the family of the *phyllidiæ* belong the genus *umbrella,* vulgarly called chinese parasol, a very rare shell of which we have four beautiful specimens; the *patella* of which we have seventy-five species and more than three hundred specimens, mostly exotics; the *chiton,* whose shell is formed of several small pieces so adapted as to admit of the animal's rolling itself into a ball in the manner of the woodlouse; we have twenty-five species.

The last order of univalve mollusca (*pteropoda*) is divided into five genera. We shall only mention the *hyalæa* and the *clio:* we have seven species of the first; the most common is the tridentata (*anomia tridentata,* Forsk.) which is seen in great quantities swimming on the surface of the water between the tropics. The *clio borealis* is so abundant in the North seas, that it forms the principal food of the whales and other large cetaceæ.

Here ends the series of univalve: we shall now examine those whose shell consists of two pieces, and which are named *conchifera.*

The first family, the *tubicola,* comprehends the genera *aspergillum, fistulana* and *teredo.* The two first come from the Moluccas, and are very rare. We have two species. The teredinæ, which form their tube in the interior of submarine wood, are natives of the same country; but they thrive as well in our seas, particularly on the Dutch coasts, where they do much mischief to the dikes. The the family of the *pholadaria* have the same habits, but they penetrate into the hardest substances, such as limestone-rocks and pebbles. They form two genera ; the *pholas*, which has a white and often a very large shell, and the *gastrochæna,* of which the shell is small and rather rare. We have but two species of this last.

The *solenaceæ* live under the sand : we shall particularize among them, the *solen,* vulgarly called razor shell, of which there are nineteen species in the collection; the *panopæa,* a fossil shell found at Placentia in Italy: the *anatinæ,* very fragile and much esteemed shells. The *anat. subrostrata,* the largest of the six species we possess, comes from the seas of New Holland.

The family of the *mactracea* consists of seven genera ; the most numerous is the *mactra,* of which we have twenty-seven species. The *m. triangularis* and the *m. spengleri* are very rare shells. The genus *crassatella* was known only in a fossil state when captain Baudin's expedition procured six species from New Holland. The genera *ungulina* and *solemya* are very rare: of the first we have only one species, and two of the second, one from the South seas and the other from the Mediterranean. This last specimen, which served M. de Lamarck as a type of the genus, was found on the sea shore at Hieres by M. Ch' Lacépède.

The genera *corbula* and *pandora* compose the family of the *corbuleæ*. We have ten species of corbula, of which four are fossil. The three species of pandora known (*tellina inæquivalvis*, Lin.) are found at St. Malo.

The *lithophagi* comprehend the genera *saxicava, petricola,* and *venerupis;* they inhabit stones which they have bored, as may be perceived by the specimens of the genus petricola.

Ten genera comprising more than eighty species, compose the family of the *nymphaceæ*. The genus *sanguinolaria* comprehends four species, one of which, known by the name .of *sol occidens,* is a rare shell. Forty-seven species belong to the genus *tellina;* they are for the most part of a beautiful rose colour, from which circumstance they have been named *rising suns.* The *tel. foliacea* is much sought after, and the *tel. lingua felis,* as well as some other species, are remarkable for the asperity of their external surface.

The family of the *conchæ* comprehends one hundred and seventy-two species, divided into sea and river shells. These last compose three genera; the *cyclas*, which are very small shells, inhabiting our ponds and rivers; the *cyrena*, more especially found in India and America; and the *galathea*, a very valuable shell from Ceylon. The sea or marine concha, of which there are more than a hundred and sixty species, are distributed into four genera, the *cyprina*, the *cytherea*, the *venus*, and the *venericardia*. The venericardia is known only in a fossil state; it is very common at Grignon and throughout Champagne. There are more than seventy species of the genera cytherea and venus, all very remarkable from their form and the variety of their colours. The rarest among them are the *v. erycina*, the *v. dionæ*, the *v. plicata*, the *v. lamellata*, which are all foreign; and the *v. decussata*, known in Provence under the name of *clovis*.

The *cardiacea* consist of more than one hundred and fifty species, divided into five genera, of which the *cardium* is the most numerous. It is seldom that the two valves of the *cardium costatum* are found belonging to the same individual: the specimen in the Museum is of an uncommon size. The *c. æolicum, cardissa, junoniæ,* and *unedo* inhabit the seas of hot countries: they are not very common, and are remarkable for their form and colour. The *isocardiæ* are very singular shells in the form of a heart, the extremities of which are pointed and twisted like ram's horns. The *is. cor* inhabits the Mediterranean; another, the *is. moltkiana*, a very rare species, was brought by M. Dussumier, from the Chinese seas; a third, the *is. semi-sulcata*, comes from New Holland; and a fourth, the *is. arietina*, is found in a fossil state in Italy, whence it was brought by M. Cuvier.

The four genera, *cucullæa, arca, pectunculus* and *nucula,* compose the family of the *arcaceæ.* There are only two species of the genus *cucullæa* known. The *c. auriculifera,* which is recent and rare, inhabits the Indian seas: the other, found in the fossil state near Beauvais, was presented by M. Lucas, jun^r. We have twenty-nine species of the genus *arca;* we shall point out as among the rarest the *arc. semi-torta,* the *ovata* and the *tortuosa.* The shells which belong to the genus *nucula* are very small and brilliant: we have seven species, one of which, the *arca nucleus,* Lin., is common on the coasts of France.

Next to the *arcaceæ* are the *trigoniæ;* to which belongs the *trig. pectinata,* a very valuable shell, from its rarity, and its being the only recent species known; it was brought from New Holland by Peron. Many others are found in the same strata with the ammonites.

The family of the *nayades* is altogether composed of river shells. Of the four genera into which it is divided, the most numerous is the *unio*, of which there are forty-six species in the collection, nearly all inhabiting the lakes and rivers of North America. We are indebted for them to MM. Michaux, Lesueur and Milbert. The *unio pictorum* and the *littoralis* are found in the Seine. The two species of the genus *hyria* come from Ceylon. Amongst the nine species of the genus *anodonta* the *a. cignorum* lives in our ponds, and is so named because it is eaten by swans.

The *etheriæ*, the *chamæ*, and the *dicerates* compose the family of the *chamaceæ*. We have but one species of *etheria*, the *elliptica*, from the Persian gulph, which is the rarest, largest, and most beautiful of the genus

The second and last order of the conchifera is that of the *monomyaria*, which is divided into seven families.

1st. That of the *tridacneæ*, formed of the two genera *tridacna* and *hippopus*. We have six species of the tridacna, one of which, the *chama gigas*, Lin., is sometimes more than 6 feet in circumference, and weighs upwards of six hundred pounds. Peron mentions having seen at Timor, specimens of this shell which two men could with difficulty move. There is but one species of the *hippopus* in the collection.

2d. The *mytilaceæ* comprising the genera *pinna, modiola* and *mytilus.* The animals of this family, as also those of the next, attach themselves to marine substances by means of filaments called *byssus,* which being very delicate and silky in two species inhabiting the Mediterranean, are wrought into gloves and other articles, and when mixed with wool and woven into cloth they give it a sort of golden lustre. The *pinna squamosa* is sometimes found three feet long. Amongst the *modiolæ,* of which we have six species, we shall notice the *m. tulipa,* which is of a pearly white, tinged with red, and the *m. lithophaga,* which penetrates into calcareous rocks. There are twenty-seven species of *mytilus* in the collection: that from Japan, that from the straights of Magellan, and the opal muscle, are much valued on account of their rarity and the beauty of their colour.

3d. The family of the *malleaceæ,* comprehending the genera *crenatula, perna, malleus, avicula* and *meleagrina.* We have four species of *crenatula,* found in the seas of New Holland; and ten of *perna*, the largest of which is the *p. maxilla-ris*, found fossil in Italy and America. The genus *malleus* derives its name from its form, which is like that of a hammer or the letter T: these shells are remarkable for their size and the small-ness of the cavity in which the animal is lodged. The white species is very scarce; we have only two specimens from the Molucca islands. The shells belonging to the genus *avicula* have one side only of the valves very much extended, and when open they appear like two wings borne on a long tail. The largest of the eleven species in the collection is the *a. macroptera :* of the ligaments of these shells are made the jewels known by the name of peacocks' eyes, a specimen of which is here seen. The species of the genus *meleagrina* are not very numerous, but they deserve to be noticed, as one of them is the pearl oyster (*mytilus margaritiferus*, Lin.), which is taken at the depth of many fathoms in the Persian gulph. The pearls are produced by an exudation from the interior of the shell, which is itself wrought into sundry small utensils, ge-nerally known under the name of mother of pearl. One of these shells is remarkable for the great number of pearls adhering to it. This spe-cies, or one nearly allied to it, is also found on the coast of Cumana.

4th. The family of the *pectinideæ* comprehending seven genera. The first, *pedum*, is composed of only one very valuable species from the Indian seas. We have forty-six species of the genus *pecten*, all fine and brilliant shells. The most rare and beautiful amongst them are the ducal mantle, the sole, and the bishop's mantle (*p. bifrons,* Lam). The *spondyli* have still more brilliant colours; there are nineteen species in the collection, amongst which we shall notice the *radians* and the *longispina*.

5th. The family of the *ostraceæ*, composed of the genera *gryphæa, ostrea, vulsella, placuna* and *anomia*. We have sixty species of the second genus and five or six of each of the others. It was long supposed that these shells existed only in a fossil state; but we now have one, the *gryphæa angulata* which is recent and is considered as very rare and valuable. The *vulsellæ* are found in sponges. The species of *placuna*, vulgarly called the saddle oyster, is much sought after on account of its form; our specimen is of an extraordinary size. The *anomiæ* are particularly remarkable for the hole at the extremity of their lower valve, from whence issues the pedicle with which the animal fixes itself on the rocks.

6th. The six genera which compose the family

of the *rudistæ,* which are all fossil, except one lately sent us from the isle of France.

The 7th and last family, that of the *branchiopoda,* has but two genera, the *lingula* and the *terebratula.* Only one species of the first is known under the name of duck's bill; it is a very singular shell, from its two valves being supported on a long peduncle, by means of which it fixes itself upon different bodies. We have seventy-two species of the genus terebratula, sixty of which are fossil; the recent species are sought after by conchologists.

The following class is the *tuniciers,* which M. de Lamarck has placed between the conchifera and radiaria, conformably to the recent observations of Savigny, Lesueur and Desmarest. The greater number of its genera are known only by their descriptions, and we possess only about thirty species, which are preserved in bottles next to the naked mollusca. They are marine animals, without heads, and not symmetrical. Some of them are free, others are attached to fuci and various substances. They are divided into two sections, the combined *tuniciers* or botryllariæ, and the free *tuniciers* or ascidiæ. The first order contains those that are agglommerated, so that several individuals appear to be animated by a common life, and to form but one animal; the second, those that are disunited, and which have no internal communication, although they are often grouped together or lie very close to each other. We observe among those of the first section, 1st, the *pyrosoma atlantica*, discovered by Peron: it is an hollow cylinder, closed at one end, open at the other, and covered with tubercles, which are so many little animals. It is so highly phosphorescent, that where it abounds the sea appears covered with burning coals, from whence its name, which signifies body of fire. When its phosphorescence ceases, it assumes various colours. 2d. The *sinoïcum aurantiacum*, brought from New Holland by Peron, and of which M. Lamouroux has formed the genus *telesto*: a dried specimen may be seen in one of the cases of madrepores near the tubulariæ. 3d. A new genus brought by M. Delalande from the seas near the Cape. Examples of the second section will be found in the *salpæ*, which float like long ribbons and form garlands on the surface of the sea in warm climates; and in several *ascidiæ*, such as the *a. mammillaris*, which is found on our own shores, and the *a. conchilega* from the Cape.

The class of the *radiaria*, to which we shall next proceed, has been so named on account of the body's being formed in the manner of rays.

M. de Lamarck, availing himself of the labours of Peron on these animals, has divided them into two orders, the *soft radiaria* and the *echinoder-mata.* Those of the first order have been divided into two sections, the *anomala* and the *medu-saria.* They have all gelatinous bodies. There is a numerous series of them in the cabinet of anatomy. Amongst those under our inspection we shall point out in the first section the *beroe* and the *noctiluca*, which are the principal cause of the phosphorescence of the sea. The *phy-salia*, a specimen of which, placed near the echini, shows its internal and cartilaginous blad-der. Persons who have travelled beyond the tro-pics have been struck with the beautiful colours of this animal, as well as with its form, which is like the hull of a ship. In the second section we remark the *cephea rhizostoma*, Peron, whose gelatinous body, skirted with purple, is often found on the shores of the English channel.

The *radiaria* of the second order have a hard and tough skin, frequently crustaceous; they are divided into three sections, the *stelleridæ*, the *echinides* and the *fistulides.*

The *stelleridæ* or sea-stars have a flattened and circular body, whence issue five principal rays which are sometimes subdivided. These animals feed on worms and small crustaceæ. They pre-

sent a striking phenomenon in the rapid reproduction of such parts as they have lost. In very
hot weather two or three days suffice for completing the process; and what is much more extraordinary, if one of the rays is severed from
the body, it soon after becomes a star-fish similar
to the one from which it was separated. By
considering the form and the disposition of the
rays, the characters of the four genera which
compose this family have been determined: these
are the *comatula, euryale, ophiura*, and *asterias*.
The *comatulæ* are contained in six cases; we have
seven species, four of which were brought from
the South seas by Peron and Lesueur. The *c. solaris* is the largest, and measures 12 inches in
diameter. From one specimen of the *c. rotularis*
we may see the attitude which these animals
adopt in order to seize their prey; suspending
themselves to the fuci or madrepores by means
of their dorsal rays, and extending the others, they
catch the small crabs which happen to swim
within their reach. Various species of *euryale*,
vulgarly called medusas' heads, occupy the four
next cases; they are remarkable for the great
number of divisions in their principal rays: we
have four species of them. The rays of the
ophiuræ are sometimes smooth and sometimes
rough; there are seven species and a great num-

ber of varieties in the collection. There are thirty-seven species of *asterias* which occupy twenty-seven frames. The lobes of some of them are so very short as hardly to surpass the disk, as is seen in the *ast. discoidea, rosacea* and *calcar;* but they are long in the greater number. The most common species on our coasts is the *a. rubens.* We shall also point out the *a. helianthus* and the *a. echinites.*

The *echinides,* otherwise called urchins, have a calcareous shell covered with long spines, which are sometimes articulated, and fixed upon moveable tubercles. This shell is pierced with a great number of small holes, from each of which issues a tube or retractile sucker, through which the animal breathes; they are regularly set, and have been named *ambulacra;* sometimes they extend all round the body and sometimes only over a part: their number, their form, and the respective position of the two intestinal canals, serve to characterize the eleven genera of this section. The very complicated apparatus of the mouth of the urchins has been named lanthorn. We may here see a specimen of the whole apparatus as well as of its several parts. The echinides feed on small shell fish. There are one hundred and seven species in the collection: among which we shall particularly notice the *scutella latissima;*

the *ananchites striata,* which is found in a fossil state in the chalk at Meudon, near Paris; the *echinus melo,* which is sometimes more than 8 inches in diameter; the *echinus esculentus* and the *lividus,* which are eaten by the inhabitants of Mediterranean coasts; the *ech. variegatus, radiatus* and *atratus,* which have variegated colours, and the *ech. tuberculatus,* whose spines are implanted on large round tubercles; the *cidarites imperialis, verticillaris, calamarius, baculosus,* whose spines are very remarkable; and the *cid. tribuloïdes,* which has close to its mouth a short spine in the form of a club: this remark acquires a degree of interest from the great number of similar bodies which are found in a fossil state.

The *fistulides,* which compose the third section, have a tough fleshy skin: they have been divided into five genera; twenty species of them are placed next to the radiaria and to the mollusca preserved in spirits. We shall only notice the *sipunculus edulis,* which lives buried under the sand in the Indian seas.

The class we have yet to examine is that of the polypi, the last in the animal kingdom; they are complex animals, some of them naked, others protected by an envelope which they construct and inhabit in common, and which is extremely remarkable.

The general aspect of these habitations, which are called *polypi*, and the interior form and distribution of the cells, are so diversified that they have furnished the grounds of a division into orders and genera. We have a very numerous series of them: but it may be proper to say something of the animal itself, before we speak of the characters of the orders.

The animals we have hitherto considered, are provided with an interior system of organs more or less complicated; their size, and the consistency of their flesh or envelopes, have admitted of their being preserved dried or in spirits; but so excessively small, soft, and glutinous are the polypi, that the instant they are taken out of the element in which they live, they dissolve into a watery substance, in which it is with the greatest difficulty that any trace of their organization can be seen.

However it has been observed through the transparency of the water, that their body is of a cylindrical form, like a tube, shut at one end, and furnished at the other with moveable tentaculæ, by means of which the animal seizes its food, rejecting the superfluous parts by the help of the same tentaculæ and by the same aperture.

These tubes considered separately are so many distinct animals, but as they are in a manner grafted together by the lower extremity, and partake of the same nutriment, they may be said to constitute but one complex animal.

Although the polypi are the most simple in their organization of all living beings, their vital phenomena, when first described by Trembley, excited to the highest degree the interest and astonishment of naturalists. We shall here mention some of the observations made on the naked or soft polypi, and more particularly on a species of the genus *hydra*, which is common in our stagnant waters. The *hydræ* propagate themselves by means of *gemmæ*. A small bud grows on the tube, lengthens and becomes a new animal, which either remains fixed on the first, or falls off, and in turn becomes the parent stalk on which new individuals branch in the same manner. They also multiply by being severed, and however small the parts into which an hydra is cut, they soon become complete animals. The *hydra viridis*, the form of which is that of a bag, can be turned inside out without the least inconvenience to the life or habits of the animal, the exterior surface, now become the interior, performing the function of digestion.

The second order of the polypi is composed of those which have a calcareous or horny envelope, the result of an exudation from their body, by which they are protected, as the mollusca by their shells. As these tubes, which by their assemblage compose what is generally called a polypus, have been moulded upon the animal itself, they give us an idea of its form ; but this notion being too imperfect for naturalists, M. Savigny lately undertook to examine several species of this order, and discovered in their organization the most curious and extraordinary phenomena.

A very remarkable circumstance amongst the numberless animals which unite to construct a polypus, is the regularity and order with which they work and branch their cells, so as to obtain for the whole set a general form or outline, which, differing in the various species, is uniform in each. Some are wrought into rounded masses, others into blades or leaves, the shape of a cup, or that of a shrub. Nothing can be more curious than to observe, at low water, on the shores of the tropical islands, the polypi so diversified in their form and colour, with their surface in perpetual motion from the activity of the small animals half protruded from their cells.

The polypi act a very important part in the creation from the prodigious facility with which they multiply. To them is due the existence, extent, and rapid increase of those shoals which are so frequently met with and are so dangerous to navigators in the equatorial seas, and more particularly in the latitude of New Guinea. They fill up straits, unite islands, which rest upon heaps of these polypi; and, as Peron says, whilst man is, with the utmost labour and difficulty, endeavouring to build on the surface of the earth fabricks which at once betray his weakness and the power of time, a small insect, of whose existence he was not aware, erects in in the depths of the ocean those astonishing monuments of a power that bids defiance to ages. Geologists frequently meet with whole strata formed of the remains of polypi, so that the history of this class of animals presents the most curious problems, and is connected with the earliest revolutions of the globe.

There are two sorts of polypi, those which are formed of one substance only, with the cells running through the whole mass of the common receptacle; and those which consist of two substances, one forming the axis and the other its bark or envelope. The animals belonging to this second sort are lodged betwixt the bark and the axis. They are called *dendroides* from their similitude to small trees, which has induced several naturalists, and Linnæus himself, to think that they partake of the nature of plants as well as of animals. But Bernard de Jussieu, Ellis, and others convinced themselves that these envelopes are destitute of organization, and serve only to protect the delicate animals which construct them; and their opinion, adopted by M. de Lamarck, is now generally received.

The collection in the Museum is composed of one thousand specimens belonging to five hundred and fifty species. It was begun in 1795 with those procured from the Dutch cabinet, to which were added the beautiful collections made by Maugé, Peron, and Lesueur in the West Indies and the South seas, and some no less curious specimens presented by other travellers. They are arranged in the lower part of the chests of drawers, and divided into five orders according to the method of M. de Lamarck.

The first is that of the floating polypi, which are nearly allied to the radiaria. They are united into an elongated, fleshy, living mass, adhering to a calcareous or cartilaginous axis. Several species of the genera *veretillum, funiculina,* and *pennatula* are preserved in spirits; and dried duplicates of them are placed in the frames with the axis of the *virgularia australis,* procured by M. Leschenault from the island of Bali. By the side of the urchins we see the osseous trunk of a *pennatula* upwards of 5 feet long, from the

Brasilian coast, presented to the Museum by M. Langsdorff. Next to it is placed the *encrinus* or medusa-head, one of the rarest specimens in the cabinet; it was caught in the gulph of Mexico at a considerable depth; there is only one other known in Europe; its form is that of a tree with the branches horizontally spread. Fossil *encrini* are abundant in the strata of secondary formation. The name of fossil palm-tree is given to the entire animal, and to its separate parts according to their form, those of trochites and star-stone.

The second order, the *tubiferi*, has been established after the observations of M. Savigny We have but two species of the order *lobularia*, all the others belonging to that order are wanting in the collection. The *l. digitata* was found in the Atlantic, and the *l. palmata* in the Mediterranean.

The third order, that of the polypi with a solid envelope, is by far the most numerous, and our collection is very rich. We shall begin with those of a small size placed in the frames, and afterwards examine the larger specimens arranged in the chests below. M. de Lamarck has subdivided this order into seven sections: the polypi *incrustati, corticati, lamelliferi, foraminosi, reticulati, vaginiformes,* and *fluviatiles.*

The genera *alcyonium, spongia, flabellaria,* and *penicillus* compose the first section. We have only one species of the *penicillus*, the *p. ca-pitatus*, from the American seas.

Amongst the twenty-eight species of the genus *alcyonium*, which are contained in five frames, is the *a. purpureum*, from the seas of New Holland.

Sixty-five species of the genus *spongia* are arranged in nineteen frames; they all differ in their form and texture. The animals which inhabit them are still unknown; but it has been observed, that, when in their element, the sponges have a glutinous envelope or pulp which is irritable, and by comparing them with the alcyonia, there is no doubt that they are formed by very small, pellucid animals. When taken out of the sea this glutinous pulp becomes brittle and soon falls off; the main body of the sponge is still flexible because it is composed of elastic fibres, and the numerous holes or vacuities in its texture admit and retain the water. We may have a still better idea of the manner in which a sponge is formed by comparing it with the *gorgonia*, a genus of the following section, of which we have twenty-six species, occupying twelve frames.

The *gorgonia* is composed of a horny axis covered with a glutinous and calcareous paste, which becomes crisp by desication: if we imagine this axis reduced to the thinness of a horny thread and ramified with the branches crossing each other and uniting by the end *ad infinitum*, we shall form an exact idea of a sponge. The holes, which are seen in every direction, let in the water to the animals, which occupy the interior.

The genera *corallium, melitœa, isis, antipathes*, and *corallina* compose, with the *gorgonia*, the second section, or the *polypi corticati*. Of the melitæa we have but four species; they are very various in their colour and occupy nine frames. Several specimens of the white and of the red *corallium* are placed in one frame; some of them have their marine envelope, others are without it, polished, or even carved to shew the different forms that are given to that production. Six frames are filled with seventeen species of the genus *corallina*, and one contains five species of the genus *anthipathes;* their appearance is much like that of branches of heath, cypress, etc. The axis of the genus *isis*, is articulated and composed of horny and calcareous substances. One of the five species in the collection, *isis encrinula*, from the seas of New Holland, is very rare.

The *lamelliferi* are placed in the lower part of the chests with the *foraminosi* and *fluviatiles*.

In the section of the *reticulati*, we may remark the genera *cellepora* and *discopora*, whose fibres are excessively delicate, and the *retepora cellulosa*, vulgarly known under the name of neptune's ruffles, several varieties of which were brought home by Peron and Lesueur.

Amongst the *vaginiformes* are the *polyphysa australis*, the *sertulariæ* procured by the same travellers from New Holland, and the *acetabulum marinum*, Tournef., very remarkable on account of its form, which is that of a funnel supported by a long and thread-like pedicle, round which many young polypi shooting give it the appearance of a cluster of mushrooms.

Among the rarest and most curious species in the lower part of the chests we shall cite, of the *incrustati*, the *alcyonium cidaris* from the Mediterranean; the *a. cuspidiferum*, which resembles an assemblage of stalactites; the *a. arboreum* from the Indian seas; the *a. vesparium* from the same latitudes, whose form is that of a wasp's nest; a specimen of it, sawn length ways, shews its interior structure; the *tethia asbestella*, procured by Bougainville, at the entrance of the Rio de la Plata, and several specie of sponge, such as the *penicillata, flabelliformis, perfoliata, pella, calix,* and *mesenterica*, brought from the South seas by Peron and Lesueur; the *spongia basta* from the Indian seas, the *s. lacunosa* and *bursaria;* and the *licheniformis* which includes a great number of varieties: of the *corticati*, the *gorgonia pinnata*, the *g. laxispina* from the American seas, and the *g. flammea*, a native of the Indian seas, so remarkable for its scarlet hue: in the third section that of the *lamelliferi*, a beautiful specimen of the *oculina flabelliformis*, a very rare species from India; and many fine madrepores, such as the *palmata*, *plantaginæa*, *corymbosa* and *cervicornis*, some from the Indian, others from the American seas, the *seriatopora subulata* from India, the *pavonia agaricites*, and the *lactuca* from America, the *astrea punctifera*, the *caryophyllia truncularis*, *sinuosa*, and *fasciculata*, the *fungia limacina* and *agariciformis*, the *meandrina cerebriformis* and *labyrinthica* from the American seas: among the *foraminosi*, the *millepora complanata* and *alcicornis*, and the *tubipora musica*, which was observed by Peron on the shores of Timor; this polypus exists in half rounded masses of a vivid red colour, on which the animals expand their fringed tentaculæ of a glossy green, and these orbs floating on the surface of the waters have the appearance of verdant sods upon a coral ground. Lastly, among the *fluviatiles* we may remark a beautiful specimen of the *alcyonella stagnarum*, found by M. Robineau in a pond in Burgundy.

CHAPTER III.

CABINET OF COMPARATIVE ANATOMY.

The zoological department has presented us with the remains of animals so prepared, that we may recognize at the first view all the species, and observe the slighter differences which distinguish them, as well as the more striking characters which mark the groupes designated by the names of classes, orders, families and genera; but the external forms are only a developement of the internal organization, and our general notions are always derived from the indications afforded by these outward forms. Therefore, to establish a general classification in zoology, we are obliged to call comparative anatomy to our aid, in order to ascertain its principles: it is anatomy which determines the relation of the different species, shews the affinities and demonstrates the respective importance of the organs to the general form of the animal, its food, manners and habits. In short, anatomy enables us to compare living ani mals with the remains found in a fossil state, belonging to races that are extinct, but which existed at separate and very remote periods. But the plan of classing animals according to their organization was attended with insuperable difficulties, as details were wanting to resolve the most important questions, and as the labours of several anatomists only made more evident the necessity of a complete system of comparative anatomy.

Anatomy was for the first time methodically applied to zoology in the King's Garden; and the association of the anatomical works of Daubenton with the general views and descriptions of Buffon, gave celebrity to their joint labours. It was however confined to quadrupeds, and illustrated only such of them as Daubenton had been enabled to dissect with the assistance of Mertrud.

A new career was opened for the study of the natural sciences when, at the organization of the Museum, a chair was instituted for the anatomy of animals. Mertrud, who had been thirty years demonstrator of anatomy in the King's Garden, was named to fill it; but as his declining age did not permit him to deliver his lectures, M. Cuvier, at his request, was appointed his substitute in 1795.

M. Cuvier, desirous of embracing anatomy in its greatest extent, undertook to form a collection which should not only present the skeletons of all animals, as well as their soft parts preserved in spirits, but also comparative series of the organs of the same nature taken from each animal. He began by arranging all that the Museum possessed, setting apart what was most important for study, as also what was intended to serve as a basis for the general collection, such as the skeletons prepared by Daubenton and Mertrud, to which were added those made at the menagerie of Versailles for the Royal Academy of Sciences, by Perrault, La Hire and Duverney. After the publication of Buffon's work these skeletons were stowed in the upper part of the building, which has been since converted into the library. They were mostly broken and in very bad condition, except a few large ones, such as the camel, the African elephant and the rhinoceros, which had been prepared by Mertrud and Vicq-d'Azyr in 1793.

The closets of the cabinet contained only a few skeletons, sculls, and preparations in spirits; some very incorrect imitations in wax from the human body, by Zumbo; the bones of the ear and injections by Duverney and Hunault; and other imitations in wax of several parts of the human body, executed and presented to the Museum by Pinson

M. Cuvier was not deterred by the immensity of the enterprise, the difficulties that were to attend it, nor the time required for its execution. He set about it in 1796, upon as large a scale as if he had actually had every desirable resource at his disposal. He instructed and formed assistants; and having inflamed the zeal of his pupils by his lectures, he soon found amongst them, men of distinguished merit who zealously aided in his dissections. Such was the activity of their labours, that as early as 1806 students were admitted to the cabinet of anatomy. The series were then numerous, they have increased every year, and they are now so extensive as to form incomparably the richest collection in existence.

All the objects with a very few exceptions, were prepared at the Museum, and were procured from the animals of the menagerie, or purchased in markets or ports, or furnished by travellers. They were reduced to their present order by M. Cuvier himself, or under his direction by M. Rousseau and other anatomists; and M. Laurillard, a pupil of M. Cuvier's, is charged with the care of their preservation and the guardianship of the cabinet.

The first room on the ground floor contains

the skeletons of the horse, the ass, the zebra and the quaccha, also those of the American tapir, of the common hog, of the *dicotyles*, Cuv., and of a new species of tapir, sent from the East Indies by MM. Diard and Duvaucel. In the next room we see the skeletons of the large carnivorous animals, those of the *pachydermata* and of the *cetacea*, such as the male and female elephants from India, which lived in the menagerie, and the female of the African species, which was prepared by Duverney: these last shew the error of several travellers, who have maintained that the elephant has no joint in the leg like other quadrupeds. There are also a skeleton of the hippopotamus, one of the rhinoceros from the Cape, recently brought by M. Delalande, and one of the species from Senegal, sent by M. Roger, governor of that settlement. There are also six other skeletons of the rhinoceros: one prepared by Mertrud, three new species from the island of Java, and two others procured at Sumatra by MM. Diard and Duvaucel, and a complete skeleton of a cameleopard more than 14 feet high, sent to Europe by colonel Gordon. The skeletons of the carnivorous animals, such as the bear, dog, wolf, hyæna, lion, tiger, panther and the seal, are arranged upon the shelves on one side of the room: and on the other, several species of dolphin, the rarest of which is the dolphin of the Ganges, sent by M. Wallick, director of the East India company's botanical garden at Calcutta. We see amongst them also the *delphinus globiceps*, which we owe to M. Lemaout, who selected it from a number of the same species run aground on the coast of Britany; and lastly, the skeleton of a large seal (*manatus*), and that of the dugong (*halicore*, Illig.), lately sent from Sumatra by MM. Diard and Duvaucel. In the middle of the rooms are supported by iron bars, three whales, procured at the Cape by M. Delalande. The horny substance called whalebone, which lines the interior face and the edges of the upper jaw, when the animal is alive, has been preserved in its proper place. The extremity is divided into numerous filaments, which serve as a net, by means of which the whale catches the small fish and mollusca on which it feeds. This substance is much employed in the arts, and is produced only by the largest and the smallest of the three species in our possession. On each side of the window, at the upper end of the room, we see the head of a whale and that of a cachalot (*physeter macrocephalus*, Cuv.), each measuring 14 feet in length : an entire skeleton of this last, more than 60 feet long, is seen in the adjoining court. The substance known under the name of *sper-*

maceti is taken from the cavity in the head of this animal, and *ambergris* is a concretion formed in its intestines.

To the left of this large room, and parallel with it, are three others filled with skeletons of the ruminating quadrupeds. In the first are those of the ox, sheep, goat and antelope, the horns of which are hollow and permanent; in the second, those of the stag tribe, or animals with deciduous horns; and in the third, those of the dromedary, camel, lama, and vicunna, which are destitute of horns.

In retracing our steps and crossing the room containing the whales, we enter another occupied by human skeletons of different ages and nations: among them we remark that of an Italian with one additional lumbar vertebra; that of an ancient Egyptian, prepared from a mummy, on which are observed a great number of fractures perfectly cured; the skeleton of a Boschisman female, known in Paris by the name of the Hottentot Venus, with a cast of her standing by that of the celebrated dwarf of Stanislaus king of Poland; and also a model in wax of the skeleton of a woman named *Supiot,* whose bones had become so soft that they were all distorted. A series of fœtuses, shews the growth from the first month of conception to the birth. On the shelves we see human skulls of different ages, from one day old to a hundred years, some of them remarkable for the singularity of their form, and chosen for the most part from amongst the prodigious quantity preserved in the catacombs.

From the walls of the staircase leading to the first floor are suspended many heads of the horse, the stag, the dolphin, the hippopotamus, and several species of the ox tribe of different ages and varieties.

The first room above stairs is devoted to a series of entire heads of vertebrated animals—a great number of those of the human species, Europeans, Tartars, Chinese, New Zealanders, Negroes, Hottentots and of several American nations; all the monkeys, amongst which are an an old and a young oran-outang, the examination of which led to the opinion, that the pongo of the island of Borneo is of the same genus and perhaps of the same species with the oran-outang; several of the *simia lar* lately sent over from India by MM. Diard and Duvaucel; a considerable number of all the carnivorous animals, amongst which are those of several species of seal; every known species of the *edentata,* and almost all the *rodentia:* some in the first of these families belong to newly discovered species. Amongst the *pachydermata* we remark

the heads of the *phascochœrus africanus* and of
the *sus larvatus;* three elephants' heads, one of
which is sawn vertically to shew its internal
structure; and four of the rhinoceros belonging
to three different species.   Here are every genus
and a great number of species of the *ruminantia*,
with three cameleopards, one of which is very
young, and many buffaloes.   Near them is the
skull, found in an Egyptian tomb, of the *bos apis*
which was an object of worship.   Lastly, the
heads of many *cetacea*, amongst which those of
the *manatus*, of the *halicore*, Illig., and of the
monodon, deserve to be noticed.

The second room on the same floor contains,
on the right, the remainder of the series of heads;
namely, those of birds, reptiles and fishes.   We
remark among those of the reptiles, three of the
crocodile of the Ganges, sent by M. Wallick.   In
collecting these heads attention has been paid to
the difference of age to shew their gradual developement.   The remainder of this room is destined to the study of bones, separately considered;
on one side of it and in the middle are many
glass-cases containing those entering into the
formation of the head; their number is very considerable, especially in the fish, of which a large
variety is exposed to view.   The bones of the
foot are also placed in several boxes, and classed

according to their nature, so that we see in one frame those of the heel, etc., of all the animals. There are similar series of the large bones and of the vertebræ in the two adjoining rooms, where we may compare those of the thigh, the arm, etc.

In the third room are the skeletons of the small quadrupeds, comprising those of almost all the monkey tribe, those of the *simia troglodites*, Lin., of the pongo, several of the *simia lar*, and that of the *galeopithecus*; a most complete series of the carnivorous animals, all the species of kanguroo and didelphis known; the *rodentia*, the beavers, etc.; every known genus of the *edentata*, amongst which is the *myrmecophaga jubata*, sent from Guyana by M. Martin; the *m. tamandua*, procured by M. Gaimard, and the *m. capensis*, by M. Delalande; lastly, the skeletons of the *echidna* and *ornithorhynchus*. Above the cases are affixed to the wall, the horns of the ruminantia, and on both tables are methodically arranged a complete series of teeth, from man to the horse.

In the fourth room we see the skeletons of birds. Those which most deserve our attention are the African and American ostriches, the Indian casuary and the emu of New Holland; the skeleton of an *ibis*, taken out of the tomb of a mummy, which M. Geoffroy procured in Egypt, and the skeletons of the humming birds. The two last cases contain the tortoises, amongst which is a very large specimen of the sea and also one of the Indian land tortoise, the largest known at this day. The series of teeth, beginning with those of the horse and terminating with those of fishes, is here continued and shewn in small boxes placed on the tables. Above the cases we see the skeletons of four large crocodiles, and near the one sent by M. Wallick are bracelets which were found in the stomach of this animal, and which must have belonged to an Indian woman.

The skeletons of the reptiles, such as lizards, serpents, toads, frogs, and salamanders, with those of all the genera and of a great number of species of fish, occupy the cases in the fifth room. On the top of the front cases we see the skeleton of a *boa constrictor*, 15 feet long, brought from Java by M. Leschenault, those of a shark and of a sword-fish from the Mediterranean, and on each side a series of snouts of the saw-fish, jaws of several species of shark, the ray, etc. On the tables in this room are the dried larynx and hyoïd bones of quadrupeds; amongst which is that of the *simia seniculus*, which is inflated like a bladder: it is this peculiar form which gives the voice of this animal the great extent and extraordinary sound for which it is remarkable.

The five rooms we have been inspecting contain all that relates to osteology, and may be considered as the first part of the collection. The second part, although it occupies a smaller space, is not less interesting nor less numerous than the first.

The sixth room is devoted to myology. In the centre is a cast of the human body deprived of the skin, and on which the muscles are painted of their natural colour. The cases on one side contain small flayed figures in wax of human arms and legs: on the other are two small statues of horses and the limbs of many quadrupeds, and in the remaining cases the dissected muscles of several animals preserved in spirits. Here we may study the myology of all the *mammalia,* and that of the birds, reptiles, and fishes.

The seventh room contains the organs of sensation: the larynx and trachea of birds are also seen on the tables of this room. The cases contain a quantity of flagons, in which are preserved in spirits a series of brains and eyes, which for the most part are dissected; also the bones of the ear of all the animals from man to the reptiles and fishes. We also see here well prepared specimens of skins, furs, feathers, scales, nails, and hoofs; others of the tongue, nostrils, and different preparations of the nervous system;

and a few heads of savages with their tatooed skin dried on them.

Preparations of the viscera in general, but more especially of those belonging to the function of digestion, are placed in the eighth room. In one of the two large glass frames is a model in wax of a child twelve years old, with the breast and abdomen laid open to shew the relative situation of the viscera and of the intestines; and in the other the anatomy of the hen, exhibiting the several periods of the formation of the egg, as well as the internal organs of the fowl.

The ninth room is dedicated to the organs of circulation, and to those of different secretions. It contains a series of hearts of the mammalia, reptiles and fishes, some injected preparations, a great number of dissected tongues and larynxes; the glands belonging to several parts of the body, swimming bladders, and the organs of generation; next to which are some very delicate preparations of foetuses belonging to viviparous and oviparous animals. There are besides, on the table in this room, injected and dried viscera, which shew the degree of fineness to which the vessels arrive, that serve to carry on the circulation of the fluids.

The tenth and last room contains a series of monstrosities and foetuses of different ages. It may be here observed, that variations in nature are quite as frequent amongst the inferior animals as in the human species; thus the opinion that they are the result of the mother's imagination falls to the ground, as it cannot be supposed that inferior animals, as the rabbit, etc., amongst which monsters are not uncommon, are susceptible of these impressions. This room contains also preparations of different orders of mollusca, articulated animals and zoophytes; comprising all those which M. Cuvier has had engraved for his work on the anatomy of mollusca. On the tables are twenty-four frames containing a series of preparations of shell-fish in wax, made at Naples under the direction of Poli, and procured by professor Hermann of Strasburg. Some very interesting and delicate preparations of the hard parts of crustacea and insects, made by M. Straüs, should be seen here; but for want of space they are deposited in the fourth room, next to the frames which contain the teeth of fishes.

To this rapid description of the cabinet of comparative anatomy, and of the order of its arrangement, we shall add a numerical list of the preparations, which in December 1822 amounted to 11,486.

OSSEOUS AND DRIED PREPARATIONS:

Human skeletons   41
    16 of fœtuses.
      2 young subjects.
      2 French adults.
      6 European do.
      3 Egyptian mummies.
      1 Gouanche mummy.
     11 Negroes and Hottentots.

Human skulls   174
    60 shewing the differences of age.
    50 of varied forms.
    64 of different nations.

Skeletons of mammalia   360
———— birds   500
———— reptiles   165
———— fishes   480

Heads of all sorts of animals 1,040
——— with the bones separate   300

Bones classed according to their species   590
——— of the hand and foot   300

Preparations of teeth   870
—————— the bones of the ear   462
—————— hair, feathers, nails, etc.   240
—————— intestines, dried and injected   58
—————— the larynx   200
—————— the branchia of fish   42
—————— the hyoid   135
—————— external envelopes of crustacea and insects 230
—————— wax and plaster casts of the muscles and
           viscera of man and animals.   46

                                   ————
                                6,231

SOFT PREPARATIONS PRESERVED IN SPIRITS:

CHAPTER IV.

THE MENAGERIE.

WHEN Louis XIV fixed his residence at Versailles, the academy of sciences requested him to establish a menagerie in the magnificent park belonging to his palace. His consent being obtained, a great number of rare animals were soon assembled in a large space fitted for their reception, and drawn by Perault with a correctness hitherto unknown in France.

The menagerie continued to be enriched under the reigns of Louis XV and XVI. It was there that Buffon and Daubenton saw most of the foreign animals they have described from their own observations; and the fifteen first volumes of the Natural History, which they published together, owe to the menagerie of Versailles most of the descriptions and original remarks which render it a fundamental work in zoology.

The unfortunate Louis XVI being obliged to quit Versailles, the menagerie was neglected, and several of the animals perished for want of food. Those which remained in 1792 were offered to M. de Saint-Pierre, intendant of the garden, with a view to their dissection. He refused to accept them on such terms, and presented a memorial to the government on the necessity of adding a menagerie to the Garden of Plants (1). This memorial produced the desired effect; and six months subsequent to the new organization of the establishment, the animals were removed to the Museum (2). At the same time, by a decree of the corporation of

(1) This memorial is printed in the 12th vol. of M. de Saint-Pierre's works, page 635 to 669. The author urges the motives which should determine the government to adopt the measure, and answers the objections of those who regarded it merely as wasteful magnificence. He shews that the establishment, being designed for instruction in natural history, ought to represent a picture of the three kingdoms; that the study of zoology absolutely requires living subjects; that the idea formed of them from their skins and skeletons is as incomplete as that formed of vegetation by turning over the leaves of an herbarium ; that there should exist means of preserving the animals sent as presents by foreign governments ; that several wild animals might one day be usefully domesticated, if methods were contrived for rearing and multiplying them ; that by the crossing of different breeds new ones might be obtained to the increase of agricultural riches; that most of the animals in our farm yards were thus reclaimed from a state of nature. In short he suggests some very wise plans, which no doubt had great influence on the new organization of the Museum.

(2) There were but five of them : a very tame lion, the *antilope bubalis,* the *antilope corinna,* the *equus quaccha,* and the *columba coronata.* The lion in particular attracted the attention of the public from

Paris, all itinerant menageries were forbidden, and those who gained their livelihood by exhibiting animals were obliged to send them to the Museum, which was at once possessed of a considerable number. Some were placed in temporary dwellings, others in the groves, and the plan of a menagerie was immediately laid out. But it was only by degrees, and as circumstances would permit, that the necessary ground was obtained, and the enclosure did not attain its actual extent until the year 1822.

In the historical part of this work we gave an account of the buildings which were constructed, and of the acquisitions by which the menagerie had been enriched. We shall not return to this subject, but shall limit ourselves to describing its present state, and the order which reigns in it. As collections of animals are liable to so much fluctuation from death and other causes, we shall not mention all the animals which are at present in the menagerie, but only the most remarkable species, and those which we hope to preserve. Neither shall we point out the places where they are to be found, as

his attachment to a dog with which he constantly played. It was the same lion of which the librarian, M. Toscan, has given a very interesting history : see *l'Ami de la Nature*, page 15-47. In the same work, published in 1801, we find, page 265, some observations on the animals then existing in the menagerie.

they are removed from one park to another according to circumstances: and as a label is placed over the gate of the park or lodge of each animal, designating its name, the country from which it comes, and the name of the person who gave or sent it to the Museum.

The menagerie is 220 fathoms in length from East to West, or from the esplanade in front of the amphitheatre to the terrace along the quay; its greatest width from North to South is 110 fathoms, and it communicates with the garden by four principal entrances, one on the West, two on the South, one of which is in the centre of the chesnut walk and the other at the extremity, and the fourth on the North near the cabinet of comparative anatomy. These gates are open to the public every day from 11 to 6 in the summer, and from 11 to 3 in the winter. No person is admitted into the interior of the parks or lodges, unless he is conducted by a member of the administration.

The space destined for tame animals, which walk about at liberty, is divided into fourteen parks or enclosures, six to the west and eight to the east of the edifice called the rotunda. These parks, round which the public can walk, are subdivided into compartments, each terminated by one side of a building into which the

G. Cathelineau Del
E. Aubert Sculp
Durand Imp.

animal retires at will in the day time, and is shut up during the night. At the extremity of these parks and near the river is the building for the wild beasts.

On entering the menagerie at the door near the amphitheatre, we find an alley to the right which leads us all round it, and in front another alley which crosses it, winding round the parks, and passing between the rotunda and the aviary. Taking this path we see 1st, on each side, the African sheep with a large tail, and the *morvant* with very long legs. 2d. The *camelus alpaca,* a very remarkable animal for the length and fineness of its wool; it was almost unknown at the time when it was presented to the Museum by M. Pouydebat, a merchant of Bourdeaux. 3d. Male and female goats from Tartary, and also one which MM. Diard and Duvaucel sent from India, of the true race, the wool of which is used in making the Cashmere shawls. 4th. The goats from upper Egypt, to which the projection of their under jaw gives a very singular appearance; and those of Napaul, which are remarkable for a very important character in zoology, that of having the curved forehead of the sheep. 5th. Some goats which scarcely differ from our European species, but which may give rise to a new breed.

Observation having proved, that under certain circumstances animals, especially goats, loose a great part of their hair, and that the wool underneath is then developed in greater quantity, choice was made of a couple most abundantly furnished with wool, and from them a new domestic breed was obtained, covered with a down very similar to that of the goat of Cashmere. It is easy to conceive the advantages to our manufactures that might be derived from their multiplication.

We next arrive at an enclosure which extends nearly as far as the aviary; it is divided into five compartments, in the middle of which stands a large circular fabrick, having its roof thatched with reeds, which serves as a stable. In the first compartment is a basin where all the smaller species of aquatic birds are assembled; here are also different species of tortoises, which either remain in the water or creep about on the grass. The second, third, and fourth compartments are occupied by a great number of long-legged birds and gallinaceous fowls. We shall particularly notice amongst them the European crane, the carunculated crane from the Cape, presented by M. Taunay, the crowned crane, and the *mycteria argala*, a species of stork, the feathers of which are in great request for ladies'

G. Cathelineau Pinx.
E. Aubert Sculp.
Pacand Imp.

ornaments. The ostriches occupy the last compartment. Large trees shade the whole extent of this park, and the numerous birds that walk about it in the day retire at night into the thatched hut.

To the right of the park just mentioned is another with three divisions, towards the extremity of which is a building imitating a ruin. It was long inhabited by the wild goat (*capra ægagrus*); other animals have been successively placed there. In the western compartment is a basin for the larger aquatic birds.

To the south of this park, which is the lowest part of the menagerie, we see another more elongated, extending from the green-house to the rotunda, sloping towards the north and divided into five compartments. In the middle is a small picturesque building with four pavilions, each of which serves as a retreat to a species of deer. The *cervus axis* has been kept and has propagated here for several years.

The winding walks which encircle these parks end at the rotunda and the aviary. Beyond, we find nine other parks constructed upon the same plan. In the middle of the first, opposite to the rotunda, we see a shed encircled with wooden pillars, in which is a mule produced from an ass and a female zebra. This animal is striped like the zebra, particularly on the legs and thighs.

The eight following parks are inhabited by different species of sheep and deer. We shall principally remark amongst them the male and female of the sheep of Astracan (*ovis aries varietas*), which were presented to the Museum by the late Duke of Richelieu; as they have propagated, we have hopes of domesticating them; the wool of the young is a valuable object of commerce. The male and female great deer of Canada (*cervus strongyloceros*, Schr.), sent us by M. Milbert; the stags of Louisiana and of Bengal, the last of which, supposed to be the hippelaphus of Aristotle, was given by M. de Montbron; and several other species of deer, one of which is white.

In the furthest park, and during summer only, we see the guepard (*felis jubata*), common to Asia and Africa. This animal, which the Asiatics train to hunting, resembles the panther in the elegance of its form and the agility of its movements; it is as tame and mild as the most familiar dog; it was given to the Museum by M. Lecoupé, governor of Senegal.

Having made the tour of the different parks, we return to the rotunda.

In this edifice, which has five large pavilions, we find a young elephant, sent from India by

Grande Rotonde dans la Ménagerie.　　　The Rotunda in the Menagery.

M. Leschenault; five dromedaries, three of which were born in the menagerie; the male and female bison (*bos bison*, Lin.); the buffalo, and several small animals, such as the *pecari, dicotyles labiatus*, and *d. torquatus*, Cuv., etc.

On leaving the rotunda we proceed to the cages where are kept the monkeys and the birds of prey, and to the aviary.

Last year the cages of the monkeys formed a continuation with those of the birds of prey; a road has been cut through, which is to lead round the menagerie. This new arrangement has made the place for the monkeys much too small; but it is in contemplation to have a more convenient lodge constructed for them.

A great number of monkeys have existed in the menagerie, and many have had young, of which several are still living. The most remarkable species at present are, the *simia leucophæa*, Fr. Cuv., *s. sphinx, s. pileata*, Shaw, *s. silenus*, Lin., and the *s. cynomolgos*.

On the other side of the new path is a small gallery, with glazed doors, which are kept open in fine weather, for such small quadrupeds as require heat, such as the *viverra mungo* and the *viv. javanica*, two American squirrels, a *dasyurus*, a *phalangista*, an *arctomys empetra*, two species of *dasypus*, etc.

Next come the birds of prey : here are vultures of different species, one of which (*vultur papa*) was presented to the Museum by the duke of Orleans; the *vultur barbatus*, which, next to the condor, is the largest bird of prey known ; the *falco ecaudatus* of Senegal and several American owls.

On turning to the left we arrive in front of the aviary, which is an inclosure planted with shrubs, with a building in the rear facing the south, and divided into compartments, for foreign birds. As this enclosure is destined for the propagation of rare and wild birds, the public is not admitted into it. It contains at this moment the golden, silvery, and common pheasants, some foreign species of gallinaceous birds, such as the *crax alector* and *galeata*, the *penelope marail*, and many curious species of poultry.

Going round the aviary we return to the extremity of the menagerie to see the carnivorous animals. In the historical account it was observed, that all those which we had possessed since 1794 had been lodged in an old building at the end of the chesnut walk, and that a new one had been erected in 1817, into which they were removed in 1821. This building, of a simple and regular style of architecture, contains twenty-one lodges which have a

Loges des Animaux féroces. Dens for the Wild Beasts.

164

southern aspect; behind is a gallery lighted from the top, sufficiently large to admit of two persons walking in it without danger, to see the animals in winter when the outside shutters are closed. It is also from this gallery that the animals are fed and their apartments cleaned, by removing them from the lodge where they pass the night, to that adjoining. There are now in this building, lions and lionesses, one of which has a dog living with it; the jaguar (*felis onca*); two species of jackall (*canis anthus* and *aureus*); several black bears; the spotted and striped hyæna (*canis hyæna* and *crocuta*); foxes, and wolves, one of which is no less sensible to the notice of those who approach its den than the most affectionate dog to the caresses of its master.

The lions are from mount Atlas, and were presented to the government by the emperor of Morocco and the dey of Algiers. One of the lionesses has had young several times, but the whelps have never survived dentition. The two jackalls were sent, the one from India by M. Leschenault, the other from Senegal. This last is a new species, remarkable for its slender make and tapering head. Although these animals are of different species they have produced young.

The jaguar is a native of South America. Some years ago we had other individuals of the same species, and were thus enabled to distinguish them from the leopard and the panther which existed at the same time.

The menagerie having successively possessed a great number of foreign animals which have been dissected, has given rise to the most important researches in comparative anatomy; it has enriched the collections with many new species, and the facility which it has afforded of observing the animals during their lives has produced results still more interesting. It has furnished the means of discerning between constant and accidental characters, and of solving the important problem of the distinction of species. It has enabled the zoologist to study the instinct, intelligence and habits of animals; the influence of education, confinement, domesticity, and change of nourishment; the phenomena relative to their gestation, to the care which they take of their young, and to the developement and propagation of certain qualities, which in process of time constitute peculiar races.

This institution has given rise to two important works. The first, entitled *la Ménagerie du Museum*, or a description of the animals which have lived or still exist there, by MM. Lacépède, Cuvier and Geoffroy, with figures drawn from nature by M. Marechal, was published in 1804,

in folio and in 12°, by M. Miger, who engraved
the plates. The second by MM. Geoffroy Saint-
Hilaire and F. Cuvier was begun in 1819, and
will be continued as opportunities occur of observing new species. Forty numbers of this
work, containing two hundred and forty figures,
are published. The figures drawn on stone are
coloured with the greatest care from living animals. The text contains not only a scientific description, but also a full account of all that has
been observed in the habits and economy of each
animal during its residence in the menagerie.

To give an idea of the utility of this establishment to natural history, we shall add a list of the
remarkable animals which have existed in it,
and which have been described and drawn, and
subsequently placed in the galleries of zoology
and comparative anatomy.

An asterisk designates those which were
not at all or imperfectly known at the time
they were received in the menagerie, and the
letter L those which are living at the present
moment (1).

(1) We shall omit the names of those animals which are generally
known.

MAMMALIA.

Simia mona, *Schr.*
—— petaurista, *Gmel.*
L. —— cephus, *Linn.*
* —— entellus, *Dufresne.*
—— rubra, *Gmel.*
L. —— fuliginosa, *Geoff.*
L. —— æthyops, *Linn.*
L. —— faunus, *Linn.*
* —— griseus, *F. Cuv.*
* —— pygerythra, *F. Cuv.*
L. —— sabæa, *Linn.*
L.* —— rhesus, *Geoff.*
L. —— cynica.
L. —— cynomolgos.
L.* —— nemestrina, *Linn.*
L. —— silenus, *Linn.*
—— inuus, *Linn.*
L. —— sphinx.
* —— cynocephalos, *F. Cuv.*
—— hamadryas, *Linn.*
* —— porcaria, *Bodd.*
L.* —— leucophæa, *F. Cuv.*
L. —— mormon, *Linn.*
L. —— paniscus, *Linn.*
* —— ateles niger, *F. Cuv.*
L. —— capucina, *Linn.*
—— apella, *Linn.*
L.* —— fatuellus, *Linn.*
—— cebus hypoleucus, *Geoff.*
* —— —— robustus, *New.*
—— sciurea, *Linn.*
—— rosalia, *Linn.*
L. —— jacchus, *Linn.*
—— midas ursulus, *Geoff.*
L.* —— trivirgata, *Humboldt.*
* Lemur ruber, *Péron.*
L.* —— —— murinus.
—— —— mongos, *Linn.*
L.* —— —— albifrons (mas. and fem.), *Geoff.*
L.* —— —— nigrifrons, *Geoff.*
—— —— catta, *Linn.*

L. Ursus arctos, *Linn.*
L. —— americanus, *Linn.*
—— maritimus, *Pall.*
L. —— lotor, *Linn.*
L. Viverra nasua, *Linn.*
L.* ——— narica, *Linn.*
L.* ——— caudivolvula, *Gmel.*
* ——— vittata, *Gmel.*
L. ——— ichneumon, *Linn.*
L.* ——— mungos, *Linn.*
L.* ——— javanica, *Desmarest.*
——— genetta.
* ——— tetradactyla.
* ——— paradoxurus typus, *F. Cuv.*
L. ——— civetta.
L.* ——— zibetha.
L.* Canis aureus, *Linn.*
L.* —— anthus, *F. Cuv.*
* —— argenteus, *Geoff.*
* —— cinereo argenteus, *Geoff.*
L. —— hyæna.
—— crocuta.
L. Felis leo, *Linn.*
—— tigris, *Linn.*
L.* —— onca, *Linn.*
* —— leopardus, *Linn.*
L*. —— jubata, *Linn.*
—— pardus, *Linn.*
—— serval.
—— discolor, *Linn.*
—— melas, *Péron.*
* —— mitis, *F. Cuv.*
—— canadensis, *Geoff.*
Phoca vitulina.
* —— nova species.
Dasyurus maugeii, *Geoff.*
L. Didelphis virginiana, *Linn.*
* ——— cancrivora, *Linn.*
——— gigantea, *Gmel.*
Phalangista Cookii.
* Phascolomis ursina, *Geoff.*
L.* Sciurus alpinus, *F. Cuv.*
L. ——— cinereus, *Linn.*
L. ——— capistratus, *Bosc.*

* Sciurus arobatis, *Linn.*
L.* Arctomys empetra, *Pall.*
L. Cavia aguti, *Linn.*
L. —— paca, *Linn.*
  —— —— var.
  Castor fiber, *Linn.*
  ——— galliæ, *Geoff.*
  Mus cricetus, *Linn.*
L. Dasypus sex-cinctus.
L.* ——— nova species.
L. Elephas indicus, *Linn.*
L. Dicotyles torquatus, *G. Cuv.*
L.* ——— labiatus, *G. Cuv.*
  Equus quaccha.
  ——— zebra.
  ——— ——— (hybrid).
  Camelus bactrianus.
L. ——— dromedarius.
  ——— llacma, *Linn.*
L.* ——— alpaca.
L.* Cervus strongiloceros, *Schr.*
  * ——— alce, *Linn.*
  ——— bengalensis, mas. *F. Cuv.*
  ——— ——— ——— fem. *F. Cuv.*
  ——— axis, *Linn.*
L.* ——— virginianus, *Linn.*
  * Moschus napu.
  * Antilope gazella, *Gmel.*
  ——— kevella, *Gmel.*
  ——— corinna, *Linn.*
  ——— gnu, *Gmel.*
  ——— bubalus, *Linn.*
L.* ——— grimmia, *Gmel.*
L. Bos bubalis, *Linn.*
L.* — bison, *Linn.*
L. — taurus, var.
L. Capra-hircus, varietas asiatica.
L.* ——— ———, ———.
L. ——— ægagrus.
L. Ovis ammon.
L*. — aries varietas.

AVES.

L. Vultur fulvus.
L. ——— cinereus.
L. ——— monachus.
L. ——— papa.
L. ——— percnopterus.
L. ——— aura.
L. ——— barbatus.
L. Falco chrysaetos.
L. —— albicaudatus.
L. —— leucocephalus.
L. Strix virginiana.
L. —— nyctea.
L. Coracias tibicen.
L. Loxia oryzivora.
L. Psittacus (of each division).
    Corythaix paulina, *Temm.*
L. Crax alector.
L. —— galeata, *Temm.*
L. Penelope marail.
    ————— cumanensis.
L. Phasianus nycthemerus.
L. ———— pictus.
L. Tetrao borealis.
    ——— umbellus.
    Columba coronata.
    ———— nicobarica.
    ———— cruentata.
L. Struthio camelus.
    ——— rhea.
    ——— casuarius.
    ——— ———— Novæ Hollandiæ.
    Ardea pavonina.
L. ——— carunculata.
L. ——— crumenifera.
L. Larus marinus.
L. —— argentatus.
    Pelecanus onocrotalus.
L. Anas canadensis.
L. —— ægyptiaca.
    ——→ sponsa.

AMPHIBIA reptilia.   PISCES.

    Testudo græca.
L. ——— radiata.
L.* ——— cafra.
   * ——— nova species (Guadeloupe).
   * ——————— ——— (Brazil).
     ——— indica.
     ——— angulata.
     ——— geometrica.
     ——— europæa
     ——— picta.
     ——— centrata.
L. ——— serpentina.
     ——— viridis.
     ——— imbricata.
   * ——— nova species.
   Emys punctata.
     —— longicollis.
     —— pensylvanica.
L.* —— nova species.
L. —— variegata.
   Lacerta africana.
   Crotalus horridus.
   Proteus anguinus.
   Gymnotus electricus.
   Silurus callichtys.

CHAPTER V.

THE LIBRARY.

THE library which was annexed to the Museum at the time of the new organization, was placed in the last room of the building; the space it occupied having become necessary for the display of the collections, and being too confined for the proper arrangement of the books, we have already expressed in the first part of this work a wish that it might be transported to the the large edifice in the centre of the menagerie, which is conveniently situated in the neighbourhood of the amphitheatre, where the courses are delivered. An unexpected misfortune, the loss of M. Van Spaendonck, has obviated the necessity of this measure, by leaving at the disposition of the administration the lodgings he occupied, where it is now proposed to place the library.

Works upon natural history being alone essential to the library of the Museum, it was to be expected that the collection in that branch would be complete; but circumstances since its formation having prevented our procuring foreign works, there are yet many gaps to fill up; that they will soon disappear, and that the library may answer to the collections connected with it, is to be earnestly hoped: it contains at present fifteen thousand volumes.

As the generality of the printed works are to be met with in most public libraries, we shall mention only some manuscripts, accompanied with original designs, and the magnificent and unrivalled collection of paintings upon vellum.

Amongst the manuscripts is the work of Plumier, entitled, *Botanicum Americanum, seu Historia Plantarum in Americanis insulis nascentium; auctore R. P. Car. Plumier, ab anno* 1689 *usque ad annum* 1697, 8 vols. folio. It consists of one thousand two hundred and twenty figures of plants, some of which are coloured, accompanied with a descriptive text. There is besides a folio volume of American birds painted from nature; the drawings are not very finished, but are exceedingly accurate.

Amongst the plants figured in the manuscript of Plumier, five hundred and forty-nine have been published by himself or by Burmann, and the remainder have been since noticed by botanists who have visited the West India islands; there are still several unedited, which it would be interesting to be acquainted with (1).

2d. Tournefort's descriptions and designs of several plants observed by him in different countries, and a part of his correspondence during his travels in the Levant.

3d. Commerson's relation of his voyage, one volume folio; and what is exceedingly valuable, the original designs of objects deserving of notice in the countries he visited.

These drawings, to the number of five hundred and thirty of zoological and six hundred and ten of botanical subjects, are upon sheets of different form according to the dimensions of the objects represented; as far as possible, they are of the natural size and are accompanied with details of the characteristic distinctions; their arrangement is in the order of the natural families.

The descriptions of animals have not been bound together on account of the difference of size; they fill a separate portfolio of about five hundred pages.

M. de Lacépède in his History of fishes and

(1) At the foundation of the library, Government had decided that Plumier's works should be removed thither from the Royal library; a part only have been sent, and it is greatly to be desired that this precious collection should be united and placed in the Museum.

reptiles, M. Cuvier in his *Regne animal,* M. Geoffroy St. Hilaire in the Annals of the Museum, and M. de Jussieu in his *Genera plantarum,* have frequently made use of the designs of Commerson. As the collection of this celebrated traveller has been given to the Museum, we possess the greater number of the objects he has figured.

4th. *Description of the plants and animals of Java and the Philippines,* with plates, by Noroña, a Spanish physician and naturalist.

This manuscript, which might form 2 vols. 4°, was given to the Museum by the academy of sciences.

On his way home from Java, where he made a long stay, Noroña stopped at the isle of France, and died there in 1788.

M. de Cossigny, who had the charge of his manuscripts, presented them to the academy of sciences, on condition that M. de la Billardiere should arrange and publish them. The academy judging that the work contained many new and curious facts accepted it; but the publication, begun in 1790, was arrested the following year, when the editor embarked in search of La Peyrouse. It would be interesting to extract from Noroña's work what may have escaped the attention of naturalists that have since visited the same countries.

5th. *Oology, or a description of the nests and eggs of a great number of European birds, with an account of their manners and habits,* by the abbé Manesse, member of the academies of St. Petersburg, Erfurt, etc., 2 vol. 4°, of three hundred and fifty pages each, with one volume of excellent designs representing several species of birds and one hundred and ninety-eight species of eggs.

The author does not insist upon the characters of species already described and figured in works on ornithology, but he presents the most curious particulars of their habits, emigration, education of their young, etc. He had sent the collection of eggs figured by him, to the Museum, and these, amounting to one thousand and ninety-four, are now deposited in the galleries of zoology.

6th. Chinese paintings on rolls of paper consisting of a single sheet; one of which, 24 feet in length, presents the figures of several fish not well known; four others represent the city of Canton and Chinese edifices : there are also three folio parcels of portraits of men and women, and figures of plants and animals.

7th. We may cite as a curiosity, a Chinese manuscript in eight volumes small folio, with figures. The correctness of the writing is wonderful, and the designs bear a most striking resemblance to our copper-plate engravings. It is a work on anatomy, translated from French into Chinese, and sent from Pekin to the academy of sciences in 1723 by father D. Parennin.

The collection of paintings on vellum, of which we have given the history in the former part of this work, when removed from the Royal library to that of the Museum, consisted of sixty volumes folio, at present it forms eighty-four; sixty of plants, twenty-two of the various branches of zoology, and two of comparative anatomy.

The drawings amount to 4,750; viz. 3,500 of plants, 166 of mammiferous animals, 460 of birds, 38 of reptiles, 118 of fishes, 130 of crustacea and shell fish, 100 of insects, 26 of radiaria and polypi, and 212 of comparative anatomy. They are arranged according to the natural families, and the names of the objects are placed on the back of the volume. A great number of these drawings have been engraved, but those that represent familiar objects are still curious, as furnishing examples for the different lectures, or as models for students in the art of painting objects of natural history.

As there are at present six artists attached to the Museum, each professor causes to be painted such objects as have not yet been figured, and these designs have the advantage over those formerly executed, of presenting in detail all the characteristic parts.

The oldest artist that contributed to the collection, Nicolas Robert, who worked for Gaston of Orleans, and for whom subsequently the situation of painter to the king's cabinet was created, was never surpassed in the style he adopted. His paintings in water-colours are remarkably delicate, of great fidelity, and though now a hundred and fifty years old, they have lost none of their brilliancy of colouring: there are five hundred of them in the portfolios. The designs of Aubriet, more numerous than those of Robert, are less highly finished, but are very true to nature. He accompanied Tournefort in his voyage to the Levant, and drew on the spot many new plants. By publishing these designs with descriptions formed upon the herbarium of Tournefort, M. Desfontaines has made known those plants that were only indicated by a corollary (1).

During the old age of Aubriet, and after his death, less attention was paid to the collection; but it received new lustre at the nomination of M. Van Spaendonck. The works of this cele-

(1) The work of M. Desfontaines at first inserted in our Annals, has since been separately published under the title of *a Selection of Plants from the corollary of Tournefort*, 1 vol. 4to. Paris, 1808.

brated painter differ in style from those of Robert; they have not the finish which is the result of patience, but we remark in them that boldness of pencil and harmony of tint that characterize the superior artist, and they may be said to be perfect models of flower painting. They are not numerous; for M. Van Spaendonck, whom the king often called upon for large paintings, devolved this part of his duties upon the distinguished artist who has since been entrusted with the continuance of the work (1).

M. Marechal, who at the new organization of the Museum was chosen by the administration to paint mammiferous animals and birds, surpassed in this branch all his predecessors. Of his paintings, we possess 66 of mammalia, 8o of anatomy, and 3o of other objects of natural history.

M. Oudinot has painted with great talent a vast number of insects, crustacea and shells.

(1) As examples of M. Van Spaendonck's works, see the *Palava Malvæfolia, Pavonia Spinifex,* and *Hibiscus Palustris;* three plants of the family of the Malvaceæ, found in the same volume.

APPENDIX.

N° I.

CHANGES

MADE IN THE MUSEUM DURING THE PUBLICATION OF THE PRESENT WORK.

Having given some idea of the origin of the Museum, its actual state, and the collections assembled in it, we may subjoin that it is becoming richer every day, in consequence of its organization and the influence it exercises on the progress of the natural sciences. It would be satisfactory to present in proof of this assertion, a table of the riches it has acquired since the period at which the publication of the present work was commenced. But in casting our eye over what has passed during this interval, we feel ourselves affected with sentiments of profound grief; in the space of one month, two of our most celebrated professors have been taken from us. Their names will always constitute the glory of the establishment, and the impetus they have given to it will not be weakened; but to their pupils there is yet no consolation for their loss, and to their colleagues is recalled every day the charm of their society. It is not for us to pronounce their eulogium, the academies of which they were members have already paid this tribute to their memory; to us belongs the duty of expressing our regret, and adding a few words on the services rendered by them to the Museum.

M. Van Spaendonck has fixed a limit to the art of flower painting, which is not to be surpassed; in his paintings we see nature herself expressed; never in this style have the richness of composition, the beauty of colouring, the exactness of detail been carried to such a pitch. But the works which for fifty years have established his reputation are not what most demand our gratitude. He was not less distinguished as a professor than as a painter; and it is to the artists formed by him, that we owe the elegance of the ornaments that constitute the superiority of many of the manufactures of France, and especially the perfection of the figures, that add so much to the beauty and utility of our works on natural history. The progress made in this art by his lectures is not less due to his character than to his talent; never was there a more affectionate, a more attentive master, he strove that each of his scholars should rival himself. During the latter years of his life he did not execute any large paintings, but employed his time in making models, that might gradually smooth away the difficulties to his pupils. The object of his labours was the school he has formed at the Museum; and whilst admired in foreign countries, the only enjoyment he desired was the favourable opinion of his colleagues and students.

At the age of seventy-six he still retained all his faculties, and would have rendered us still greater services had he not been suddenly carried off, on the 11th of May 1822. His chair has been suppressed, and the iconographical course is given jointly by MM. Redouté and Huet, the one taking the vegetable, the other the animal kingdom.

Twenty days had scarcely elapsed when M. Haüy, from whom his death had been concealed, followed him to the tomb. M. Haüy had professed twenty years at the Museum; he is the founder of the French school, and the methods of other countries have been modified by his; perhaps the last age has not witnessed a more remarkable discovery, and one more completely the work of its author, it may be applied to every mineral species, it places mineralogy in the rank of the *exact* sciences, in determining by a rigorous measure the form of the primitive nucleus, and by subsequent calculations all the resulting secondary forms (1). Once published, this theory could never be lost, but it would have presented many difficulties, had not the professor by the clearness and elegance of his explanations removed such a drawback.

He not only confined himself to giving his lectures, but he united at his own house those who were fond of the study, displayed to them his collection, and superintended their labours; his instructions were proportioned to the capacity of the student; and many foreigners who

(1) In our description of the rooms containing the minerals, we mentioned that the first armory contained wooden models, serving to render sensible to the eye and at the same time to explain the crystalline structure. We may add here, that these models are the work of M. Belœuf, at present residing in the garden. This artist executes with the most rigorous precision all the varieties of form determined and described by M. Haüy. He sells at a moderate price complete sets with additional models, representing the principal results of the mechanical division of crystals and the gradation of decrease giving rise to them.

had come to Paris to study under him, have now spread his doctrine over the world.

M. Haüy was carried off at the moment his reputation was universally established; he had the good fortune to finish the structure of which he had laid the foundation. Not long before his death his crystallography had appeared, and the manuscript of the second edition of his treatise of mineralogy was in the printer's hands.

Forty years of his life were spent in forming a complete series of crystals; this collection is singular, it is the type of his great work, and is labelled by himself. It would be desirable that the Museum should obtain possession of it as an object of study, and a monument of the greatness and generality of the discovery.

M. Haüy has been replaced at the Museum by M. Brongniart, for several years his substitute at the faculty of sciences. This nomination, made by the king in accordance with the unanimous choice of the academy of sciences and the professors of the Museum, leaves no doubt that our institution will continue to preserve the just reputation it has acquired.

The description we have given of the collections of the Museum carry us to the close of 1822, but it would be incomplete if we did not notice the new riches acquired, and the changes lately made in the distribution of the objects.

At present the collection of fishes fills one part of the old room and that which contained the library. The cases emptied are now occupied by the reptiles, the species are in the order we have indicated.

At the entrance to the cabinet on the landing place of the staircase, above the basalts, have been placed two fragments of columns of the temple of Serapis at Puozzola. These fragments have been pierced by pholades consequent to a most singular geological phenomenon (1). There is also to be seen a case enclosing a calcareous block from mount Bolca, divided into laminæ, on each of which are seen fossil fishes.

The objects which were formerly in the five armories to the left of the room containing the rocks have been distributed in the collection of minerals and rocks, beside the species of which they form the matter. The saloon of rocks contains only at present the two collections that give it its name, viz, the methodical and geographical series.

(1) This phenomenon consists in the soil which supported the temple of Serapis, after having been for a lapse of time buried under the waters of the sea, again becoming dry.

The former occupies the shelves that stretch along the room to the right of the entrance; it has been considerably augmented and completely classed by the professor of geology. In this new classification the rocks are arranged in natural series; the name of each series is taken from the predominant component principle. The meteoric stones form the continuation.

The geographic collections at present occupy the cabinets of which the upper part contains the methodical collection, and also those to the left on entering; it has been much enlarged; 1st. By a series of rocks of the Indian peninsula and island of Ceylon, brought home by M. Leschenault. 2d. By a complete series of earths, constituting the soil of England, presented by M. Greenough, formerly president of the geological society of London. 3d. By a series of the rocks of the high Alps and Switzerland, presented by counsellor Escher of Zurich, correspondent of the Museum, and whose loss the sciences and humanity at present deplore.

The collection of fossil bones has been augmented by the addition of some interesting fragments, among which we cite the head of a hippopotamus, presented to the Museum by the grand duke of Tuscany, and a human skeleton enclosed in a calcareous aggregate formed of recent sea sand, and containing some terrestrial and marine shells. This skeleton was sent from Guadeloupe by M. l'Herminier, on the demand of his excellency the minister for naval affairs; it is more complete than that in the museum of London.

M. Brongniart has deemed it proper to make some arrangements in the distribution of the mineral species in the cabinets; thus several specimens are no longer in the place pointed out in our description of the gallery; but it will be easy to find them out, as in general the series of species remain the same. M. Brongniart having confined himself to the re-establishment of the methodical order in the varieties of each of those formerly placed on the shelves. These varieties are now grouped in vertical lines. Many duplicates have been removed from the cabinets to form the commencement of a new collection for study, disposed in a set of drawers in the middle of the room containing the metals. In this collection, which is destined for the lectures of the professor, the order of the specimens corresponds exactly with that of his demonstrations; but considering the small number of duplicates, the series would have been very incomplete had not

M. Brongniart generously filled up the voids with specimens from his private cabinet.

Amongst the objects more lately placed in the hall of mineralogy, we owe the following to M. Leschenault : 1st. Superb specimens of moon-stone, found by him in its gangue. 2d. Some corundums also in their gangue and of large size. We may mention also a ferruginous and siliceous pudding stone named *cascalho dos diamantes*, or pebble of diamonds, and which is the gangue of this precious gem in Brazil, whence it was brought by M. Auguste de Saint-Hilaire : it was not known in Europe.

The galleries of zoology and botany are considerably enriched, as will be seen from what we are about to say concerning the acquisitions due to the zeal of our travellers and correspondents.

M. Leschenault de la Tour and M. Auguste de Saint-Hilaire returned a few months ago ; the former, who during a stay of six years in the peninsula of India and in the island of Ceylon had sent us many specimens, has brought home a collection composed of objects from the three kingdoms. We have spoken of the minerals ; we may now cite among the mammifera the bear of the mountains of the Gates, two apes of Ceylon, the *paradoxurus typus*, which was wanting in the cabinet, and also some fishes and reptiles of the isle of Bourbon (1).

M. Auguste de Saint-Hilaire has for six years been travelling throughout Brazil and the settlements of Paraguay, from the 12th to the 34th degree ; he has analyzed and described on the spot all the plants he could collect (2). He has taken notes upon the animals, and has brought home one of the most considerable and curious collections of botany and zoology that ever arrived at our Museum (3).

(1) The *paradoxurus typus*, Fr. Cuv., was brought alive, also a small red maki of Madagascar, and six species of tortoise.

(2) M. Auguste de Saint-Hilaire has finished his manuscript of the history of the plants of Brazil. It would be important for the progress of botany that this work were published, with the engravings representing the most interesting new genera and species.

(3) To shew how much this collection has enriched us, we shall here transcribe what the professors have said in their report to the academy of sciences on the voyage of M. Auguste Saint Hilaire.

The collection contains : 1st. one hundred and twenty-nine individuals of the mammifera forming forty-eight species, of which thirteen were not in the Museum. Among these are two bats ; a new *simia seniculus*; the aguarachay, a species of jackall known only by the description of Azzara ; a porcupine with a prehensile tail, and a new species of the *rodentia* named *moco*.

M. Duvaucel, who continues his researches in India, has just sent us the skeleton of a very large elephant, a gangetic dolphin more than 6 feet long, and a great number of birds, amongst which forty-three species are unknown in the cabinet; we expect from him a collection of fishes amounting to five hundred species and two thousand individuals.

We have received from M. Lesueur the greater number of the fishes and mollusca described by him in the Journal of Sciences of Philadelphia; and from M. Milbert, some fishes taken in the lakes of the United States, and which were unknown to us.

Lastly, M. Dussumier, on his return from India, gave us a gazelle of Bassora, a species of dolphin, and twenty-eight species of birds not in the cabinet.

N° II.

LIST of the PRINCIPAL PERSONS employed in the MUSEUM in january 1823, and of their most important publications.

In the following list we shall follow the order of the chairs and places established by the decree for the organization of the Museum in 1795. We shall not particularize the memoirs inserted in the transactions of academies, in periodical works, or in dictionaries of science, which would exceed the limits of this article.

MINERALOGY.

Professor.—M. Brongniart (Alexander), born at Paris the 5th of Fe-

---

2d. Two thousand and five birds, forming four hundred and fifty-one species, of which one hundred and fifty-six were not in the Museum: the greater number of these make us better acquainted with the species described by Azzara. Amongst them we remark the *chaja*, akin to the *kamichi*, a species of *rhynchus*; the white swan with a black neck, of Paraguay; the *psittacus hyacynthus*, of which there exist but one or two species in European cabinets; the crowned eagle, several species of tangara, and the *guirayetapa* or little cock of Azzara.

3d. Twenty-one reptiles, amongst which is a new species of *lachesis*.

4th. About sixteen thousand well preserved insects, of which M. Latreille judges there are eight hundred unknown.

5th. A herbal composed of about thirty thousand specimens, forming nearly seven thousand species of plants in good preservation; two thirds of which M. Desfontaines judges to be new, and which will furnish new genera, and perhaps new families.

bruary 1770 ; member of the academy of sciences in 1815, principal engineer to the royal board of mines, named to the chair of the Museum in August 1822.

Elementary treatise on mineralogy applied to the arts, 2 vols. 8°, 1807 ; Natural history of fossil crustacea, and particularly of the trilobites, 1 vol. 4°, 1822.

Assistant.—M. de la Fosse (Gabriel), born at Saint Quentin in 1796. M. de la Fosse has not yet published any work. M. Haüy thus speaks of him in the introduction to his treatise on natural philosophy, p. 52 :

« We have great reason to felicitate ourselves on the assistance of » M. de la Fosse, who has ably seconded us in the experiments designed » to prove the truth of the new facts we are about to lay before the » public.... He has aided in the composition of several articles.... By the » treatise on crystallography, which is about to appear, the public will » be able to judge of the success with which he has cultivated that im- » portant branch of mineralogy. »

GENERAL CHYMISTRY.

Professor.—M. Laugier (Andrew), born at Paris in 1770 ; member of the academy of medicine, attached to the Museum with the title of Assistant Naturalist, charged with the analysis of bodies in June 1803, and appointed substitute to M. Fourcroy in his lectures two years after, named professor on the 17th of February 1810.

Numerous memoirs in the Transactions of the institute, the Academy of sciences, the Annals of chymistry and of the Museum.

Assistant.—M. Dubois (Anthony Charles), born at Paris in 1776 ; attached to the chymical laboratory in 1796.

CHYMISTRY APPLIED TO THE ARTS.

Professor.—M. Vauquelin (Nicholas Lewis), born at Hibertot, near Pont-l'Évèque, in 1763 ; named to the academy of sciences in 1792, director of the school of pharmacy, elected professor at the Museum in 1804.

Numerous memoirs in the Transactions of the institute and the academy of sciences, in the Annals of the Museum and those of chymistry, in the Journal of the mines and in the Bulletin of the philomattic society.

Assistant.—M. Chevreul (Michael Eugene), born at Angers in 1786 ; named to the Museum in 1809.

Author of several memoirs in the Annals of the Museum, of the chymical part of the Dictionary of natural sciences, of a work on the unctuous bodies, and of several memoirs in the Annals of the Museum.

BOTANY.

Professor.—M. Desfontaines (René Louiche), borne in 1752, at Tremblay ; member of the faculty of medicine in 1782, and of the academy of sciences in 1783, professor of the faculty of sciences of the university, named professor in the king's garden in 1786.

Flora atlantica, 2 vols. 4°, with 260 plates, 1796 ; Description of the botanic garden of the Museum, 1 vol. 8°, two editions, 1804 and 1815 ; Selection of plants from the corollary of the institutes of Tournefort, 1 vol. 4°, with figures, 1808 ; History of the trees and shrubs which may be cultivated in the open air in France, 2 vols. 8°, 1809 ; Description of a great number of new genera and other memoirs in the Annals of the Museum and in the Transactions of the academy of sciences and the institute ; of the last, that on the comparative organization of monocotyledon and dicotyledon plants, printed in 1797, is in the third volume of the Memoirs of the institute.

Assistant.—M. Deleuze (Joseph Philip Francis), born in 1753, at Sisteron (Lower Alps) ; secretary of the association of professors for the publication of the Annals of the Museum, named assistant naturalist in February 1795.

Translation of Darwin's Loves of the plants, 1 vol. 12°, 1799 ; Thomson's Seasons, two editions, 1801-6, 8° and 12° ; *Eudoxe,* or Discourses on the study of sciences, belles-lettres and philosophy, 2 vols. 8°, 1810 ; Critical history of animal magnetism, 2 vol. 8°, two editions, 1813-19 ; Defense of animal magnetism, 1 vol. 8°, 1819 ; several historical panegyricks, and some memoirs in the Annals of the Museum.

RURAL BOTANY.

Professor.—M. de Jussieu (Anthony Laurence), born at Lyons in 1748 ; doctor of the faculty of medicine in 1772, member of the academy of sciences in 1773, of the royal society of medicine in 1776, professor at the school of medicine in 1804, substitute to M. Lemonier at the king's garden from 1770 to 1787, named demonstrator at the death of his uncle Bernard in 1777, professor at the new organization of the Museum.

Genera plantarum secundum ordines naturales disposita. 1 vol. 8°, Paris,

1789; numerous memoirs in the Transactions of the academy of sciences and in the Annals of the Museum, amongst the latter are to be found the characters of many natural families and the monogrophies of several others ; many articles in the Dictionary of natural sciences, of which he is one of the principal conductors.

AGRICULTURE.

Professor.—M. Thouin (Andrew), born in the king's garden in 1747 ; named chief gardener in 1768, member of the academy of sciences in 1786, professor of agriculture at the Museum since the creation of the chair at the new organization.

His numerous writings, all relative to the principles or practice of agriculture, are inserted in the Memoirs of the academy of sciences of the institute and of the society of agriculture, in the Dictionary of natural history, printed by Deterville, in the new edition of Rozier's Course of agriculture, and in the Annals of the Museum ; in which he has given a description of the garden of seeds, of that of agriculture, that of fruit-trees, etc. He has published separately a treatise on grafts, 1 vol 4°, Paris, 1820. The manuscript syllabus of his lectures are exposed in the library of the Museum, with permission to copy them.

Assistant.—M. Leclerc (Oscar), born at Paris in 1798 ; attached to the Museum in 1818.

Chief Gardener.—M. Thouin (John), brother to the professor, born at the king's garden in 1756.

ZOOLOGY.

Mammiferous Animals and Birds.

Professor.—M. Geoffroy Saint-Hilaire (Stephen), born at Étampes, in 1772 ; member of the institute 1807, professor of the faculty of sciences of the university, professor of the Museum at the new organization.

Anatomical philosophy, 2 vol. 8°, 1818-22 ; a great number of memoirs on zoology and comparative anatomy in the Transactions of the institute and of the academy of sciences, in the Annals of the Museum, and in the Description of Egypt.

Assistant for the Preparation of Animals.—M. Delalande, who has made several voyages and added considerably to the riches of the Museum.

Reptiles and Fishes.

Professor.—M. de Lacépède (Bernard Germain Stephen), born at Agen, in 1756; peer of France, grand cross of the legion of honour, member of the institute at its creation; named keeper and demonstrator at the king's garden in 1785, and professor at the institution of the third chair of zoology in 1795.

Essay on electricity, 2 vol. 8°, 1781; Natural philosophy, general and particular, 2 vol. 12°, 1782; Poetry of music, 2 vol. 12°, 1787; History of oviparous quadrupeds and of serpents, forming a continuation of Buffon's Natural history, 2 vol. 4°, 1788, reprinted in 8° and 12°; Natural history of fishes, 5 vol. 4°, 1798, reprinted in 8° and 12°, 1803; Natural history of cetaceous animals, 1 vol. 4°, 1804; the Menagerie of the Museum, in conjunction with MM. Geoffroy and G. Cuvier, 1 vol. folio, reprinted 12°; several memoirs in the Transactions of the academy of sciences, the Dictionary of natural sciences, and the Annals of the Museum.

Substitute.—M. Dumeril (Andrew Marie-Constant), born at Amiens, in 1774; member of the institute in 1813, secretary to the medical division of the royal society of medicine.

Digest and abridgment of the two first volumes of M. Cuvier's Lectures on comparative anatomy; Analytic zoology, 1 vol. 8°, 1806; Elements of natural history, two editions, the first 1 vol. 8°, 1804, the second, 2 vols. 8°, 1807; several zoological memoirs in the Methodical Cyclopædia, in the Register of the philomatic society, etc. He is author of the articles on entomology in the Dictionary of natural sciences.

Assistant.—M. Valenciennes (Achilles), born in Paris in 1794; nominated to succeed his father in the Museum in 1812.

Several memoirs of his are inserted in the Annals of the Museum, and he has aided M. de Humboldt in the publication of his zoological observations. Under the direction of the professors of zoology and comparative anatomy, he has been charged, since the enlargement of the cabinet, with arranging and naming the collection of vertebrated animals.

Animals without vertebræ.

Professor.—M. de Lamarck, (John Baptist Peter Anthony), born at Bazantin, near Bapaume, in 1744; member of the academy of sciences in 1779, attached to the king's garden with the title of Botanist of the Cabinet in 1789, professor of zoology for invertebrated animals at the creation of the chair.

French Flora, 3 vol. 8°, 1778; new edition with additions by M. Decandolle, 5 vol. 8°, 1795; Hydrogeology, 1 vol. 8°, 1801; Researches on the causes of the principal natural phenomena, 2 vol. 8°; Zoological philosophy, 2 vol. 8°, 1809; Analytic system of the positive knowledge of man, 1 vol. 8°, 1820; System of invertebrated animals, 1 vol. 8°, 1801; the first volume of the Dictionary of botany and the *Illustrationes generum*, forming a part of the Cyclopædia; Natural history of the invertebrated animals, 7 vol. 8°, 1822; a great variety of memoirs among those of the institute, the academy of sciences and the Museum, in the Journal of natural history, in that of natural philosophy, etc.

Adjunct Professor.—M. Latreille (Peter Andrew), born at Brive in 1762; elected to the academy of sciences in 1814, attached to the Museum in 1797.

He has named and classed the entomological collection. Natural history of salamanders of France, 1 vol. 1800; Natural and general history of ants, with a collection of memoirs, 1 vol. 8°, 1822; *Genera crustaceorum et insectorum*, 4 vol. 8°, 1809 and the following years; Natural history of reptiles, forming a continuation to the Buffon of Castel, 4 vol. 18°; Natural history of crustacea and insects, forming a continuation to the Buffon of Sonnini, 14 vol. 8°; the third volume, or entomological part of Cuvier's *Regne animal*; many memoirs among those of the academy of sciences and of the Museum, and the principal articles on entomology in the Dictionary of natural history by Deterville.

Assistant Naturalist and Chief of the zoological Laboratories.—M. Dufresne (Lewis), born at Champien in 1752; named in June 1793.

Author of the article *Taxidermie* in the Dictionary of natural history, and of a memoir in the Annals of the Museum.

HUMAN ANATOMY.

Professor.—M. Portal (Anthony), born at Gaillac in 1742; doctor of the faculty of medicine of Montpellier in 1765, professor of anatomy at the college of France in 1768, member of the academy of sciences in 1769; honorary president of the academy of medicine, first physician to H. M., professor at the king's garden since 1778.

The works of M. Portal are too numerous to be mentioned in this place, we shall barely cite the following:—History of anatomy and surgery, etc. 6 vol. 8°, 1770-77; Observations on the nature and treatment of madness, 1 vol. 12°, 1779; Observations on mephitic vapours, on cases of drowning, asphyxia, madness, etc. 1 vol. 8°, 1791; Instructions respecting the treatment of asphyxia and drowning, etc. 1 vol. 12°, many times reprinted by order of government; Observations on the nature and treatment of *phthisis pulmonalis,* first edit. 1 vol. 8°, 1792, second edit. 2 vol. 8°, with notes from German and Italian authors; Observations on the nature and treatment of rickets, etc. 1 vol. 8°, 1796; Collection of memoirs on the nature and treatment of several maladies, etc. 2 vol. 8°, 1800; a great number of memoirs in the Transactions of the academy of sciences and the Institute, in the Annals of the Museum, and in scientifick journals.

Assistant.—M. Martin (John Paul), born at Cahussac in 1788; named to the Museum in 1809.

COMPARATIVE ANATOMY.

Professor.—M. Cuvier (George), born at Montbeliard in 1769; counsellor of state, member of the royal council of public instruction, professor of natural history at the college of France, named member of the Institute at its creation, perpetual secretary of the class of natural sciences in 1803, one of the two perpetual secretaries of the academy of sciences, member of the French academy in 1818, named assistant to M. Mertrud in 1795, professor in 1802.

Elementary description of the animal kingdom, 1 vol. 8°, 1798; Lectures on comparative anatomy digested by MM. Dumeril and Duvernoy, 5 vol. 8°; Historical report on the progress of the natural sciences since 1789, and on their actual state, 1 vol. 8°, 1810; Researches on fossil bones, first edit. 4 vol. 4°, 1812, second edit. 5 vol. 4°, 1822; Memoirs on the history and anatomy of mollusca, 1 vol. 4°, 1817; the *Règne animal,* 4 vol. 8°, 1817; Collection of historical panegyrics of the members of the academy of sciences, 2 vol. 8°, 1819; Report on the state of public instruction in Holland and Italy, 2 vol. 8°; Numerous memoirs among those of the academy of sciences and of the Museum; and several articles in the Dictionary of natural sciences and in the Universal biography.

Assistant.—M. Rousseau (Simon Peter), born at Belleville near Paris in 1756; attached to the Museum in 1795.

He prepared the greater part of the skeletons in the cabinet of comparative anatomy.

Keeper of the galleries of comparative anatomy.—M. Laurillard (Charles Leopold), born at Montbeliard in 1784; named in March 1812.

He has assisted M. Cuvier in his researches and furnished a great number of zoological and anatomical drawings, several of which are engraved in the works we have just cited.

GEOLOGY.

Professor.—M. Cordier (Peter Lewis Anthony), born at Abbeville in 1777; inspector at the royal college of mines, named professor in 1819.

Memoirs on volcanic productions, 1 vol. 4°, Paris, 1815; Memoirs on the coal-mines of France, 1 vol. 8°, 1815; memoirs in the *Journal de physique*, in the *Journal des mines*, in the Annals of natural philosophy and chymistry, in the Description of Egypt, and in the Annals of the Museum.

Assistant.—M. Regley (Francis Theophilus Marie), born at Paris in 1777; named to the Museum in 1812.

DRAWING MASTERS.

For plants.—Redouté (Peter Joseph), born at Saint-Hubert (Ardennes) in 1759; named to the chair in 1823. He has been attached to the Museum ever since the new organization.

The beautiful engravings of plants accompanying the works of l'Heritier and Ventenat were executed after the designs of this artist; he has contributed much to the perfection of coloured engravings, and has himself published some considerable works of botanical Iconography. Lilies, 8 vol. fol.; Plants of the family of Cacti, 2 vol. fol.; Roses, 3 vol. fol. and 4°; more than 400 of his drawings on vellum are found in the Museum.

For animals.—M. Hüet (Nicholas), born at Paris in 1770; attached to the Museum in 1804, named to the chair of iconography jointly with M. Redouté.

Besides the branch with which he was specially charged (the drawing of worms, insects and shells), he has furnished figures of quadrupeds and birds, and has executed a great number of anatomical drawings for MM. Geoffroy and Cuvier.

M. Redouté (Henry Joseph), born at Saint-Hubert in 1762, brother to the former.

He accompanied the expedition to Egypt, and executed more than 60 drawings for the great work which was the result of it; many of his designs are found in the portfolios of the Museum.

M. Dewailly (Peter Francis), born at Paris in 1775; professor of drawing at the royal conservatory of arts and trades; after a successful competition he was named painter of the Museum at the death of M. Marechal in 1803, and has added to the collection of drawings such living animals as were designated by the professor of zoology.

M. Bessa (Pancracius), born at Paris in 1772 ; named painter to the Museum in March 1823.

His principal works are : the drawings for the 50 last numbers of Duhamel's *Traité des arbres;* those in the first 76 numbers *de l'Herbier de l'amateur;* one half of the figures comprised in the first 20 numbers of Ferussac's works on shells ; a considerable number of his drawings are met with in the work on Egypt and many other productions, in which he has displayed great ability. In 1816 H. R. H. the duchess of Berry named him her painter of flowers, and since December 1820 he has had the honour of teaching her the art.

M. Meunier (John Baptist), born at Orleans in 1785; named painter to the Museum in April 1823.

Has executed a great number of drawings for the work on Egypt ; those on entomology in Olivier's work and in M. Latreille's Genera ; the drawings of crustacea and trilobites in the work MM. Brongniart and Desmarest published jointly; and several no less important drawings all very remarkable for their accurate representation of nature.

M. Toscan (George), born at Grenoble in 1756; appointed librarian in 1794.

The Friend of nature, or interresting observations on different objects of nature and art, 1 vol. 8°, 1800 ; Translation of Spallanzani's work, intitled Travels in Sicily and in several parts of the Appennines, 6 vol. 8° ; principal author of articles on natural history in the Philosophical and litterary *Decade.*

KEEPER OF THE MENAGERIE.

M. Cuvier (Frederic), born at Montbeliard in 1775; inspector of the academy of Paris, named keeper of the Menagerie in 1805.

Natural history of mammalia, conjointly with M. Geoffroy Saint-Hilaire, of which 40 numbers with 6 plates each have appeared; Of the Teeth of mammalia considered as zoological characters, 1 vol. 8°, 1822; several memoirs in the Annals of the Museum; the Zoology of mammalia in the Dictionary of natural sciences.

KEEPERS OF THE GALLERIES OF NATURAL HISTORY.

M. Lucas (John Francis), born in the king's garden in 1747; appointed to the place he now occupies at the new organization.

M. Lucas (John Andrew Henry), son of the preceding, born at the king's garden in 1780; associated with his father on the 12th of February 1799.

Methodical tables of mineral species, 2 vol. 8°, 1806-13; several articles on mineralogy in Deterville's Dictionary of natural history.

ADMINISTRATION.

M. Thouin (James), born at Paris in 1751; secretary and cashier.

MILITARY GUARD.

The service of the establishment is confided to a company of non-commissioned officers, at present under the command of M. Gouvion Saint-Cyr.

N° III.

TRAVELLING NATURALISTS.

M. Leschenault de la Tour, attached as botanist to the expedition of captain Baudin, resided some time in Java, whence he brought home many specimens of zoology, and a considerable herbarium. Sent to Pondicherry in 1817, he has traversed part of the peninsula of India, and transmitted three collections in every branch of natural history: the first, in 1818, contained the richest collection of fish and crustacea ever received; the second in 1819, and the third in 1820, presented many new objects in zoology and botany, and several living animals; among which was the young elephant now in the menagerie. Each of the collections was accompanied with explanatory catalogues and memoirs relative to the productions and culture of the country. On his way home he stopped at Ceylon, where he remained some time, and whence he brought a very rich collection. After a few months stay in Paris, M. Leschenault embarked for South America, where he is to visit all the French settlements.

M. Milbert sailed for New York in 1814; he has made eighteen remittances of quadrupeds, birds, reptiles and fishes; he has also sent us several living animals which had never been in the menagerie, and which are still seen there.

M. Lesueur, attached as painter of natural history to captain Baudin's expedition. During the voyage he became intimately connected with Peron, and gave himself up entirely to the study of zoology. He went to the United States in 1814, and has made us two remittances of birds and fishes, and communicated several memoirs for the Annals of the Museum.

M. de Saint-Hilaire (Augustus) went to Rio Janeiro in 1816; he has traversed several provinces of Brazil, and made four remittances of quadrupeds, birds, shell-fish and insects. Botany being his principal object, it is in that branch that his collection is most extensive. He has returned this year with a considerable addition to what he had already sent. His memoirs on several families of plants, inserted in the Annals before and after his departure, prove that he possesses equally the talent of observation and that of philosophic induction. We are convinced that he will thoroughly make known the Flora of Brazil.

M. Diard went to the East Indies in 1816. As soon as the affairs which called him thither were terminated, he gave himself up entirely to his taste for zoology and anatomy, which he had studied under M. Cuvier, having assisted him in researches on the developement of the fœtus, and on the eggs of quadrupeds. Being joined by M. Duvaucel at Calcutta, they proceeded together in their researches, visited Chandernagor, Sumatra, and several islands of the Indian archipelago, and sent home the first Cashmere goat seen in France, and many new objects of zoology. They then separated to embrace a wider range. M. Diard proceeded to Java, whence he has sent us a considerable remittance, and thence to Cochin-China.

M. Duvaucel sailed in 1817 for Bengal, where he met M. Diard, as we have already mentioned. Besides their joint remittances, M. Duvaucel has made one from Sumatra, in which were many precious objects we had not been able to procure. To his collections were added notes and interesting descriptions. He is at present at Chandernagor. It was not till after MM. Diard and Duvaucel had transmitted their rich collections that government thought proper to reimburse a part of their expenses, and to enable them to prosecute their researches.

M. Plée sailed for St. Thomas in 1820, whence he proceeded to Martinique and to the United States. He has sent us three miscellaneous collections.

M. Savigny was ordered to Senegal in 1820, by the minister for naval affairs, who had demanded from the administration of the Museum a person skilled in agriculture and botany : he is come back with a considerable collection.

M. Fontanier went first to the borders of the Black sea, whither he was sent by the minister of the Interior. We have some reason to believe that he is now visiting mount Caucasus, his last communications being dated from Teflis.

N° IV.

CORRESPONDENTS OF THE MUSEUM.

1. M. A. de Humboldt, named at his departure for America in 1798. We have spoken in the first part of this work of the services he has rendered to the Museum.

2. M. Bonpland, at present at Buenos-Ayres, named at the same time with M. de Humboldt; on his return from America he commenced the publication of the plants they had collected together and given to the Museum.

3. M. Baillon, at Abbeville, has enriched the zoological collection with the greater part of the birds of Europe, and especially the water-fowls which he has particularly studied : he has also presented mammalia and fishes. His father, formerly a correspondent of the king's garden, had sent a great number of objects

for the zoological collection, and many notes of his are quoted by Buffon in the History of birds.

4. M. Bory Saint-Vincent, named at the departure of captain Baudin, whom he accompanied, stopped at the isle of France, and thence proceeded to visit the principal islands of Eastern Africa. He has inserted in our Annals some memoirs on the confervæ, and has published several works on natural history.

5. M. Pichon, zoologist at Boulogne-sur-mer, named in 1802, has sent several birds to the Museum.

6. M. Faujas (Alexander), son to the late professor of geology, at present at Guadeloupe, named in March 1803, has communicated observations on the fossils of Guadeloupe.

7. M. Leonhard, professor of mineralogy at Heidelberg, named in 1808, has sent several specimens of minerals for the cabinet. He has published a great number of works on mineralogy and lately an excellent treatise on that science.

8. M. Marcel de Serres, professor of mineralogy at Montpellier, named in 1809, before setting out for Germany: he has transmitted from that country interesting objects for the collections of mineralogy and zoology, and inserted in the Annals several memoirs on the anatomy of insects.

9. M. Troost (Gerard), surgeon in the Dutch navy, named in 1810, has sent minerals for the cabinet.

10. M. Correa de Serra, member of the king's council and of the council of finances at Lisbon, named correspondent in 1811, at his departure from France for the United States : he has inserted in our Annals some excellent memoirs on carpology, and transmitted from Philadelphia several objects for the cabinet.

11. M. Taunay, at Rio Janeiro, named at his departure for Brazil in 1815, has transmitted a collection of birds in 1816.

12. M. Saint-Yves, surgeon in the navy, has remitted several parcels of seeds from the East Indies.

13. M. Leach (W. E.), of the royal society of London, late keeper of zoological collections at the British museum, named in 1818, has made additions to the cabinet in every branch.

14. M. Mitchell, professor of natural history at New-York, named in January 1821, has sent many objects of natural history from North America.

15. M. Wallick, superintendent of the botanical garden at Calcutta, named in January 1821, has sent us many seeds and skeletons of animals.

16. M. Macleay (W. S.), member of the Linnæan society at London, named in April 1821, author of a work entitled *Horæ entomo-logicæ*, has presented a great number of insects, and an orni-thorhynchus preserved in spirits.

17. M. Dorbigny, at La Rochelle, named in May 1821, has sent many specimens of mollusca, crustacea and fossils, and inserted several memoirs in the Annals of the Museum.

18. Rev. W. Buckland, professor of mineralogy and of geology at Oxford, named in June 1821, has furnished some beautiful specimens of minerals and fossil bones.

19. M. Durand de Villegegue, at Martinique, named in July 1821, has sent home bulbs of some very beautiful plants of the family of the lilies.

20. M. Maraschini (Peter), mineralogist, at Schio in the Vicentino, named in August 1822.

21. M. Sylveira Caldeira, in Brazil, named in November 1822.

22. M. Greenough (G. B.), late president of the geological society of London, named in March 1823.

N° V.

DIRECTIONS

TO PERSONS WHO VISIT THE ESTABLISHMENT.

Nothing is paid in the King's Garden for admittance, nor for objects ceded.

General instruction being the aim, strangers have the same privileges in the institution with natives ; but to obtain the objects they are de-sirous of examining or of possessing, they must address themselves to the person who has the charge of them.

THE GARDENS.

The gardens are daily open to the public from six o'clock in the morning till night. When the doors of the establishment are shut, no one is admitted but those who dwell in the Museum, or such as come to visit them; for which purpose sentries are placed at the three principal gates, and a patrol warns those that have remained in the garden, when it is shut. During the night the same survey in the interior is made, and sentries are placed near those parts of the Museum where it is thought necessary for the safety of the collections.

The several schools or enclosures devoted to a particular branch of horticulture can never be entered, except at the time when the gardeners are at work in them, by obtaining the permission of the professor, M. Thouin, or of the chief gardener, which is seldom or never refused.

Those who from a taste for botany are desirous of visiting the hot and green-houses, in which the plants of warmer climates are reared, must also apply to the professor or chief gardener for admittance; this is rather less easily granted, as the cultivation of the delicate plants kept in them requires some regularity in the heat, and the constant attention of the gardeners who have the charge of them, and as they are set so close for want of space, that not more than two or three people can be admitted at once, without danger to the plants. As few persons, but those who have some knowledge of the science make such applications, there is no example of a refusal. It must be observed however, that children are not admitted.

The botanical garden, in which all the known plants are systematically arranged and labelled, is open from the time the lectures on botany begin, until the end of autumn, every day from six o'clock, except during the two hours when the gardeners take their meal. No ticket of admittance is here required.

At the time when the lectures on horticulture are given, those who have been inscribed as attending them, are provided with a ticket of admittance into the garden for fruit-trees, kitchen herbs and agriculture, where they can observe the several modes of cultivation, and the experiments that have been prepared by M. Thouin for the elucidation of his discourses.

Each year, from December to February, a general distribution of the seeds that have been gathered in the gardens, and of those sent by the numerous correspondents of the Museum, or purposely procured from other countries by the board of agriculture, is made to cultivators, nursery-men and amateurs; first throughout France, and then to those of foreign countries.

To partake in this yearly distribution, requires only a written application, accompanied with the list of such seeds as are desired, to the board of the professors, or even to M. Thouin, professor of agriculture, who never fails of directing the parcel in its proper time. In the same manner young plants and trees are obtained, at the proper periods of the year, viz. in March, before the shooting of the leaves, and after the slight frosts of November or December, when they are totally fallen.

At all seasons young specimens of the tropical plants can be distributed; but these are only granted to amateurs or cultivators, who are provided with the necessary implements, and know how to rear them.

THE GALLERIES OF BOTANY, ZOOLOGY, MINERALOGY, AND COMPARATIVE ANATOMY.

The galleries of botany are open only to botanists, and such as wish to ascertain the identity of a plant, its synonym, etc. The public in general would gather but little instruction from the mere sight of herbals closely wrapt in paper, and uniformly placed in cases over which curtains are constantly drawn. Therefore, if it be wished to take a survey of these galleries, an order must be obtained from the professor of botany. But any one who has made botany his study, and who has recourse to the collections to complete his knowledge, is readily admitted by making himself known to the professor.

The galleries of natural history are open to the public at large on Tuesday and Friday of each week, from three to six o'clock in the summer, and from three till night in winter. Every Monday, Wednesday, and Saturday, from eleven to two o'clock, students and visitors are admitted; the first by means of a card, delivered at the office of the administration to those who have had their names registered at the lecture room in a book, kept for that purpose by the assistant naturalist. This card cannot be used by any other but the person whose name it bears.

Casual visitors get introduced with another ticket, which serves them but once, and admits several persons at the same time; with these tickets they can visit also the galleries of anatomy and the menagerie. The professors only distribute them: they are never refused when asked or written for.

The cabinet of comparative anatomy is not yet rendered public, but with a ticket it is to be seen every day of the week, Sunday excepted, from eleven to two o'clock.

Students, wishing to make comparative anatomy a part of their acquirements, can, by being introduced to M. Cuvier, obtain permission for a regular and daily frequentation of the cabinet.

THE MENAGERIE.

The gates of the menagerie are open all the year, from eleven to six o'clock in summer, and to three in winter. There the public can walk round the enclosures, and become thoroughly acquainted not only with the animals, but also with the trees and shrubs, as a label bearing the latin and French names is affixed in front of each park and suspended to each tree.

Some accidents having proved fatal to the visitors, in the inside of the dens or habitations of the animals, the keepers have orders to admit no person unaccompanied by one of the professors or assistant naturalists.

THE LABORATORIES.

There are in the Museum proper places where the skins and remains of animals are prepared and stuffed, in order to be placed in the galleries.

These laboratories are not open to the public; but such is the liberality of this institution, that those who wish to become acquainted with the processes are admitted to assist and act as if they belonged to the Museum, after being presented by one of the professors to M. Dufresne, who has the general direction of the zoological department, or to M. Rousseau, for comparative anatomy. The same may be said of the laboratory of chymistry, by applying to the professors, or to M. Dubois, the assistant naturalist.

It was intended to give here a short account of the methods used in the Museum for preparing and stuffing the skins of animals; but since this work was begun, a gentleman, a friend to the editor, has been at the trouble of translating into English M. Dufresne's article on *Taxidermy, or the Art of collecting, preparing, and mounting objects of natural history*, which was first published in Deterville's Dictionary of natural history. The small 12° vol. in which all the desirable details on the subject are to be found, has had two editions. As it can be procured, and is sold in London, at MM. Longman, Hurst, and Co.'s, Paternoster-row, we refer to it our readers who are desirous of becoming thoroughly acquainted with the mode adopted for the preservation of animal remains.

THE LIBRARY.

The library is open to the public every day from eleven to two (Sunday and Thursday excepted) during the summer, and only three days of the week in winter. No book or painting can be removed from the establishment without an order from the administration ; but to all is afforded the facility of studying and even making copies of the drawings in the library.

PUBLIC LECTURES.

The lectures begin in May. The order in which they are delivered, the days and hours when they take place, and the object of them are announced on printed advertisements, posted up in town and in the Museum.

Every person is admitted to the lectures.

REFERENCES TO THE PLANS.

FIRST PLAN—1640.

A. Street of the *Jardin du Roi*.

B. Lane, called the *Petit-Gentilly*.

C. Patouillet's enclosure, where kitchen plants are cultivated.

D. River *Bièvre*.

E. Kitchen-gardens and wood-yards.

F. Marshes extending to the *rue de Seine*.

G. Site of the hotel *Vauvray*.

H. Lands and houses of the *Nouveaux-Convertis*.

I. Houses bordering the hospital of *la Pitié*.

N° 1. Entrance.

2. Amphitheatre, where the lectures were given.

3. The intendant's dwelling-house.

4. Gallery in which was arranged the *materia medica*.

5. A court or yard.

6. Garden of the Indian plants.

7. ——— with compartments for medicinal plants.

8. ——— for kitchen plants, and of which Tournefort made a nursery.

9. The orangery with a garden.

10. Beds for plants of the south of France.

11. Steps ascending to the hills.

12. Botanical garden.

13. An orchard planted in quincunx.

14. Hot-beds for delicate plants.

15. Uncultivated spot whence sand was taken for the alleys,

16. Grove planted with wild trees.

17. Terrace commanding the marshy grounds.

18. Pavilion inhabited by professor Winslow, and where he died.

19. Hillock planted with mountain-trees.

20. Larger hill with winding paths, first planted with mountain-trees, transformed into a vineyard under Chirac, and afterwards planted with evergreen-trees.

SECOND PLAN—1788.

A. Street of the *Jardin du Roi*.

B. Buffon's street.

C. Boulevard of the Salpêtrière.

D. Quay Saint-Bernard.

E. Wood-yards and kitchen-gardens.

F. Seine street.

G. Grounds belonging to the *Nouveaux-Convertis.*

H. Houses bordering the hospital of *la Pitié.*

N° 1. Dotted line shewing the former extent of the garden.

2. Galleries of natural history.

3. A chapel near the principal entrance.

4. A building commenced by Buffon, and finished since.

5. Dwelling-house for the intendants.

6. The amphitheatre.

7. The orangery with its garden.

8. The spot where a new orangery was to be constructed.

9. The old hot-houses.

10. Dufay's hot houses.

11. A new hot-house built by Buffon.

12. The larger hill with a labyrinth and a pavilion.

13. The small hill.

14. A lane which gave entrance into the garden from the *rue de Seine.*

15. Hotel *de Magny,* the entrance to which is in the same street.

16. The gardens of that hotel.

17. New amphitheatre.

18. Two buildings at the sides of the amphitheatre.

19. Limits of the enclosure between Magny's garden and the small hill.

20. Hot-beds for seeds.

21. Botanical garden.

22. An ancient parterre.

23. The nursery.

24. Irregular plantation of trees, in the centre of which is a coffee house.

25. The two principal avenues planted with lime-trees.

26. Avenue of horse-chesnut-trees.

27. Avenue of Buffon.

28. A basin sunk to the level of the river.

29. New parterre.

30. The square for fruit-trees.

31. Square for economical plants.

32. Two other squares, an addition to the nursery.

33. Four squares for trees of the four seasons.

34. Avenue of Canadian poplars.

35. ———— of oriental planes.

36. ———— of Virginian catalpas.

37. ———— of Judas trees.

38. ———— of Virginian tulip-trees.

N° 39. Avenue of European larches.
    40. ——— of American maples.
    41. ——— of aylanthuses.
    42. An iron railing on the boulevard and Buffon's street.
    43. A terrace and gate on the quay.
    44. A green-house transformed into dens for wild beasts.

THIRD PLAN—1721-23.

    A. Street of the *Jardin du Roi*.
    B. Buffon's street.
    C. Square of the king's garden.
    D. Quay Saint-Bernard.
    E. Seine street.
    F. Square of *la Pitié*.
N° 1. Galleries of natural history.
    2. ———— of botany, office and laboratories for zoology.
    3. ———— of comparative anatomy.
    4. Amphitheatre and laboratories for chymistry.
    5. Gate of entrance near the galleries.
    6. —— near the river.
    7. —— in the *rue de Seine*.
    8. —— *rue de Buffon*.
    9. A descent into the quarries.
   10. Additional rooms of the cabinet of anatomy.
   11. The blacksmith's workshop.
   12. The joiner's shop.
   13. Workrooms for the painters of the Museum.
   14. A coffee-house.
   15. Small grove of foreign and indigenous trees.
   16. Garden for annual flowers.
   17. ——— for perennial plants.
   18. Seed-beds for trees and shrubs.
   19. A shed for artificial soils.
   20. Avenue of tulip-trees,
   21. Winter groves.
   22. Avenue of larches.
   23. Autumnal grove.
   24. Avenue of Virginian maples.
   25. Summer grove.
   26. Avenue of aylanthuses.
   27. Spring grove.
   28. Medicinal indigenous plants.
   29. ——— foreign plants.
   30. Foreign vivacious plants.

Nº 31. Flowers for ornament.
  32. Trees and shrubs.
  33. Aquatic plants.
  34. Ancient orangery.
  35. Vaulted green-house.
  36. Cold green-house.
  37. Green-house for the botanical garden.
  38. Hot-houses (Buffon, Baudin and Philibert).
  39. Frames for rare plants.
  40. Green-house for shrubs.
  41. ———————— for succulent plants.
  42. ————————— for mesembryanthema.
  43. ———————— for Indian plants.
  44. ———————— for cacti.
  45. The new green-house.
  46. Workrooms, seed laboratory and lodgings for gardeners.
  47. Frames for bulbous plants.
  48. The garden of the orangery.
  49. The botanical garden.
  50. The garden for fruit-trees.
  51. ———————— for cereal plants and pot-herbs.
  52. ———————— for practical agriculture.
  53. Avenue of kælreuterias and medlars.
  54. ———— of oriental planes.
  55. ———— of Virginian catalpas.
  56. ———— of Judas trees.
  57. ———— of horse-chesnut-trees.
  58. ———— of lime-trees.
  59. The terrace of the green-house.
  60. The seed-garden.
  61. Garden for the naturalization of plants.
  62. Place where the trees of the green-house are exposed in summer.
  63. The hillock planted with ever-green-trees.
  64. The labyrinth.
  65. The cedar of Lebanon.
  66. Daubenton's tomb.
  67. A dairy.
  68. Slope leading to the lower part of the garden.
  69. Bee-hives.
  70. The nursery.
  71. The rotunda for large herbivorous animals.
  72. Park and hut for the zebra.
  73. ———————— for the goats.
  74. ———————— for several species of stags.

N° 75. Ancient dens for the wild beasts.
76. New dens for carnivorous animals.
77. Fosses in which the bears were kept.
78. Park and hut for small ruminating animals.
79. —— for foreign sheep and goats, the alpaca, etc.
80. —— for European deer.
81. —— for the wild goats and buffalo.
82. —— for the ostriches, cassiowaries and fowls.
83. —— and basin for water birds.
84. ————— for foreign ducks.
85. The pheasant walk.
86. Aviary for birds of prey.
87. Cages for the monkeys and parrots.
88. Garden for experiments.
89. Hut and park for the elk or great deer of Canada.
90. The library and lodgings for the professor of zoology and the keepers of the cabinet.
91. The guard-house.
92. Lodgings for the professors of chymistry applied to the arts and of botany.
93. Lodgings for the professor of botany
94. ———————————— zoology (mammalia).
95. ———————————— human anatomy.
96. ———————————— comparative anatomy.
97. ————— for two assistant naturalists.
98. ————— for the assistant naturalist for anatomy.
99. ——————————————— botany.
100. ————————————————— zoology and the keeper of the menagerie.
101. ———— for the professors of mineralogy and agriculture.
102. ———— for the professor of geology.
103. A house and land lately added to the menagerie.
104. Lodgings for several workmen.
105. ———— for the professor of general chymistry.
106. Newly acquired grounds.
107.⎫
108.⎬ Houses let till the plan of the garden is completed.
109.⎭
110. Ten acres of private property.
111. Gates of entrance to the menagerie.

FINIS.

ERRATA.

P.   9, l. 22, for *appointed* read *assigned*.

   11, l. 10, for *correspondance* read *correspondence*.

   12, l. 15, for *repetedly* read *repeatedly*.

   21, l. 3, for *had excited* read *excited*.

   47, l. 23, for *and animals* read *of animals*.

   65, l. 14, for *concieved* read *conceived*.

   67, l. 10, for *what* read *with*.

   95, l. 19, for *casuaries* read *cassiowaries*.

   96, l. 20, after *set* read *of*.

  108, l. 8, for *Somerat* read *Sonnerat*.

  123, l. 1, for *polypuses* read *polypi*.

  128, l. 20, dele *in philosophy*.

   — l. 27, after *perception* insert *to*.

  140, note, for *glands* read *acorns*.

  154, l. 20, for *animals* read *animal*.

  169, note, before *virgilia* dele *the*.

  181, l. 14, before *seed* insert *the*.

  196, l. 9, dele *it*.

  205, note, for *lemon* read *melon*.

  206, l. 11, for *Auguste* read *Augustus*.

  208, note, for *at* read *of*.

  227, l. 9, for *and* read *et*.

  244, l. 11, for *polypis* read *polypi*.

  261, l. 4, fo *called* read *classed*.

  262, l. 27, for *diferent* read *different*.

  264, l. 14, for *fellow* read *follow*.

  278, l. 28, for *Harz* read *Hartz*.

  357, l. 27, for *bufaga* read *buphaga*.

  434, l. 3, read *bichir*.

  438, l. 12, for *Baillan* read *Baillon*.

  449, l. 24, for *Xyphius* read *Xyphias*.

  463, l. 28, for *scarabæus* read *scarabæi*.

  491, l. 28, for *metamorphoses* read *metamorphosis*.

  504, l. 6, for *bases* read *basis*.

  512, l. 15, for *from isles* read *from the isles*.